Fixed Broadband
Wireless Communications

Fixed Broadband Wireless Communications

Principles and Practical Applications

Dr. Douglas H. Morais

Adroit Wireless Strategies
San Mateo, California
www.adroitwireless.com

PRENTICE
HALL
PTR

PRENTICE HALL
Professional Technical Reference
Upper Saddle River, New Jersey 07458
www.phptr.com

Library of Congress Cataloging-in-Publication Data

Morais, Douglas H.
  Fixed broadband wireless communications : principles and practical
applications / Douglas H. Morais.
     p. cm.
  ISBN 0-13-009367-X
  1. Wireless communication systems. 2. Broadband communication
systems. I. Title.

  TK5103.2M67 2004
  621.382'1--dc22

                                        2003023686

Editorial/production supervision: Jessica Balch (Pine Tree
   Composition, Inc.)
Cover design director: Jerry Votta
Cover design: Nina Scuderi
Art director: Gail Cocker-Bogusz

Manufacturing buyer: Maura Zaldivar
Publisher: Bernard Goodwin
Editorial assistant: Michelle Vincenti
Marketing manager: Dan DePasquale
Full-service production manager: Anne R. Garcia

© 2004 Pearson Education, Inc.
Publishing as Prentice Hall Professional Technical Reference
Upper Saddle River, NJ 07458

PRENTICE
HALL
PTR

Prentice Hall PTR offers excellent discounts on this book when ordered in quantity
for bulk purchases or special sales. For more information, please contact:

U.S. Corporate and Government Sales
1-800-382-3419
corpsales@pearsontechgroup.com

For sales outside of the U.S., please contact:

International Sales
1-317-581-3793
international@pearsontechgroup.com

Other company and product names mentioned herein are the trademarks
or registered trademarks of their respective owners.

Printed in the United States of America
First Printing

ISBN 0-13-009367-X

Pearson Education Ltd., *London*
Pearson Education Australia Pty, Limited, *Sydney*
Pearson Education Singapore, Pte. Ltd.
Pearson Education North Asia Ltd., *Hong Kong*
Pearson Education Canada, Ltd., *Toronto*
Pearson Educación de Mexico, S.A. de C.V.
Pearson Education–Japan, *Tokyo*
Pearson Education Malaysia, Pte. Ltd.

*To my wife Christiane,
and our children,
Stephan and Natalie*

Contents

Preface   xiii

Acknowledgments   xv

Chapter 1 Introduction   1

    1.1   Fixed Broadband Wireless: A Definition   1
    1.2   Brief History of Fixed Wireless   1
    1.3   Primary Fixed Wireless Applications   3
        1.3.1   Mobile Cell Site Connection   3
        1.3.2   Broadband Access   4
        1.3.3   Public Network Interoffice Links   4
        1.3.4   Private Network Links   4
    1.4   Alternative Technologies   4
        1.4.1   Fiber Optic Technology   5
        1.4.2   T1/E1 Carrier   5
        1.4.3   DSL   5
        1.4.4   Cable Modem Technology   6
        1.4.5   Internet Satellite Transmission   6
    1.5   Coverage Methods   6

Chapter 2 Fixed Wireless Payload: Digital Data   9

    2.1   Introduction   9
    2.2   Circuit Switched Data   9
        2.2.1   Pulse Code Modulation   9
        2.2.2   Fundamentals of Time Division Multiplexing   13
        2.2.3   The 24-Channel Primary PCM System   15
        2.2.4   The 30-Channel Primary PCM System   15
        2.2.5   Line Codes   16
        2.2.6   Synchronous/Asynchronous Digital Multiplexing   19
        2.2.7   The Plesiochronous Digital Hierarchy (PDH): North American   19

2.2.8   *The Plesiochronous Digital Hierarchy (PDH): European   20*
2.2.9   *Limitations of PDH   20*
2.2.10   *The Benefits of a Synchronous Digital Multiplexing Hierarchy   21*
2.2.11   *Synchronous Digital Multiplexing Hierarchies   21*
2.3   Packet Switched Data   24
2.3.1   *TCP/IP   25*
2.3.2   *Ethernet Access Scheme   27*
2.3.3   *Asynchronous Transfer Mode (ATM)   29*
2.4   Digitized Video   33
2.5   The Service Channel   34
2.6   Wayside Channels   34

Chapter 3 Mathematical Tools for Digital Transmission Analysis   35

3.1   Introduction   35
3.2   Spectral Analysis of Non-periodic Functions   35
3.2.1   *The Fourier Transform   36*
3.2.2   *Linear System Response   38*
3.2.3   *Energy and Power Analysis   39*
3.3   Statistical Methods   41
3.3.1   *The Cumulative Distribution Function and the Probability Density Function   41*
3.3.2   *The Average Value, the Mean Square Value and the Variance of a Random Variable   43*
3.3.3   *The Gaussian Probability Density Function   44*
3.3.4   *The Rayleigh Probability Density Function   46*
3.3.5   *Thermal Noise   46*
3.3.6   *Noise Filtering and Noise Bandwidth   48*

Chapter 4 Fixed Wireless Digital Modulation: The Basic Principles   51

4.1   Introduction   51
4.2   Principles of Baseband Data Transmission   51
4.2.1   *Baseband Filtering   51*
4.2.2   *Probability of Error of a Binary (2-Level) PAM Signal in the Presence of White Gaussian Noise   58*
4.2.3   *Probability of Error of Multilevel PAM Signals in the Presence of White Gaussian Noise   63*
4.3   Linear Modulation Systems   66
4.3.1   *Double-Sideband Suppressed Carrier (DSBSC) Modulation   66*
4.3.2   *Binary Phase Shift Keying (BPSK)   72*
4.3.3   *Quadrature Amplitude Modulation (QAM)   75*
4.3.4   *Quaternary Phase Shift Keying (QPSK)   76*
4.3.5   *Offset Quaternary Phase Shift Keying (OQPSK)   81*
4.3.6   *High-Order 2^{2n}-QAM   81*
4.3.7   *High-Order 2^{2n+1}-QAM   89*

        4.3.8   8-PSK   95

        4.3.9   Modem Realization Techniques   96

                4.3.9.1   Line Code to NRZ / NRZ to Line Code Conversion   96

                4.3.9.2   Scrambling/Descrambling   97

                4.3.9.3   Carrier Recovery   98

                4.3.9.4   Timing Recovery   105

                4.3.9.5   Differential Encoding/Decoding   105

                4.3.9.6   Differential Phase Shift Keying (DPSK)   107

                4.3.9.7   Differential Quadrature Phase Shift Keying (DQPSK)   109

  4.4   Nonlinear Modulation, 4-Level FSK   112

  4.5   Transmission Frequency Components   121

        4.5.1   Transmitter Upconverter and Receiver Downconverter   123

        4.5.2   Transmitter RF Power Amplifier and Output Bandpass Filter   123

        4.5.3   Power Amplifier Predistorter   124

        4.5.4   The Receiver Front End   124

Chapter 5 The Fixed Wireless Path   127

  5.1   Introduction   127

  5.2   Antennas   128

        5.2.1   Introduction   128

        5.2.2   Antenna Characteristics   128

        5.2.3   Typical Broadband Wireless Antennas   130

  5.3   Free Space Propagation   134

  5.4   Received Signal Level   136

  5.5   Fading Phenomena   137

        5.5.1   Atmospheric Effects   137

                5.5.1.1   Refraction   138

                5.5.1.2   Reflection   142

                5.5.1.3   Rain Attenuation and Atmospheric Absorption   142

        5.5.2   Terrain Effects   146

                5.5.2.1   Terrain Reflection   146

                5.5.2.2   Fresnel Zones   147

                5.5.2.3   Diffraction   149

                5.5.2.4   Path Clearance Criteria   150

        5.5.3   Signal Strength versus Frequency Effects   154

                5.5.3.1   Flat Fading   154

                5.5.3.2   Frequency Selective Fading   154

                5.5.3.3   Multipath Fading Channel Model   157

        5.5.4   Crosspolarization Discrimination Degradation Due to Fading   161

  5.6   External Interference   162

  5.7   Outage and Unavailability   163

  5.8   Outage Analysis   164

        5.8.1   Rain Outage Analysis   164

        5.8.2   Multipath Fading and Interference Induced Outage Analysis   170

5.9   Diversity Techniques for Improved Availability   178
     5.9.1   Space Diversity   178
     5.9.2   Angle Diversity   182
     5.9.3   Frequency Diversity   183

Chapter 6 Link Performance Optimization in the Presence of Path Anomalies and Implementation Imperfections   189

6.1   Introduction   189
6.2   Forward Error Correction Coding   190
     6.2.1   Introduction   190
     6.2.2   Block Codes   191
         6.2.2.1   BCH Codes   193
         6.2.2.2   Reed-Solomon (RS) Codes   193
     6.2.3   Convolution Codes   195
     6.2.4   Trellis Coded Modulation (TCM)   207
     6.2.5   Code Interleaving   218
     6.2.6   Concatenated Codes   219
         6.2.6.1   Serial Concatenated Codes   219
         6.2.6.2   Iteratively Decoded Concatenated Codes
                  (Turbo Codes)   220
6.3   Adaptive Equalization   225
     6.3.1   Introduction   225
     6.3.2   Frequency Domain Equalization   225
     6.3.3   Time Domain Equalization   227
         6.3.3.1   Introduction   227
         6.3.3.2   Adaptive Baseband Equalization Fundamentals   228
         6.3.3.3   QAM Adaptive Baseband Equalization   233
         6.3.3.4   Initialization Methods   235
6.4   Cross-Polarization Interference Cancellation (XPIC)   236
6.5   Automatic Transmitter Power Control (ATPC)   237

Chapter 7 Point-to-Point Systems in Licensed Frequency Bands   241

7.1   Introduction   241
7.2   Spectrum Management Authorities   241
7.3   Frequency Band Channel Arrangements and Their Multihop Application   243
7.4   Popular Frequency Bands   247
7.5   Single- and Multichannel System Architectures   248
7.6   Typical Radio Terminals   251
     7.6.1   Witcom WitLink-2000, 16 x E1, 18 GHz, 4-CPFSK Radio Terminal   252
     7.6.2   Harris Constellation 3 x DS3, 6 GHz, 4-D 128-QAM Trellis Coded
             Radio Terminal   257
     7.6.3   Stratex Networks Altium MX 311, 2 x OC-3, 23 GHz, 256-QAM
             Radio Terminal   261

Chapter 8 Point-to-Point Systems in Unlicensed Frequency Bands 265

8.1 Introduction 265
8.2 Spread Spectrum Techniques 265
    8.2.1 Direct Sequence Spread Spectrum (DSSS) 266
        8.2.1.1 The Basic Principles 266
        8.2.1.2 Interference 271
        8.2.1.3 QPSK-Derived DSSS 273
        8.2.1.4 Pseudonoise (PN) Sequences 275
        8.2.1.5 Performance in a Multipath Environment 276
    8.2.2 IEEE 802.11b's (Wi-Fi's) Complementary Code Keying (CCK) 277
    8.2.3 Frequency Hopping Spread Spectrum (FHSS) 282
8.3 Unlicensed Frequency Bands 284
8.4 Typical Radio Terminal: The Plessey Broadband Wireless MDR 2400/5800 SR, 4 T1/E1 plus 10BaseT, 2.4/5.8 GHz, DSSS CCK Unit 288

Chapter 9 Fixed Broadband Wireless Access Systems 291

9.1 Introduction 291
9.2 Multiple Access Schemes 292
    9.2.1 Frequency Division Multiple Access (FDMA) 292
    9.2.2 Time Division Multiple Access (TDMA) 293
    9.2.3 Direct Sequence-Code Division Multiple Access (DS-CDMA) 295
    9.2.4 Frequency Hopping-Code Division Multiple Access (FH-CDMA) 297
9.3 Transmission Signal Duplexing 297
    9.3.1 Frequency Division Duplexing (FDD) 297
    9.3.2 Frequency Switched Division Duplexing (FSDD) 297
    9.3.3 Time Division Duplexing (TDD) 297
9.4 Medium Access Control (MAC) 298
9.5 Dynamic Bandwidth Allocation (DBA) 300
9.6 Adaptive Modulation 302
9.7 Non-Line-of Sight (NLOS) Techniques 302
    9.7.1 Orthogonal Frequency Division Multiplexing (OFDM) 303
    9.7.2 DS-CDMA with RAKE Receiver 312
    9.7.3 Spatial Processing Techniques 313
        9.7.3.1 Spatial Diversity (SD) 313
        9.7.3.2 Multiple Input, Multiple Output (MIMO) Transmission Scheme 314
        9.7.3.3 Adaptive Beamforming 315
9.8 Mesh Networks 318
9.9 Air Interface Standards 320
    9.9.1 IEEE 802.16 Standard, 10–66 GHz Systems (WiMAX Supported) 321
    9.9.2 IEEE 802.16a Standard, 2–11 GHz Systems (WiMAX Supported) 326
    9.9.3 IEEE 802.16e Planned Standard for Combined Fixed and Mobile Operation 329

9.9.4   *ETSI/BRAN HIPERACCESS, Systems above 11GHz*   *329*
9.9.5   *ETSI/BRAN HIPERMAN, Systems below 11GHz
         (WiMAX Supported)*   *330*
9.9.6   *IEEE 802.11a Standard, 5 GHz Systems (Wi-Fi Supported)*   *330*
9.9.7   *IEEE 802.11g Standard, 2.4 GHz Systems (Wi-Fi Supported)*   *331*
9.10   Popular Frequency Bands   331
9.11   Two Commercial PMP Systems   333
9.11.1   *The Alvarion BreezeACCESS VL, 54 Mb/s, 5.8 GHz, NLOS, OFDM,
          TDM/TDMA, PMP System*   *335*
9.11.2   *The Alvarion WALKair 3000, 26 GHz, LOS, Multicarrier
          TDM/TDMA PMP System*   *336*

Appendix   Helpful Mathematical Identities   341

Abbreviations and Acronyms   343

Index   349

PREFACE

The worldwide market for fixed broadband wireless systems, though substantial and growing, is a subset of a much larger market for broadband communications transmission systems. This broader market encompasses such technologies as digital subscriber line (DSL), coaxial cable systems, optical fiber systems, satellite systems and third generation (3G) mobile systems. The field of communication transmission is one of ever increasing complexity and sophistication. As a result, it is becoming more and more difficult to address in any detail advancements in specific areas while at the same time trying to cover the broader arena. Many excellent current texts deal with communications transmission and often touch on fixed broadband wireless. However, the author is unaware of any up-to-date publication addressed exclusively to fixed broadband wireless that provides a comprehensive coverage of the principles involved and the practical applications thereof. This text is intended to fill that void. It should be noted, however, that, despite this focus, much of the theoretical material presented is applicable to the emerging area of truly broadband portable and mobile wireless communications.

The material provided here is directed to industry professionals as well as academics. On the industry side, it should prove valuable to engineering managers, system engineers, design engineers, technicians and anyone who would benefit from a rounded understanding of fixed wireless technology in order to more effectively execute their job. On the academic side, it should provide the university student with a useful introduction to the theory of fixed wireless communication and the application thereof. It is assumed that the reader has had some previous exposure to the elementary concepts of electronics and communications. That said, the material is intentionally not rigorous so as to be friendly to a wide audience and in keeping with a desire to convey a somewhat high-level view of the technology. Mathematics, though clearly necessary for any meaningful study of the subject at hand, has been intentionally minimized. However, references are provided to allow the more inquisitive reader to explore the fundamentals in further detail. Numerous real-world examples are provided where appropriate to reinforce the practical side of the subject matter.

Chapter 1 is introductory in nature. It deals with the evolution of fixed broadband wireless, its primary areas of application, and its place in today's communications transmission landscape.

Chapter 2 describes the payload transported by fixed broadband wireless systems. This payload is all digital, typically originating from voice, video, or data sources, or a combination thereof. Though the term *broadband* is somewhat ill defined, here it is used to imply a payload bit rate of at least that of a T1 stream (i.e., 1.5 Megabits per second). Circuit switched signal formats such as PDH and SONET/SDH as well as packet switched formats such as IP, ATM, and Ethernet are reviewed.

Chapter 3 reviews some mathematical tools important for digital transmission analysis presented in later chapters.

Chapter 4 presents some of the fundamentals of digital communication theory. It then addresses, at an introductory level, modulation techniques, such as QPSK and QAM, that have proven to be the most popular in fixed broadband wireless systems.

Chapter 5 introduces the subject of microwave propagation. It describes the structure of a microwave link and the various propagation phenomena that impact system performance. It discusses how to compute link availability in the presence of these phenomena and considers diversity techniques to improve availability.

Chapter 6 addresses design techniques used to optimize performance of nondiversity (single transmitter/receiver) links in the presence of implementation imperfections and the propagation phenomena described in Chapter 5. Techniques covered include forward error correction ones such as Reed-Solomon coding, convolution coding, turbo coding, and trellis coded modulation. Also addressed are time domain and frequency domain adaptive equalization, cross-polarization interference cancellation, and automatic transmitter power control.

Chapter 7 describes point-to-point systems that operate in licensed frequency bands and that are therefore likely protected from harmful levels of direct, same frequency interference. These systems constitute by far the largest segment of the fixed broadband wireless market. Real-world examples of such systems are provided.

Chapter 8 also describes point-to-point systems but those that operate in unlicensed bands and are therefore subject to unregulated levels of direct interference. It discusses the use of spread spectrum techniques, including Wi-Fi's CCK modulation, to mitigate potential interference problems and provides an overview of a commercial product using such a technique.

Chapter 9 addresses point-to-multipoint and mesh structured systems in licensed and unlicensed bands. These systems are used primarily for "last mile" access. Though relatively new, such systems are receiving much attention and are projected to be deployed in significantly increasing quantities over the next several years. There are many state-of-the-art techniques incorporated in them, and this chapter covers the key ones, including OFDM, DS-CDMA, MIMO transmission, and adaptive beamforming. It also gives an overview of standards developed for these systems, including IEEE's 802.16/16a and ETSI's HIPERMAN, which are WiMAX supported, and ETSI's HIPERACCESS. Finally, key features of two commercial systems are outlined.

Like most modern digital-based technologies, fixed broadband wireless communications is in a constant state of evolution. However, throughout this evolution, certain principles remain fundamental to its realization. This text strives to set out those principles, as well as the latest realization techniques, in a tractable yet informative format.

ACKNOWLEDGMENTS

I would like to express a few words of gratitude. First, my sincere thanks to Dr. Kamilo Feher for putting the idea of writing this book into my head during a conversation I had with him in Lake Tahoe and for his encouragement throughout the process. Much thanks also to Steven Hirschman, Dharmendra Lingaiah, and Bernard Sklar for their thorough reviews of the entire text and many helpful suggestions. Thanks also to Paul Kennard, Richard Laine, Dr. Carlos Belfiore, and Donald Arndt for reviewing and providing helpful suggestions on Chapters 4, 5, 6, and 8, respectively. My gratitude to Dr. Roger Marks and Ken Stanwood for reviewing the section in Chapter 9 on the IEEE 802.16/16a standard.

Special appreciation to Alvarion Limited, Harris Corporation, Plessey Broadband Wireless, Stratex Networks, and Witcom Limited for graciously providing me with detailed product information and permitting publication thereof.

Finally, my gratitude to the late David J. Hadley and to William B. Farinon for opening the door for me to the wireless communications industry.

Douglas H. Morais

CHAPTER 1

INTRODUCTION

1.1 FIXED BROADBAND WIRELESS SYSTEMS: A DEFINITION

Wireless communication in today's world is diverse and ubiquitous. Its largest and most visible segment is clearly cellular mobile radio. However, long before the advent of cellular, fixed wireless systems have been a key component of the world's telecommunications infrastructure. Further, such systems continue to evolve in features and capabilities, leading to their deployment in many different types of networks. *Fixed wireless systems*, as the nomenclature implies, refers to wireless communication systems between fixed locations. As practically all such systems provided today are digital, only digital systems will be addressed in this text. In the North American standard digitized voice multiplexing hierarchy, the primary level signal results from the multiplexing of 24 digitized voice channels. Such a signal has a bit rate of approximately 1.5 Megabits per second (Mb/s) and is often referred to as a T1 signal. The European primary level signal results from multiplexing 30 digitized voice channels, has a bit rate of approximately 2 Mb/s, and is often referred to as an E1 signal. Signals, with bit rates equal to or greater than primary level rates, are commonly referred to as *broadband* signals. The fixed wireless systems that will be addressed in this book are those capable of transporting such broadband signals.

1.2 BRIEF HISTORY OF FIXED WIRELESS

The first significant deployment of commercial fixed wireless, multichannel voice systems began soon after World War II. These systems were deployed by telephone companies that had a need for transporting voice signals over long distances (hundreds of miles) economically and without the unacceptable build up of noise that accompanied land lines in use at the time. Such systems typically transported *frequency division multiplexed* (*FDM*)[1] voice channels over *frequency modulated* (*FM*)[1] or *phase modulated* (*PM*)[1] radio systems operating in the *ultra high frequency* (*UHF*) band. Typical path lengths were tens of miles. In the 1950s AT&T deployed the now classic TD-2 system.[1] It employed FDM-FM technology, was initially capable of transporting 480 voice channels in a 20 Megahertz (MHz) bandwidth RF channel, and operated in the 4 Gigahertz (GHz) band. Alongside telephone company operated public fixed wireless networks, private fixed wireless networks (i.e., networks not part of the *public switched telephone network* [*PSTN*]), were being installed globally. Private operators included entities such as railroads, oil and gas distribution utilities, electric utilities, and state and local governments. By the late 1960s FDM-FM systems were carrying from 12 to1800 voice channels per RF channel at operating frequencies ranging from 150 MHz to close to 12 GHz.

In the 1970s pressure grew for more capacity in a given bandwidth. In response to this, *single sideband amplitude modulated (SSB-AM)* systems were introduced that enabled the transmission of up to 6000 voice channels in a 30-MHz bandwidth channel. However, even as these highly spectrally efficient systems were being refined and deployed, their death knell, along with those of most other analog-based wireless systems, was being sounded. Digital wireless communications was rapidly rising above the horizon.

In the 1960s telephone companies began digitizing voice signals at their switching offices using *pulse code modulation (PCM)*[1] (Chapter 2) and multiplexing them together using *time division multiplexing (TDM)*[1] (Chapter 2). Switching also began to go digital. However, long-distance links were still all analog. To create an all-digital network, long-distance digital transmission was required. The first solution to this problem was via long-haul digital radio. Several commercial systems were introduced in the 1970s. These systems typically used relatively simple modulation techniques such as QPSK and 8-PSK (Chapter 4). They operated primarily in frequency bands between 2 GHz and 11 GHz and at the high-capacity end transmitted 90 Mb/s (the equivalent of about 1300 voice channels) in a 30-MHz RF channel. Path lengths achievable were, like their analog counterparts, in the tens of miles. The resulting spectral efficiency, however, was such that the voice traffic handling capability of these systems was less than that of existing FDM/FM analog systems and much less than the SSB-AM systems coming into vogue. A major effort was therefore launched to achieve higher spectral efficiency and capacity.[2,3] As a result, today we have systems employing modulation techniques such as 256-QAM and 512-QAM (Chapter 4) with resulting efficiencies many multiples of that of the early systems. Depending on the application, these systems have capacities ranging from a few Mb/s to over 300 Mb/s per RF carrier.

As long-haul digital radio technology was advancing in the early 1980s, another wireless technological push began. This was the move toward higher and higher transmission frequencies. Lower frequencies were becoming congested while higher frequencies lay fallow. As a result, a number of manufacturers harnessed the technology necessary to start offering digital radio systems at 18 GHz. Soon 23 GHz systems were available and, by the early 1990s, 38-GHz systems were offered. Because of the physics of atmospheric propagation (Chapter 5), links established with these higher frequency radios are limited to a few miles. Consequently, fixed wireless systems operating at these frequencies are often referred to as *short-haul* systems. They are also often referred to as *millimeter wave* systems because their signal wavelengths are in the millimeter range. Like current long-haul radio suppliers, today's short-haul radio manufacturers offer spectrally efficient systems with capacities of over 300 Mb/s per RF carrier.

As improvements in digital radio were taking place, however, another technological revolution was taking place that would forever impact the role of microwave radio in long-distance transmission. Fiber optic transmission[4] became a commercial reality in the early 1980s and has never looked back. With the ability to handle almost immeasurably more data than microwave and at costs that have, since inception, continually trended down, fiber, by the end of the 1980s, had replaced microwave as the dominant long-haul, high-capacity technology.

Fiber optic ascendancy in long-distance transmission did not, however, spell the demise of fixed wireless as an important and growing communication medium. On the contrary, it fueled its growth, albeit in new applications. Though fiber created the digital backbone of major communication systems, its cost effectiveness was and still is tied to its transporting large amounts of data. This is because the very high in-ground installation cost of fiber optic cable is independent of data capacity. As the fiber digital backbone grew in coverage and capacity (today it transports typically tens of gigabits per second [Gb/s] of data), the need arose for tributary systems where the lower capacity requirements didn't always favor fiber over microwave from a cost-effective point of view. Further, in many parts of the world, limited communication needs or geography made deployment of fiber cable simply prohibitive. In such situations fixed wireless was often, and continues to be, the only viable long-distance solution.

1.3   PRIMARY FIXED WIRELESS APPLICATIONS

1.3.1   Mobile Cell Site Connection

Probably the largest application for fixed wireless today is the connection of mobile cell sites to an operator's *mobile telephone switching office (MTSO)*. A typical such connection is shown in Fig. 1.1, where each cell site communicates via 1 T1 with the MTSO. Fixed wireless links are particularly suited for deployment here. They can be quickly installed, can be made environmentally unobtrusive, and usually result in an operating cost less than that available from any other technology. The data handled is all digital. Even older analog mobile systems normally convert voice information to digital format for communication between cell sites and the MTSO. The MTSO in turn connects to the *public switched telephone network (PSTN)*, usually via a fiber optic link or fixed wireless, allowing mobile users worldwide access.

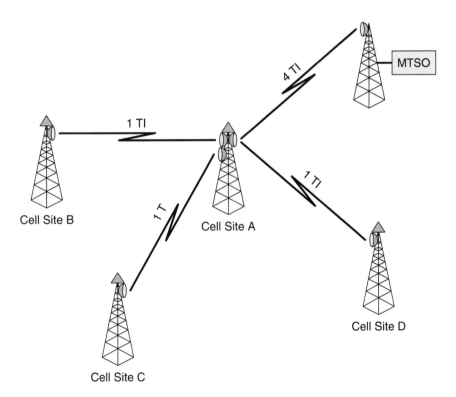

FIGURE 1.1 Typical cell site to MTSO connections.

1.3.2   Broadband Access

As data transmission via the Internet has grown over the last decade, so has the need to provide broadband access to end users. For most businesses and for many private subscribers the data transfer capability afforded by a 56 kilobits per second (kb/s) dial-up modem is simply inadequate. One of several solutions to this limitation is fixed wireless. This technology brings many advantages to the subscriber access arena, the more important of which are as follows:

- It can be rapidly deployed.
- It can provide two-way data at rates of up to tens of Mb/s.
- It can be easily expanded or reconfigured to address changing subscriber needs.
- It can distribute data transfer capability dynamically between subscribers depending on need.
- It does not rely on existing infrastructure.
- It is often, though not always, cost competitive with alternative technologies.

1.3.3   Public Network Interoffice Links

As mentioned in Section 1.2, the huge fiber optic networks now deployed owe their existence to the need for transporting vast amounts of data. However, when capacity needs don't exceed a few hundred Mb/s, telephone companies often deploy fixed wireless systems to interconnect switching offices. Such systems vary in length from a few to a few hundred miles. Geography is often a determining factor in the use of wireless. Hilly, rocky terrain usually makes fiber optic deployment prohibitive, whereas fixed wireless thrives in this environment as it facilities line-of-site between prominent protrusions.

1.3.4   Private Network Links

Like telephone-operated fixed wireless networks, private fixed wireless networks originally employed analog technology. However, as the public networks transitioned to digital, starting in the 1970s, so did the private ones. Private communication networks are usually deployed for cost, security, or reliability reasons. Fixed wireless is often an ideal medium in situations such as the following: (a) the capacity requirements are usually too low to make fiber viable, (b) security needs are easily achieved with wireless, and (c) desired reliability is attainable via a number of proven wireless techniques.

Typical operators of private networks include oil and gas distribution companies, electric utilities, railroads, state and local governments and private companies with large campuses. With the exception of those on large business campuses, these networks typically have long backbones, often hundreds of miles in length, with shorter tributaries.

1.4   ALTERNATIVE TECHNOLOGIES

For each of the aforementioned applications, there are one or more alternative technologies. Primary alternatives are from among the following:

- Fiber optic cable
- T1/E1 carrier

- Telephone company twisted pair copper wire equipped with high bit rate *digital subscriber line* (*DSL*)
- Cable TV company coaxial cable equipped with high speed cable modems
- Internet satellite transmission

1.4.1 Fiber Optic Technology

We have already touched on fiber optic technology. It is the dominant technology for public network interoffice links and broadband access to very high-density commercial buildings. Some private networks are also now being built with fiber backbones. Fiber optic technology's great advantage is its almost unlimited data bandwidth capability. Its major disadvantage is the large upfront cost associated with digging trenches and laying the cable. Wherever data transmission in the Gb/s range is required, cable is the natural choice and is usually deployed unless a right of way is not available or the terrain doesn't economically support trenching.

1.4.2 T1/E1 Carrier

T1/E1 carrier (Chapter 2) is the oldest standardized digital transmission technology in use by telephone companies and was introduced in the early 1960s. It provides two-way data transmission at 1.5/2.0 Mb/s over two twisted pairs of copper wire. It requires signal regeneration via repeaters spaced about every mile, but, with regeneration, can span distances of hundreds of miles. Service provided is reliable, but can be relatively expensive if several repeaters are required. It is an alternative technology for mobile system cell site to MTSO connection if telephone access lines are available close to the base stations. Further, despite the introduction of DSL, it is still used for commercial building access, particularly for connection to *private branch exchanges* (*PBXs*).

1.4.3 DSL

DSL is used for broadband subscriber access and comes in several varieties. Transmission is over telephone company twisted pair cable. Unlike dial-up modem connections, DSL is always on. Very common for residential use is *asymmetrical DSL* (*ADSL*).[5] As the name implies, the data capability is asymmetric. It utilizes one copper pair per subscriber. In the *downstream* direction (from the *central office* [*CO*] to the subscriber) the rate is 1.5 to 9 Mb/s, with the actual rate a function of the line condition and length. In the *upstream* direction (from subscriber to the CO) the rate is 16 to 640 kb/s, again a function of line condition and length. Even under the best conditions, the maximum distance from the central office that ADSL can be deployed is about 3½ miles. The asymmetric quality is acceptable if the user is primarily interested in Internet downloading. However, if symmetric capability is required, ADSL is clearly inadequate. One DSL solution to this problem is *high-speed DSL* (*HDSL*). HSDL has been available from the early 1990s, provides a fixed symmetrical data rate of either 1.5 Mb/s (T1) or 2 Mb/s (E1), and has been very popular with small businesses. The maximum distance that it can be deployed from the CO is about 2½ miles. One negative is that it requires two copper pairs per subscriber. To overcome this limitation, a new version is now being offered called HDSL2. It utilizes only one copper pair and supports service up to about 4½ miles from the CO.

1.4.4   Cable Modem Technology

Cable modem technology, also used for broadband subscriber access, is in some ways similar to ADSL. It provides always-on, asymmetric data to the subscriber. In theory, it can provide downstream rates of up to about 36 Mb/s in U.S. standard systems and 50 Mb/s in European standard systems. In the upstream direction the maximum data rate is about 10 Mb/s. A significant limitation to this technology, however, is that the actual access speed is a function of the number of subscribers on the system at a given time. As the number of subscribers increases, the access speed available to each subscriber decreases. Further, cable TV companies often limit the speeds to about 1.5/2.0 Mb/s. This is because (a) the practical processing speed of most PCs is well below the maximum downstream rates indicated previously, and (b) the cable system may be connected to the Internet via a line of data rate less than the theoretical downstream maximum. Unlike ADSL, cable modem performance is almost independent of its distance from the central cable office. An obvious limitation to its deployment is that it is only available where cable services exist. Cable companies have shown no interest to date in offering cable modem Internet access to areas not served with basic cable offerings.

1.4.5   Internet Satellite Transmission

Internet satellite transmission, again used for broadband subscriber access, was first applied commercially to Internet access in the mid-1990s. The original offering, called DirecPC and provided by Hughes Network Systems, is still in service. It provides one way satellite downstream transmission at a rate of 400 kb/s. Upstream transmission is achieved via dial-up modem over the public network at a maximum rate of 56 kb/s. In locations where no other high-speed Internet access offering is available, the system has found acceptance. However, when in direct competition with DSL and/or cable modem technology it has never been a favorite. A new option, however, appears to be improving the acceptance of satellite access. This new option is two-way satellite transmission that's always on. Services now being offered include downstream speeds of up to 500 kb/s and upstream speeds of up to 256 kb/s. Though still less than DSL and cable modem offerings, these speeds are believed to be more in the ballpark of acceptability. And in areas where DSL and cable modem service are not available, it represents the first real taste of two-way high-speed interconnection. Because of the long distances that the downstream and upstream signals have to traverse via the satellite, the signals suffer noticeable time delay, up to about ¼ second per direction. Depending on how the Internet is being accessed (for example, if it's being used for an interactive game) this delay could prove to be irritating.

1.5   COVERAGE METHODS

Traditionally, fixed wireless systems have been configured as a series of *point-to-point* (*PTP*) links. They tend to be multiple-link backbone systems with tributaries as required, with the exact layout defined heavily by the topography. Figure 1.2 shows a typical PTP system layout that includes a four-hop backbone between Terminal A and Terminal E, two single-hop tributaries or spurs off Repeater 1, and a two-hop

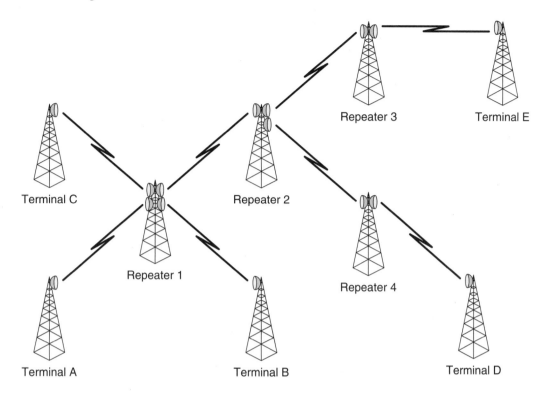

FIGURE 1.2   A point-to-point wireless system.

tributary off Repeater 2. Detailed information on PTP links is provided in Chapters 5, 7, and 8.

For certain applications, coverage architectures have evolved away from this tradi-tional PTP approach. Most prominent among newer architectures is the *point-to-multipoint* (*PMP*) configuration. Here a common base station transceiver communicates via a single antenna with a plurality of remote transceivers. This structure is essentially the same as that used in mobile cellular communications, with the exception that the re-mote stations are fixed. As a result, much of the technology developed for mobile commu-nications is directly applicable here. The main difference is that fixed systems are inherently more reliable as they suffer far less from signal fading. Figure 1.3 shows a typi-cal PMP fixed wireless cell comprising three sectors, with all remote units in a particular sector communicating with one antenna at the base station. PMP systems are covered in detail in Chapters 9.

A recently introduced and somewhat intriguing network configuration is the *mesh network*. Mesh networks are in fact comprised of PTP links but are configured in such a way that, via software control, traffic can be routed over several different paths. This al-lows routing around line of site obstructions, for example, and rerouting should a link go down. More information on these networks is also provided in Chapter 9.

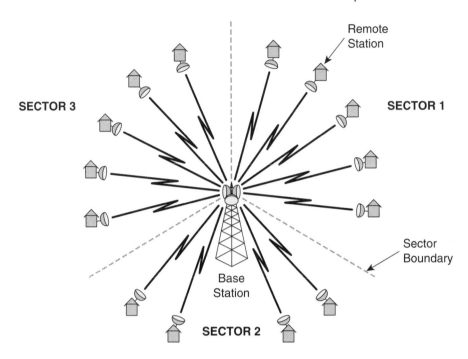

FIGURE 1.3   Point-to-multipoint three-sector wireless system.

REFERENCES

1. Members of the Technical Staff, Bell Telephone Laboratories, *Transmission Systems for Communications*, Revised Fourth Edition, Bell Telephone Laboratories, Inc., 1971.
2. Feher, K., guest editor, Special Issue on Digital Radio, *IEEE Trans. Commun.,* Vol. COM-27, Dec. 1979.
3. Hartmann, P. R., "Digital Radio Technology: Present and Future," *IEEE Communications Magazine,* Vol. 19, July 1981.
4. Winch, R. G., *Telecommunication Transmission Systems*, McGraw-Hill, 1993.
5. Maxwell, Kim, "Asymmetric Digital Subscriber Line: Interim Technology for the Next Forty Years," *IEEE Communications Magazine,* Vol. 34, Oct. 1996.

CHAPTER 2

FIXED WIRELESS PAYLOAD: DIGITAL DATA

2.1 INTRODUCTION

In this chapter we will explore the more common types of digital data transported by fixed broadband wireless systems. The source of this data can be voice, video, computer derived, and so on. However, for efficient transmission, data is always transported in a standardized form. We will examine the North American and European standard digital hierarchies. Traditional telephone communication is based on *circuit switching*, where circuits are set up and kept continually open between users until the parties end their communications and release the connection. Traditional data communication is based on *packet switching*, where the data is sent in discrete packets. Here, unlike circuit switching, a circuit is configured and left open only long enough to transport an addressed packet, then freed up to allow reconfiguration for transporting the next packet, which may be from a different source and addressed to a different destination. As fixed wireless is today being used to support both circuit switched and packet switched networks, we will review data formats transported over both types of switched networks.

2.2 CIRCUIT SWITCHED DATA

2.2.1 Pulse Code Modulation

Pulse code modulation[1] is a means of converting analog signals into digital ones. Thus, with PCM, analog voice signals and analog video (image) signals can be converted into digital format. PCM is achieved via the principles of sampling, companding, quantizing, and coding. A simplified end-to-end PCM transmission system is shown in Fig. 2.1.

    Low-pass filtering is the first step in the conversion of an analog signal into a digital one. This filtering limits the highest frequency of the signal to f_m, say. Filtering is followed by sampling. As the name implies, the latter involves the sampling of the signal amplitude at regular intervals of time, T_s say, as shown in Fig. 2.2. The sampling frequency, f_s samples per second say (note that $f_s = 1/T_s$) is set by the *sampling theorem*.[2] This theorem states that f_s must be at least twice f_m, the highest frequency of the analog signal being sampled, if the analog signal is to be retrieved at a receiver without impairment. In the typical telephony PCM system, voice signals are limited to a maximum frequency of approximately 3.4 kb/s, and the sampling rate is set at 8 kilo samples per second (ks/s).

9

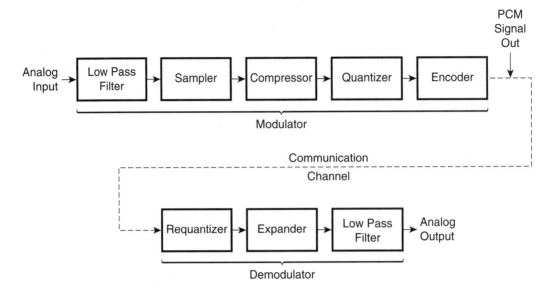

FIGURE 2.1   Simplified PCM transmission system.

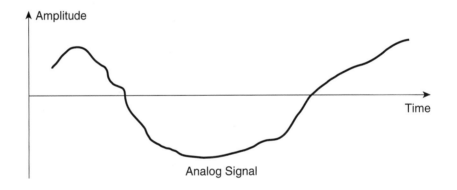

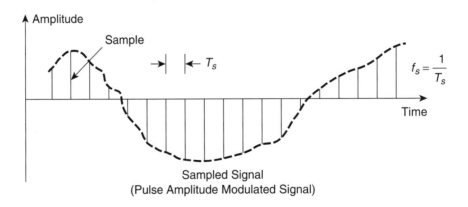

FIGURE 2.2   Sampling of analog signal.

The sampled signal is called a *pulse amplitude modulated (PAM)*[2] signal, and each signal sample has an amplitude that is one of an infinite number of levels. However, in order to digitally encode the signal sample, each amplitude level must be restricted to one within a finite set. This process is called *quantization*. Figure 2.3 illustrates the basic concepts of quantization, where, for simplicity, only eight quantization levels are shown. The number of quantization levels is always a power of 2, for a reason that will become clear as we proceed. The quantized sample assumes the value of the nearest quantization level to the signal sample. The difference in amplitude between the quantized level and the signal sample is called *quantization error*. In Fig. 2.3, the first signal sample shown has a value of 1.3 and the nearest quantization level is 1.5. Thus, the resulting quantized sample has a value of 1.5 and the quantization error is 0.2. The more quantization levels, the closer the quanitzed signal level will be to the sampled signal level and hence the less quantization error that will be introduced into the process. For voice signals it has been found that telephone company signal quality can be achieved with 256, or 2^8, quantization levels. However, for this to be realized, additional signal processing, referred to as *companding*, is required.

Intuitively, it is clear that a large varying signal ranging over all or most of the quantization levels will acquire an average signal to quantization noise ratio that is larger

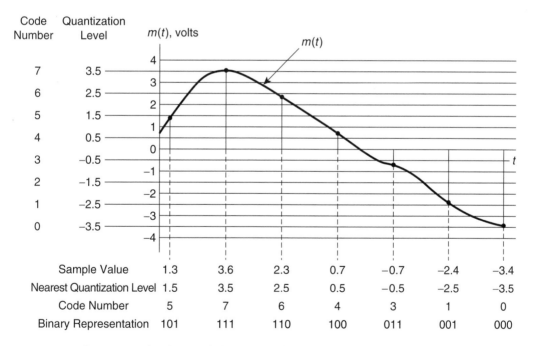

A message signal is regularly sampled. Quantization levels are indicated.
For each sample the quantized value is given and its binary representation is indicated.

Sample Value	1.3	3.6	2.3	0.7	−0.7	−2.4	−3.4
Nearest Quantization Level	1.5	3.5	2.5	0.5	−0.5	−2.5	−3.5
Code Number	5	7	6	4	3	1	0
Binary Representation	101	111	110	100	011	001	000

FIGURE 2.3 Example of quantization and binary encoding in a PCM system. (From Taub, H., and Schilling, D., *Principles of Communication Systems*, McGraw-Hill, Inc., 1971 and reproduced with the permission of the McGraw-Hill Companies.)

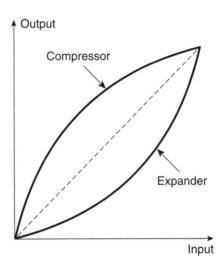

FIGURE 2.4   Compressor and
expander input–output relationship.

than that acquired by a small signal ranging over only a few quantization levels. This is
because although the average quantization error per sample is the same, the large signal
will have a larger average value. In voice telephony, signal levels vary significantly from
speaker to speaker. To minimize signal to noise degradation when low signals are transmitted, a technique called *companding*[2] is used. Companding is achieved by placing a
nonlinear input–output device called a *compressor*[2,3] ahead of a uniform step size quantizer. The nonlinearity is such as to amplify small signal levels more than large ones. The
transfer function of a typical compressor looks similar to that shown in Fig. 2.4. In effect, by amplifying the small signals more than the large ones, the compressor fools the
quantizer into believing it is processing a large signal even when it is processing a small
one. As a result, improved signal to noise for small signals is achieved. In fact, the compressor normally used for telephone company class communication results in an improvement in signal-to-noise ratio (S/N) of approximately 26 dB for systems with 256
quantization levels.

The final step in creating a digital signal is encoding. In encoding, the quantized
sample is simply converted to a binary signal that represents its level. Thus, for an eight-
level quantization system, for example, each level is represented by a three-digit binary
signal. Clearly, to convert the quantized levels to binary format with the minimum number
of bits per sample, the number of quantized levels needs to be a power of 2. In Fig. 2.3 we
see that the eight quantization levels have each been assigned a code between 0 and 7.
Each code number has a binary arithmetic representation ranging from 000 for code number 0 to 111 for code number 7. For the first quantization sample, where the level is 1.5,
we note that the corresponding code is 5 and thus the binary representation is 101.

Based on the preceding, the transmission rate of a standard voice channel can be
calculated. In the sampling process 8 ks/s are created. Each sample is quantized into one
of 256 quantization levels. To code this number of quantization levels, 8 bits of information are required for each quantized sample ($2^8 = 256$). Thus, one voice channel sampled
at 8 ks/s and generating 8 bits per sample results in a transmission rate of 64 kb/s.

For a receiver to convert PCM data back to its original format, the process is essentially the reverse of that used to create the data. First, the received binary data stream is converted to a stream of pulses of amplitude equal to the original quantized samples and of rate equal to the original sampling rate. This process is called *requantization.*[2] Of course, should there be errors in the binary data stream as a result of transmission, these errors will show up as incorrect quantization levels in the recreated stream. Next, the requantized signal is passed through an *expander.*[2,3] An expander is a nonlinear input–output device with a transfer function such that if its input were to come directly from the output of the associated compressor, then its output would be the same level as the input level of the compressor (see Fig. 2.4). As a result, its output is similar to the original PAM signal fed into the compressor. Finally, the expanded quantized signal is converted to the original analog signal by passing it through a low-pass filter of bandwidth f_m. To understand why this filtering results in the original analog signal requires a mathematical analysis of the PAM signal,[2] which shows that it consists of the original analog modulating signal plus modulated harmonics of the sampling frequency, where the lowest frequency of the modulated first harmonic exceeds f_m.

2.2.2 Fundamentals of Time Division Multiplexing

Rarely is a single PCM voice signal transmitted over a fixed wireless system. Instead, several voice signals are combined sequentially in time. This process of sequential combining is called time division multiplexing (TDM).[1,2] Figure 2.5(a) shows a simplified TDM-PCM system where four voice channels are multiplexed. It will be noticed that the system is very similar to the single channel system shown in Fig. 2.1. The only difference on the transmitter side is the input to the compressor. Instead of being PAM samples from one filtered analog signal, it is instead a repetitive sequence of PAM samples as shown in Fig. 2.5(b), each sample in the sequence being from a different filtered signal. This sequencing is made possible by sampling the different signals fast enough that all pulses can be stacked sequentially in time before the first signal in the sequence has to be sampled again. In the system shown in Fig. 2.5(a) this is achieved via a rotary switch called a *commutator.* At the receiver, the reverse process takes place. At the output of the expander the quantized samples from each individual signal are distributed to separate outputs via another rotary switch, here called a *distributor.* The separated samples are then filtered, resulting in signals similar to the original analog signals. Clearly, for the system to work, the timing actions in the receiver must be carefully synchronized with those in the transmitter. This is accomplished via timing recovery circuits. Assume that the voice signals in the system shown are individually sampled at 8 ks/s and that each sample is coded into an 8-bit word. Then, given that there are four signals being sampled sequentially, the composite PCM signal transmitted by this system has a rate of 256 kb/s. In practical systems, additional bits are normally added to the binary output of the encoder. For example, so called *framing bits* are added. These allow the identification of *frames*, the grouping of codewords that represent a full sequence of coded voice signals. *Signaling bits* are also added to allow the transmission of signaling information, such as a ring signal or an off-hook signal, associated with each channel.

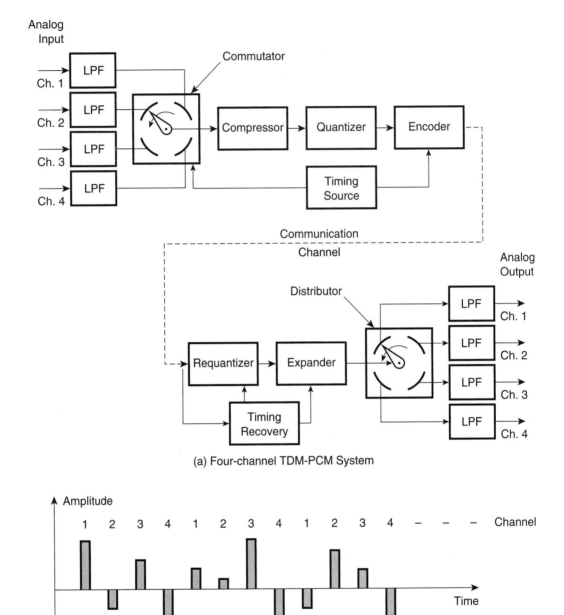

(a) Four-channel TDM-PCM System

(b) Typical Commutator Output

FIGURE 2.5   Simplified TDM-PCM transmission principles.

2.2.3   The 24-Channel Primary PCM System

In North America and Japan, the first standard level of PCM voice channel multiplexing is one where 24 channels are multiplexed. As each voice signal is sampled at a rate of 8 ks/s, the sampling interval, T_s, is 125 μs. Since all signals have to be sampled in this time, the *frame period* is also 125 μs. The frame structure is shown in Fig. 2.6. This frame contains 24 channel time slots numbered 1 to 24, one from each channel, each such slot having an 8-bit codeword, often referred to as a *byte*. Thus the channels contribute 24 bytes, or 192 bits to the frame. It will be observed that the multiplexing is done on a codeword-by-codeword basis. This is called *word-by-word interleaving,* or sometimes *sequential byte interleaving*. At the start of every frame, one additional bit is added for framing purposes. Thus a complete frame contains 193 bits. As there are 8000 frames per second, the multiplexed signal has $193 \times 8000 = 1.544$ Mb/s. If only one bit is added per frame, then how is signaling achieved? It is accomplished by robbing bits from the information data. The first 7 bits in each channel time slot are always used for voice encoding. The eighth bit is also used for voice encoding, but only in the first five out of every successive six frames. In the sixth frame, the eighth bit in each channel time slot is used for signaling associated with the channel occupying the channel time slot. Since there are 8000 frames a second but only every sixth frame carries a signaling bit per channel time slot, there are $8000/6 = 1333$ signaling bits/s per channel.

2.2.4   The 30-Channel Primary PCM System

In European PCM systems, the first standard level of voice channel multiplexing is one where 30 channels are combined. Like the North American system, the sampling rate for each voice signal is 8 ks/s and hence the frame period is also 125 μs. The frame structure is shown in Fig. 2.7. Note that, like the 24-channel system, word-by-word interleaving is

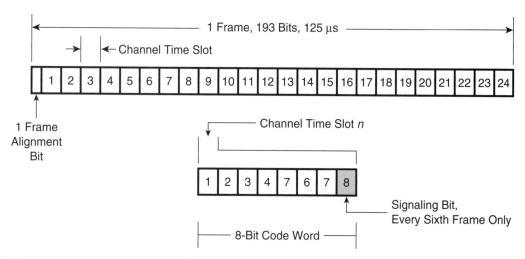

Note: Each time slot contains 8 bits.

FIGURE 2.6   24-Channel PCM frame structure.

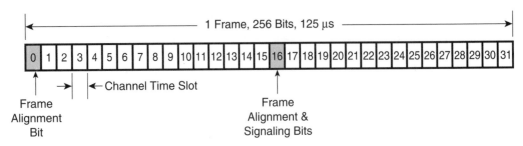

Note: Each time slot contains 8 bits.

FIGURE 2.7   30-Channel PCM frame structure.

used. Each frame contains 32 (not 30) channel time slots, each containing an 8 bit codeword. Observe that the time slots are, by convention, numbered from 0 to 31. Channel time slots 1 to 15 and 17 to 30 are used for voice channel generated codewords. The additional two channel time slots are used for framing and signaling. Specifically, channel time slot 0 is used for framing purposes and channel time slot 16 is used for both framing purposes and for transmitting signaling information for each of the 30 voice channels. The multiplexed signal has 8000 (samples) × 32 (channel time slots) × 8 (bits per time slot) = 2.048 Mb/s.

2.2.5   Line Codes

A PCM multiplexed signal consists of a bit sequence of 1s and 0s. Typically these signals are formatted as shown in Fig. 2.8. Signals with such a format are called *unipolar, non-return to zero* (*NRZ*) signals: unipolar because the signal has only two states, one of which is at zero voltage; NRZ because in bit state 1 the signal level is a constant real voltage for the entire bit period (i.e., it does not return to zero for the entire bit period). Such a signal format is not ideal for transmission as it contains a DC component that cannot pass through AC coupled media. To overcome this limitation the signal format is normally changed. There are several types of signal formats, called as a class *line codes.*[1] In this re-

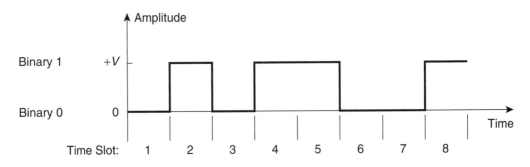

FIGURE 2.8   A unipolar NRZ signal.

view, only those codes typically found in multiplexers or other digital signal generating equipment that interface with fixed wireless systems will be reviewed. Clearly, for the wireless system designer, exact knowledge of the signal format is essential if such a signal is to be appropriately accepted at the transmit end, manipulated for transmission, and re-created at the receive end.

Alternative Mark Inversion (AMI) Code. This code is reviewed first because it's the code recommended for the 24 voice channel, 1.544 Mb/s interface. An AMI coded signal is shown in Fig. 2.9. It is seen that in this code 0s are represented by 0 voltage over the entire bit period whereas 1s are represented by alternative positive and negative pulses over the first half of the bit period. This code is also referred to as the *bipolar code*. Unfortunately, if a long sequence of 0s is present, the signal appears over this interval as no signal at all and clock recovery becomes difficult.

HDB3 Code. The term *HDB3* stands for high-density binary with a maximum of three consecutive 0s. This code is the one recommended for the 30 voice channel, 2.048 Mb/s interface. An HDB3 coded signal is also shown in Fig. 2.9. As can be seen, it is somewhat similar to the AMI code. However, it is constructed in such a way as to limit to three the number of 0s outputted from a long sequence of 0s inputted. As a result, and

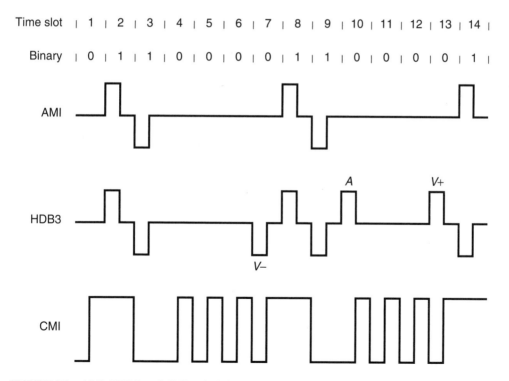

FIGURE 2.9 AMI, HDB3, and CMI coded binary signals.

unlike AMI, clock recovery in the receiver is assured. Longer sequences of more than three 0s are eliminated by the replacement of one, and in some cases two 0s by 1s according to a given set of rules. These rules are as follows:

1. A binary one is coded the same as for AMI (i.e., either a positive or negative pulse, and of opposite sign to that of the preceding pulse).
2. For strings of three 0s or less, a binary 0 is coded the same as for AMI (i.e., by a zero voltage).
3. For strings of four 0s or more, the zero voltage of the fourth zero is replaced with a pulse that violates the AMI rule.
4. Every violation pulse must change polarity with the previous violation pulse.
5. If complying with rule 4 results in a violation pulse that does not violate the AMI rule, then the zero voltage of the first zero is changed to a pulse in accordance with the AMI rule. As a result of this last change, the violation pulse will now violate the AMI rule.

In Fig. 2.9:

Violation pulse $V-$ in time slot 7 is a result of rule 3.
Violation pulse $V+$ in time slot 13 is a result of rule 4.
Pulse A in time slot 10 is a result of rule 5.

B3ZS Code. The term *B3ZS* stands for binary 3 zero substitution. This code is the one recommended for the North American 672 voice channel, 44.736 Mb/s interface. It is almost identical to the HDB3 code, the only difference being that for the B3ZS code the number of 0s in a sequence is limited to two, with the third 0 being substituted with a pulse that violates the AMI rule.

B6ZS Code. This code, the binary 6 zero substitution code, is recommended for the North American 96 voice channel, 6.312 Mb/s interface if symmetric pair wire is used. It is a modified AMI code where the number of consecutive 0s is limited to five. In it, a group of six consecutive 0s, preceded by a positive pulse (+), is replaced with 0 + –0 – +. If preceded by a negative (–) pulse, then the group of six 0s is replaced by 0 – +0 + –.

B8ZS Code. This code, also a modified AMI code, is recommended for the 6.312 Mb/s interface if coaxial cable is used. It is also recommended for the 1.544 Mb/s interface if symmetric pair wire is used. It limits the number of consecutive 0s to seven. In it, a group of eight consecutive 0s, preceded by a positive pulse (+) is replaced with 000 + –0 – +. If preceded by a negative (–) pulse, then the group of eight 0s is replaced by 000 – +0 + –.

Coded Mark Inversion (CMI) Code. A CMI coded signal is illustrated in Fig. 2.9. This code is the one specified for the European 1920 voice channel, 139.264 Mb/s interface. It is not a variant of the AMI code as were the preceding four. Here all zero bits are coded as a negative value in the first half of the bit interval and as a positive value in the second half of the bit interval. The one bits are coded as a fixed value over the entire bit interval, this value being alternated between positive and negative with succeeding bits.

2.2.6   Synchronous/Asynchronous Digital Multiplexing

The 24-channel and 30-channel PCM multiplexers described previously are *synchronous digital multiplexers*. Synchronous multiplexers combine bit streams, or *tributaries*, originating from different sources that are all synchronized off a master timing device. Thus the bit streams being combined can be interleaved with one another in a predictable and nonvariant time relationship. The term *synchronous digital hierarchy* is commonly used to describe a digital multiplexing hierarchy based on synchronous operation. Note, however, that the official name for the European synchronous digital hierarchy is also *Synchronous Digital Hierarchy*[3-5] (*SDH*), whereas the North American synchronous digital hierarchy is called the *Synchronous Optical NETwork*[3-6] (*SONET*). To avoid misunderstanding, when referring to the *synchronous digital multiplex* hierarchy in the generic sense, this text will abbreviate it to *SDM* to differentiate it from SDH.

The word *asynchronous* means "not occurring at the same time." Asynchronous digital multiplexers combine bit streams, or tributaries, from different sources that all have the same nominal bit rate but are not tied to a common timing device. As a result, the individual bit rates vary slightly relative to each other. Multiplexing these asynchronous streams clearly presents a challenge. The problem is resolved by employing *bit-by-bit interleaving* as well as *pulse stuffing*, otherwise known as *justification*. Pulse stuffing is a complex process and will not be described in any detail here. Suffice it to say, however, that prior to bit-by-bit interleaving, the individual input bit streams are all brought up to the same identical bit rate by stuffing them with additional bits. In effect, the input streams are made synchronous! The number of bits added to each stream is a variable as it depends on the stream's bit rate at the time of stuffing. The stuffed bits, alternatively called *justification bits*, are added in such a fashion that they are recognized and discarded in the demultiplexing process, leaving only the original tributaries. As a result of stuffing, the multiplexed output bit rate is a predetermined, slightly higher one, than the sum of the input bit rates.

It has become common in the telecommunications multiplexing world to replace the word *asynchronous* with *plesiochronous*, after the Greek meaning "almost synchronous," as indeed, the bit streams being mulitplexed have the same nominal rate (i.e., are "almost synchronous"). Thus the term most often used today to describe a digital multiplexing hierarchy based on asynchronous operation is *plesiochronous digital hierarchy*, or *PDH*.

2.2.7   The Plesiochronous Digital Hierarchy (PDH): North American

Table 2.1 summarizes the first four levels of the North American PDH and associated multiplexing. It should be noted that modern North American rate PDH fixed wireless systems typically provide inputs that are either multiples of T1 or multiples of T3. The T2 rate is now hardly ever used. Also note that although by the official hierarchy a T3 signal is created by multiplexing seven T2s, fixed wireless suppliers normally provide multiplexers that skip the T2 level, multiplexing 28 T1s directly up to a T3. A multiplexer that multiplexes level X signals to a level Y signal is referred to as a MX-Y multiplexer. Thus one that multiplexes 28 T1s to 1 T3 is called a M1-3 multiplexer.

TABLE 2.1   First Four Levels of North American Plesiochronous Digital Hierarchy and Associated Multiplexing

Level	Multiplexer Nomenclature	Equivalent No. of Input VCs	Input Signals	Output Bit Rate (Mb/s)	Output Nomenclature	Output Line Code
1	24 Ch. PCM	24	24 VCs	1.544	DS1/T1	AMI or B8ZS
1C	48 Ch. PCM	48	48 VCs	3.152	DS1C/T1C	AMI or B8ZS
2	M1-2	96	4XDS1	6.312	DS2/T2	B6ZS or B8ZS
2	M1C-2	96	2XDS1C	6.312	DS2/T2	B6ZS or B8ZS
3	M2-3	672	7XDS2	44.736	DS3/T3	B3ZS
3	M1-3	672	28DS1	44.736	DS3/T3	B3ZS

2.2.8   The Plesiochronous Digital Hierarchy (PDH): European

Table 2.2 summarizes the first four levels of the European PDH and associated multiplexing. European rate PDH fixed wireless systems provide inputs that are multiples of E1, E2, or E3. Also, as with North American rate systems, suppliers of European rate fixed wireless systems offer M1-3 multiplexers that skip the second level, multiplexing 16 E1s directly up to an E3.

2.2.9   Limitations of PDH

An obvious benefit of PDH is that the input tributaries don't have to be under the same master clock control. However, PDH suffers from a number of shortcomings. One very major one is that the use of justification bits at each level in the hierarchy makes it impossible to access a first-level tributary from a high-level multiplexed stream without a complete demultiplexing of the signal. For example, consider a 140 Mb/s system operating between two distant cities, with several repeaters at towns in between. In order to access a single 2 Mb/s tributary from such a system at a repeater, the 140 Mb/s signal must be demultiplexed to four 34 Mb/s lines, one of these demultiplexed to four 8 Mb/s lines, and, finally, one of these demultiplexed to four 2 Mb/s lines. Once the desired 2 Mb/s line has

TABLE 2.2   First Four Levels of European Plesiochronous Digital Hierarchy and Associated Multiplexing

Level	Multiplexer Nomenclature	Equivalent No. of Input VCs	Input Signals	Output Bit Rate (Mb/s)	Output Nomenclature	Output Line Code
1	30 Ch. PCM	30	30 VCs	2.048	E1	HDB3
2	M1-2	120	4XE1	8.448	E2	HDB3
3	M2-3	480	4XE2	34.368	E3	HDB3
3	M1-3	480	16XE1	34.368	E3	HDB3
4	M3-4	1920	4XE3	139.264	E4	CMI

been accessed, the remaining lines must be remultiplexed back up to 140 Mb/s for continuation of the transmission.

A second major shortcoming of PDH is its inadequate provision for network monitoring and control. The larger the network involved, the more this limitation is felt.

2.2.10 The Benefits of a Synchronous Digital Multiplexing Hierarchy

To overcome the limitations of PDH, the concept a standard synchronous digital multiplexing hierarchy was aggressively pursued in the 1980s. Commercial deployment of such systems began in the mid-1990s and now it is the dominant digital multiplexing hierarchy being deployed in major networks worldwide. Its benefits include the following:

- The ability to access directly lower-level tributaries in high-level signals without going through multiple demultiplexing/multiplexing processes.
- Extensive provisioning for operations, administrative, and maintenance (OA&M) functions. Note that monitoring and control fall within these functions.
- Multiplexers that accept a wide variety of tributaries, including all PDH bit rates up to 140 Mb/s.
- The provision of standard interfaces, greatly simplifying the international connection of digital networks and permitting multivendor interoperability.

2.2.11 Synchronous Digital Multiplexing Hierarchies

Synchronous digital multiplexing (SDM) hierarchies were created based on the principles of sequential byte interleaving and a four-layered network architecture. These four layers, starting from the lowest layer, are the photonic, the section, the line, and the path layer. Figure 2.10 shows a typical SDH transmission system, indicating a path, lines, and sections, with an SDH microwave radio as one section.

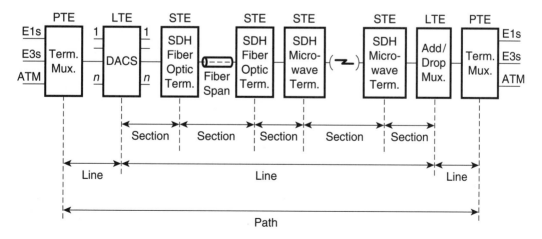

FIGURE 2.10   Typical SDH transmission system.

The *photonic layer* deals with the transport of bits across the physical media and is so named because that media is normally optical fiber. Note, however, that fixed wireless is also used in synchronous systems as the transport media. No overhead is associated with the photonic layer.

The *section layer* deals with the transport of an SDM frame over a section. A section is a single link spanning two SDM network elements, with such elements capable of accessing, generating and processing section overhead only. These network elements, referred to as *Section Terminating Equipment* (*STE*), are interconnected via a physical medium such as an optical fiber system or a fixed wireless system. An example of section terminating equipment is a line regenerator or transmission equipment containing a line regenerator such as an SDH fiber optic terminal or an SDH radio terminal.

The *line layer* deals with the transport of an SDM frame over a line. A line consists of one or more sections, including network elements at each end, with one or both capable of accessing, generating, and processing line overhead. These network elements are referred to as *Line Terminating Equipment* (*LTE*) and typically include add/drop SDM multiplexers and SDM *digital access and cross-connect system*s. (A digital access and cross-connect system [DACS] is one with *n* digital signal inputs and *n* digital signal outputs and capable, via software, of connecting any input port to any output port.)

The *path layer* deals with the transport of an SDM frame over a path. A path is one or more lines, including network elements at each end, capable of accessing, generating, and processing path overhead. These network elements are referred to as *Path Terminating Equipment* (PTE), a typical example of which is an SDH terminal multiplexer.

As indicated earlier, the North American synchronous digital multiplexing hierarchy is called SONET. Its first level is called the *Synchronous Transport Signal–Level 1* (*STS-1*) if electrical, the *Optical Carrier–Level 1* (*OC-1*) if optical, and is synchronized to the network master clock. Its frame structure, shown in Fig. 2.11, consists of a matrix of 90 columns and 9 rows, with each matrix element being an 8-bit byte. There are thus 810 bytes, or 6480 bits per frame. Bytes are transmitted row by row, from left to right and, with a frame repetition rate of 8000 frames per second, the entire frame is transmitted in 125 μs. The STS-1 signal rate is thus $6480 \times 8000 = 51.84$ Mb/s.

The SONET overhead is divided into section, line, and path layers. The first three columns of the STS-1 frame contain section and line overhead bytes. The remaining 87 columns are used to carry the STS-1 *Synchronous Payload Envelope* (*SPE*), which includes 9 bytes of path overhead. The STS-1 SPE is the container for carrying information inside a STS-1 signal and has a bit rate of 50.112 Mb/s. It can carry a wide variety of payloads, including a mixture of DS1, E1, DS1C, DS2, E3, and DS3 frames as well as ATM cells (Section 2.3.3). Thus a SONET terminal multiplexer is much more flexible than a PDH multiplexer, which only multiplexes PDH bit streams of equal rate.

Higher-level SONET signals are created by byte interleaving *N* STS-1 signals to form an STS-*N* signal. Figure 2.12 shows an STS-*N* frame structure. Thus an STS-3 signal has 270 columns and nine rows, and a bit rate of $51.84 \times 3 = 155.52$ Mb/s.

The European hierarchy is called, as previously noted, the Synchronous Digital Hierarchy (SDH). Its first level is called *Synchronous Transport Module–Level 1* (*STM-1*). It is the equivalent of and compatible with an STS-3 and thus has a bit rate of 155.52 Mb/s. However, following its original conception, a so-called Sub-STM or STM-0 level, which is in effect the STS-1 level, has been added for use in fixed wireless links and satellite

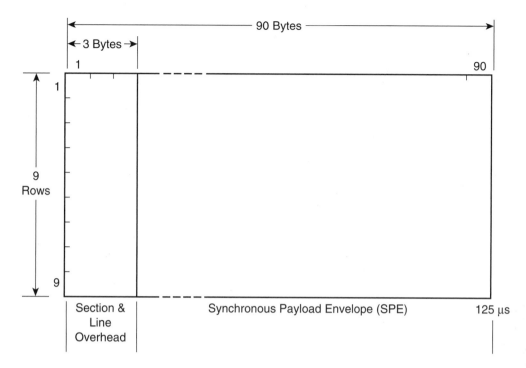

FIGURE 2.11   STS-1 frame.

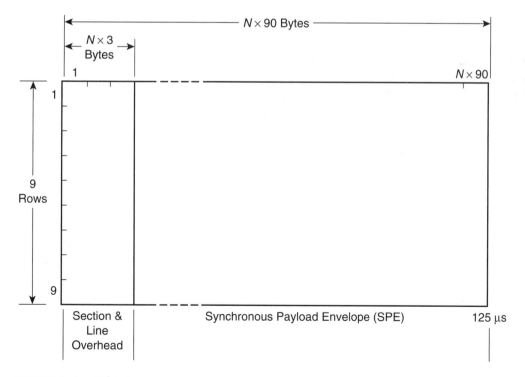

FIGURE 2.12   STS-N frame.

23

TABLE 2.3   **Synchronous Signals Transported by Fixed Wireless Systems**

| Optical Signal | Electrical Signal | | Bit Rate (Mb/s) | Line Code |
	Synchronous Transport Signal	Synchronous Transport Module		
OC-1	STS-1	STM-0	51.84	B3ZS
OC-3	STS-3	STM-1	155.52	CMI
OC-12	STS-12	STM-4	622.56	CMI

connections. The STM-1 SPE can carry, in addition to the PDH frames carried by an STS-1 SPE, the E4 (140 Mb/s) frame.

Table 2.3 shows the line rates and associated line codes for standard STS and STM signals transported by fixed wireless systems. Higher-rate systems become exceedingly difficult to transport because of the tremendous RF bandwidth that would be required. Also shown in Table 2.3 are the associated optical levels. Many synchronous rate fixed wireless systems offer an optical interface option to the external network as the most prevalent means of transporting synchronous rate data is via optical networks.

2.3   PACKET SWITCHED DATA

In packet switched networks,[5,7] data to be communicated is sent in discrete packets, each packet being a sequence of bytes. Contained within each packet is its address and the address of the intended recipient. The individual packet seeks the best available route to its destination, and after arrival, the elements of that route are available for other packets from other sources. In some packet switched networks the packets at the receiving end don't necessarily arrive in the order sent, so they must be reordered before further use. It is not the intent to investigate packet data communications and its many protocols in any detail here. However, a concise overview is in order as more and more fixed wireless systems, particularly in the form of wide area network (WAN) interconnection and broadband subscriber access links, are being used to convey such data. The demand for fixed wireless systems with packet switched data handling capability is being driven largely by the explosive growth of the Internet. The Internet communicates with the aid of a suite of software network protocols called *Transmission Control Protocol/Internet Protocol*, or *TCP/IP*. These protocols are, in essence, preestablished rules of communication. With TCP/IP, data to be transferred along with administrative information is structured into a sequence of so-called *datagrams*, each datagram being a sequence of bytes. TCP/IP works together with a physical data link layer to create an effective packet switched network for Internet communications. The data link layer protocol puts the TCP/IP datagrams into packets (it's like putting an envelope into another envelope) and is responsible for the reliable transmission of packets over the physical layer. If the datagrams are too large to fit into the packets, they are broken down into smaller units then put into the packets. The data link layer network types most often encountered with TCP/IP in the wireless broadband world are *Ethernet* and *Asynchronous Transfer Mode* (*ATM*) based ones. In the following subsections a brief introduction to TCP/IP, the Ethernet access scheme, and ATM is presented.

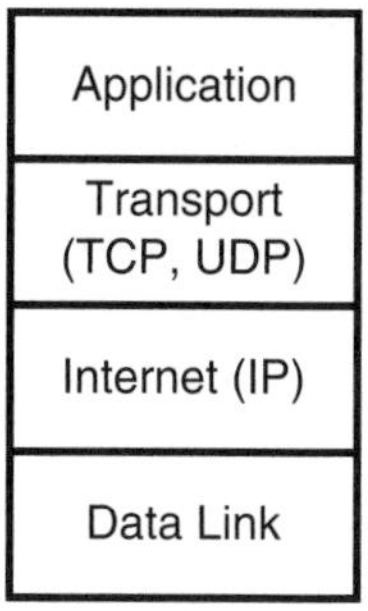

```
┌─────────────────────┐
│     Application      │
├─────────────────────┤
│     Transport        │
│     (TCP, UDP)       │
├─────────────────────┤
│    Internet (IP)     │
├─────────────────────┤
│     Data Link        │
└─────────────────────┘
```

FIGURE 2.13   The TCP/IP
protocol stack.

2.3.1   TCP/IP

The acronym TCP/IP commonly used in the broad sense and as applied here refers to a hierarchy of four protocols but derives its name from the two main ones,.namely TCP and IP. The four protocols are stacked as shown in Fig. 2.13. Following is a short description of each protocol layer:

- **Application layer**: This is where the user interfaces with the network and includes all the processes that involve user interaction. Protocols at this level include those for facilitating World Wide Web (WWW) access, e-mail, and so on. Data from this level to be sent to a remote address is passed on the next lower layer, the transport layer, in the form of a stream of 8-bit bytes.
- **Transport layer**: There are two protocols at this layer, the User Datagram Protocol (UDP) and the Transmission Control Protocol (TCP). The former does not provide reliable communication and is only used in special circumstances. TCP, on the other hand, provides reliable end-to-end communication and is used by most Internet applications. With TCP the two users at each end of the network must establish a connection before any data can be transferred between them. It is thus referred to as a *connection-oriented* protocol. It guarantees that data sent is received at the far end, and received correctly. It does this by having the far end acknowledge receipt of the data. If the sender receives no acknowledgment within a specified time frame, it re-sends the data. When TCP receives data from the application layer above for transmission to a remote address, it first splits the data into manageable blocks based on its knowledge of how large a block the network can handle. It then adds control information to the front of each data block. This control information is called a *header,* and the addition of the header to the data block is called *encapsulation*. TCP adds a 20-byte header to each data block to form a datagrams. It then passes these datagrams to the next layer below, the IP layer. A TCP datagram is shown in Fig. 2.14. When TCP receives a datagram form a remote address via IP, the opposite procedure to that involved in creating a datagram takes place (i.e., the header is removed and the data passed to the application layer above). However, before passing it above, TCP takes data that arrives out of sequence and puts it back into the order in which it was sent.
- **Internet layer**: The protocol at this layer, IP, is the core protocol of TCP/IP. Its job is to route data across and between networks. When sending datagrams, it figures

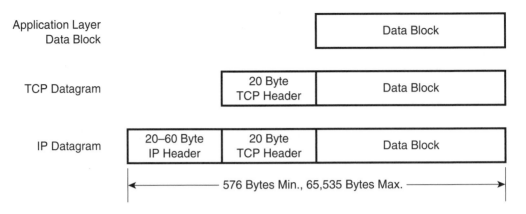

FIGURE 2.14   Encapsulation to the IP layer of TCP/IP data.

out how to get them to their destination; when receiving datagrams, it figures out to whom they belong. It is an unreliable protocol, unconcerned as to whether datagrams it sent arrive at their destination and whether datagrams it received arrive in the order sent. If a datagram arrives with any problems, it simply discards it. It leaves the quest for reliable communication to the TCP level above. The IP is defined as a *connectionless* protocol. There are, naturally, provisions to create connections per se or communication would be impossible. However, such connections are established on a datagram-by-datagram basis, with no relationship to each other. It processes each datagram as an entity in itself, independent of any other datagram that may have preceded it or may follow it. In fact, there is no information in an IP datagram to identify it as part of a sequence or as belonging to a particular task. Thus, for IP to accomplish its assigned task, each IP datagram must contain complete addressing information. An IP datagram is created by adding a header, normally 20 bytes and at most 60 bytes, to the TCP datagram as shown in Fig. 2.14. This header includes source and destination address information as well as other control information. The minimum size of an IP datagram is 576 bytes, and its maximum size is 65,535 bytes. All IP networks must be able to handle IP datagrams of at least 576 bytes in length. IP passes datagrams destined to a remote address to the next layer below, the link layer. On the receiving end, datagrams received by the IP layer from the link layer are stripped of their IP headers and passed up to the transport layer. The maximum amount of data that a link layer packet can carry is called the *maximum transfer unit* (*MTU*) of the layer. As explained later, because each IP datagram is encapsulated within a link layer packet prior to transmission to the remote address, the MTU of the link layer places a maximum on the length of an IP datagram that it can process. Should an IP datagram be larger than this maximum, it is broken up into smaller datagrams that can fit in the link layer packet. This breaking up process is called *fragmentation,* and each of the smaller datagrams created is called a *fragment.* Fig. 2.15 shows the fragmentation of an IP datagram 1000 bytes long into two fragments, in order that it may be processed by a link layer with an MTU of 576 bytes. It will be noticed that each fragment has the same basic struc-

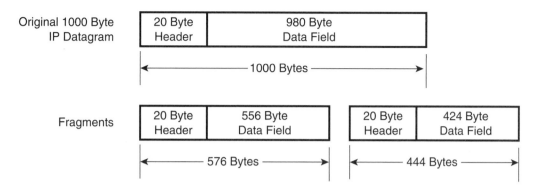

FIGURE 2.15   An example of the fragmentation of an IP datagram.

ture as an IP datagram. Information in the fragment header defines it as a fragment, not a datagram. At the destination, fragments are reassembled to the original IP datagram before being passed to the transport layer. Should one or more of the fragments fail to arrive, the datagram is assumed to be lost and nothing is passed to the transport layer. Fragmentation puts additional tasks on Internet hardware, and so it is desirable to keep it to a minimum. Obviously, it can be entirely eliminated by using IP datagrams 576 bytes long since all IP handling networks must have a MTU of at least 576. In fact, most bulk TCP/IP data transfer is done using IP datagrams 576 bytes long.

• **Data Link Layer**: The Data Link layer consists of the network hardware that effects actual communication. When transmitting IP data onto the network, it creates packets by taking IP datagrams and adding headers and, in some instances, trailers (control data trailing the datagrams) onto them, and dispatches these packets out onto the network. When data packets are received from the network, the Data Link layer strips them of their headers and trailers, if any, and passes them on to the IP layer above. As indicated previously, the data link layer network types most often encountered with TCP/IP in the wireless broadband world are Ethernet and ATM based.

2.3.2   Ethernet Access Scheme

The Ethernet access scheme is based on a technique known as *carrier-sense multiple access with collision detection*[5,7] (CSMA/CD). This is a technique in which all devices connected to the network (multiple access) listen for any transmission in progress by others (carrier sense) before starting to transmit. The instant a device wanting to transmit data senses no carrier, it begins to transmit data. However, should two or more begin transmitting at the same time and their transmissions collide, this collision is detected by the devices (collision detection), and each backs off for a different amount of time before attempting to transmit once more. Two Ethernet-based schemes whose signals are often transported by fixed wireless links are the IEEE-specified 10BaseT[3,7] (IEEE 802.3i) and 100BaseT[8] (IEEE 802.12) schemes. It is therefore of interest to understand the signal structure of these schemes.

In 10BaseT transmission, 8-bit bytes are used and packets range in size from 72 to 1526 bytes. Fig. 2.16(a) shows the 10BaseT packet format. It consists of a 22-byte header, followed by information data, and ending in a 4-byte trailer. The header commences with a 7-byte preamble used for synchronization of the receiving station's clock, followed by a 1-byte start frame delimiter used to indicate the start of the frame. This is followed by two 6-byte address codes, the first representing the packet's destination, the second representing the packet's source. Next is a 2-byte length field used to indicate the length of the information data field. Following this is the information data field, which is an integer number of bytes, from a minimum of 46 to a maximum of 1500. Note that if the actual number of data bytes to be sent is less than 46, then extra bytes are added at the end to total 46. The 4-byte trailer follows the information data field and forms the frame check sequence, which provides the error detection and correction at the bit level to ensure that

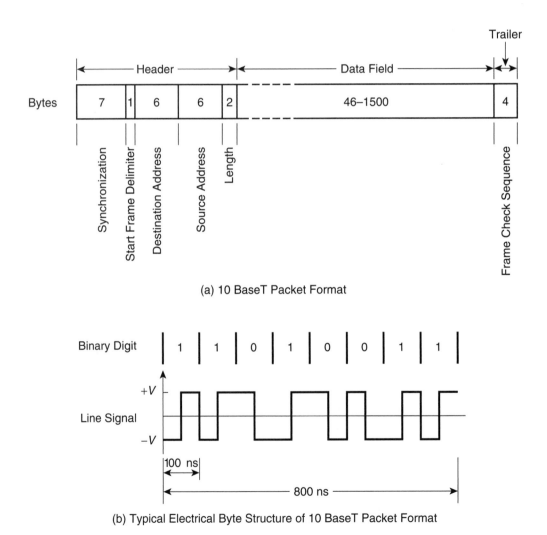

(a) 10 BaseT Packet Format

(b) Typical Electrical Byte Structure of 10 BaseT Packet Format

FIGURE 2.16   10BaseT packet format and byte electrical structure.

the data from the destination address through the information data field reaches its destination correctly. The individual packets are transmitted at rates of 10 Mb/s. Local communication (100s of meters) is via two twisted pairs of phone wires, one pair for each direction. Note that although individual packets are transmitted at a rate of 10 Mb/s, there is not a continuous and uninterrupted transmission of packets, and hence bits, as with PDH systems. Between packets there is no transmission. Fig. 2.16(b) shows one byte of a 10BaseT signal. The system encodes data using Manchester encoding. With this code each bit is divided into two halves, one positive, one negative. A negative to positive transition in the middle of the bit denotes a binary 1, while a positive to negative transition denotes a 0. Each bit occupies 100 nanoseconds, and thus a byte occupies 800 nanoseconds.

In the 100BaseT (also called *Fast Ethernet*) scheme the packet format is identical to 10BaseT, but the packet transmission rate is 100 Mb/s; thus each bit occupies 10 nanoseconds and a byte occupies 80 nanoseconds. Local communication is via either two twisted pairs or four twisted pairs.

Ethernet 10/100BaseT offers three types of connection related service: type 1, unacknowledged connectionless service; type 2, connection mode service, a connection-oriented protocol; and type 3, acknowledged connectionless service. The sending node determines the type of connection-related service.

IP datagrams are transported via Ethernet 10/100BaseT packets by encapsulating them between the packet's header and trailer to form a packet, as shown in Fig. 2.17. Since the Ethernet 10/100BaseT packet can handle up to 1500 information data bytes, it can easily accommodate the 576-byte minimum length IP datagram. However, IP can support datagrams as large as 65,535 bytes. Thus, when IP datagrams are to be transported via Ethernet 10/100BaseT, this is communicated to TCP/IP, which limits the length of its IP datagrams to no more than 1500 bytes.

It should be noted that when data packets, such as Ethernet 10BaseT or 100BaseT packets, are transported over continuous rate broadband wireless systems, the system need not be able to communicate at the individual packet rate to allow throughput without significant delay. This is because, given that the traffic is bursty, the average data rate is typically much less than the packet rate. Take, for example, a 10BaseT packet signal that sends packets on average one-tenth of the time. Since the individual packet rate is 10 Mb/s, the average data rate is 1 Mb/s. Such a packet stream could therefore be comfortably carried via a T1 (1.544 Mb/s) wireless link. This is accomplished by feeding the packets into buffer storage and clocking them out at the continuous channel rate, 1.544 Mb/s.

2.3.3 Asynchronous Transfer Mode (ATM)

Asynchronous Transfer Mode (ATM) is a data transport packet technology that is used in local area networks (LANs), wide area networks (WANs), and backbone networks. A key feature of ATM is that it uses very short, fixed-length packets called cells. An ATM cell is shown in Fig. 2.18(a). It is only 53 bytes long and includes a 5-byte header, which

22 Byte Ethernet Header	IP Datagram	4 Byte Ethernet Trailer

FIGURE 2.17   IP datagram Ethernet encapsulation.

contains identification, control priority, and routing information. The remaining 48 bytes comprise the data field. The short, fixed-length nature of the cells allows very fast switching of cells from the input to the output of a data switch. An ATM network can support and multiplex a large number of users, easily handling voice, video, and bursty data simultaneously. ATM allocates available data capacity (often referred to loosely as bandwidth) more flexibly than TDM by giving users access, if it is available, to the entire data capacity when they need it, for as long as they need it. On the other hand, should the channel be in use, a new user may have to wait before access is granted. Fig. 2.18(b) gives an example of ATM cells assigned between three users—namely, user 1, user 2, and user 3—and the interleaving of these cells onto the transmission media. User 3, whose data is video derived, requires the most bandwidth and so is assigned more cells than user 1 and user 2. User 2, whose data is voice derived, has an ongoing but low bandwidth requirement. User 1, whose data is derived from a computer terminal, has a sporadic and low bandwidth requirement. Note that there is no repetitive pattern to the assignment of cells. Thus, like an IP datagram, the header in each cell must contain an address. ATM provides for the predictable performance of the circuit technology. It accomplished this through the use of specific *Quality of Service* (*QoS*) classes for each traffic type.

ATM networking depends on the establishment of connections between end terminals across the network. No information may be transferred until such connections are established; thus, ATM technology is connection oriented. A key attribute of ATM is the fashion in which connections are established. ATM used the concept of *virtual connections* (*VCs*), also referred to as *virtual channels* as well as *virtual circuits*, and *virtual paths* (*VPs*) to accomplish these connections.

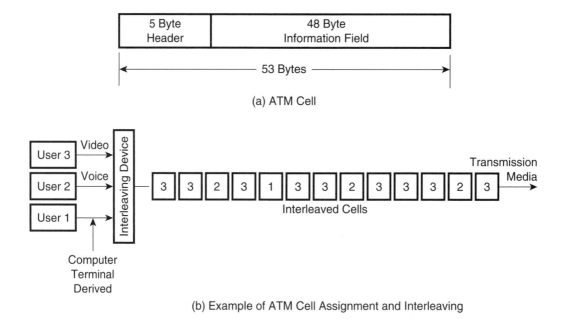

(a) ATM Cell

(b) Example of ATM Cell Assignment and Interleaving

FIGURE 2.18   ATM cell structure and cell assignment.

A VC is a connection between two communicating ATM end terminals, with this connection consisting of links between a series of interconnected switches. Since all communication between a pair of end terminals proceeds along the same VC, cell sequence is preserved. VCs allow multiple connections to be defined simultaneously over a single ATM network. VCs may be established on demand or they may be preconfigured. When established on demand and set up automatically through the signaling protocol, they are referred to as *switched virtual connections* (*SVCs*); when preconfigured, typically via network management, they are referred to as *permanent virtual connections* (*PVCs*). PVCs eliminate the need to establish a route each time an end user needs to communicate with a remote terminal.

Virtual paths are a group of VCs carried between two points in the network and, like VCs, may involve many links. VPs can be regarded as a bundling of VCs that have a common path. VPs simplify transmission since, in switching cells between the terminals of a VP, the switching equipment only has to refer to the VP information in the header rather than the entire header address.

ATM is a three-protocol layer technology, the three layers being the physical layer, the ATM layer, and the adaptation layer. Fig. 2.19 shows an example of an end-to-end IP over ATM network and the participation of the various protocol layers.

The *ATM layer* is the center layer and is concerned with the data transmission between two adjacent network nodes. At the transmit end it formats the 53 byte cells, defines the cell header content, and then multiplexes cells from individual VPs and VCs into one stream. On the receive end the process is reversed. It functions on a link-by-link basis, not an end-to-end one, and thus cell addressing is between adjacent nodes only.

The *ATM adaptation layer* (*AAL*) sits on top of the ATM layer and provides some of the most important functions of the ATM communication process. It sets out the rules for breaking up the individual data units received from the level above it (such data units are referred to as *protocol data units* [*PDUs*]) into ATM cells for transmission over the network and the rules for the reassembly of the PDUs at the remote end. Unlike the ATM process, the AAL process, as shown in Fig. 2.19, is an end-to-end one, used only by the communicating devices at each end to insert and remove data from the ATM

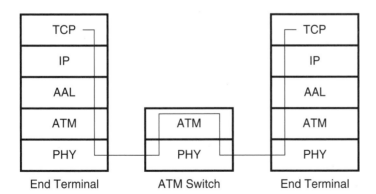

FIGURE 2.19   Protocol layers and layer participation in an IP over ATM network.

layer. AAL gives ATM the versatility to transport data originating from several different types of service, from voice- and video-derived continuous TDM data to bursty data generated by LANs, all within the same format. Most PDUs are longer than 48 bytes. Thus the first job of the AAL is usually to divide the PDUs into sections of shorter length for encapsulation in the ATM cells. There are presently five different AAL standards, namely AAL 1 through AAL 5, each designed for a different type of service. AAL 5 is referred to as the *Simple and Efficient Adaptation Layer* (*SEAL*). It is designed for data traffic and the one most often used for IP transmission. By providing for no error recovery, on the assumption that a higher-layer protocol will address this, it allows all 48 bytes of the ATM cell information field to be used for data from the higher-level PDU. Fig. 2.20 shows how AAL 5 breaks up user data from a PDU and packages it into ATM cells. In the last cell a pad is added to fill the information field, and an 8-byte trailer with control data appended to it. Further, one bit in the cell header, which is normally 0, is set to 1 allowing the end of the packet stream containing the PDU to be flagged. AAL 5 presents an interface that accepts/delivers variable length IP datagrams. However, it only accepts, transfers, and delivers IP datagrams of 9180 bytes or less. Thus, if an IP datagram is larger than 9180 bytes, IP fragments it and passes on the fragments to AAL 5.

The physical layer sits below the ATM layer, and its main function is to transport ATM cell content accurately over a single physical link. The terminals of this link can be connected via electrical cable, optical cable, wireless equipment, or a combination thereof. The physical layer is in fact divided into two sublayers, the *physical media dependent* (*PMD*) sublayer and the *transmission convergence* (*TC*) sublayer. The PMD sublayer is the lower of the two and thus at the very bottom of the ATM protocol stack. It is responsible for the actual transmission and reception of bits on the physical medium and thus is

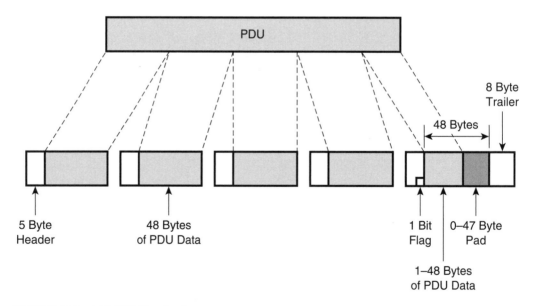

FIGURE 2.20   AAL 5 PDU packaging.

also tasked with keeping bit timing in sync with the timing source and line encoding and decoding. There are two classes of PMD sublayers, those with a frame structure and those without. PMD sublayer members with a frame structure include SONET/SDH over fiber or wireless links and PDH over cable, fiber, or wireless links. If the PMD sublayer has a transmission frame structure, the TC sublayer is responsible for placing the bits of the ATM cells received from the ATM layer in the appropriate frame structure for transport over the PMD sublayer. Before this conversion, however, it inserts idle cells to the ATM cell stream in order to adapt the rate of the ATM cells to the transmission rate of the PMD sublayer. On the receive side, the process reversed. If the PMD structure is cell based with no frames, the transmitting TC sublayer sends a continuous stream of cells, by inserting an idle cell whenever the ATM layer has not provided one. On the receiving side, the TC sublayer deletes the idle cells before passing cells up to the ATM layer. The functions of the TC sublayer are dependent on the PMD member used. A change of the PMD member requires a change to the TC sublayer.

2.4   DIGITIZED VIDEO

Like voice, TV and other video signals were originally transmitted in analog form. However, video transmission today is in the midst of transformation to digital. A digitized video signal is created by converting an analog video signal to a digital one via PCM techniques. The device that converts a video signal from analog to digital and vice versa is often referred to as a video CODEC. The major issue with digitized video transmission is that, because of the inherently much higher information content of video compared to voice, very high bit rates (compared to voice) are required if no compression of the signal rate is employed. Video signal compression is a complex process and will not be addressed here. Suffice it to say, however, that it is possible because there are redundancies in a video signal's uncompressed format and because visual perception has certain characteristics that can be used to advantage. Regarding format redundancies, video compression in essence results in the transmission of only differences from one video frame to the next, rather than the transmission of each new frame in its entirety.

The standard digital video signal, noncompressed, as defined by the International Telecommunications Union in its recommendation ITU-R 601, has a bit rate of either 216 Mb/s for 8 bit samples or 270 Mb/s for 10 bit samples. When compared to the standard 64 kb/s standard voice signal, it is seen that this noncompressed video signal requires a bit rate that's about 4000 times that of voice! For high-definition television (HDTV)[5] an even greater noncompressed bit rate of approximately 1.5 Gb/s is required. The transmission of video signals over long distances at these very high bit rates would be extremely expensive; hence great effort has been expended in video signal compression.

There are today a myriad of compression options. The most popular specifications are provided by the *Moving Picture Expert Group* (*MPEG*). MPEG is a working group of the International Standards Organization (ISO) in charge of the development of standards for coded representation of digital audio and video. CODECs built to MPEG specifications provide compressed bit rates from as low as about 10 kb/s for signals used in video conferencing and the like up to about 50 Mb/s for HDTV. CODECs are also available that provide bit rates at standard digital multiplexing hierarchy levels, such as T3, E3, and STS-1.

2.5 THE SERVICE CHANNEL

Fixed broadband wireless systems usually come with a *service channel*. The service channel, as the name implies, allows service technicians easy communication between sites. It is normally a digitized voice channel that is digitally multiplexed with the main payload and easily accessible at all equipment locations throughout the system. An alternate and often used name for the service channel is the *orderwire*.

2.6 WAYSIDE CHANNELS

Wayside channels are low-capacity channels typically added to the payload of high-capacity microwave systems. They are analogous to small side roads beside a large highway, allowing easy access to stops along the way. A typical example of a system with wayside channels would be one whose main payload would be a T3 signal (45 Mb/s) but also providing two T1 wayside channels. Wayside channels are used for a variety of purposes. These include the provision of service channels, the establishment of low-speed data links, network control, and accessing intermediate stations independent of the main payload.

REFERENCES

1. Townsend, A. R., *Digital Line-of-Sight Radio Links*, Prentice Hall International (UK) Ltd., 1988.
2. Taub, H., and Schilling, D., *Principles of Communication Systems*, McGraw-Hill, 1971.
3. Winch, R. G., *Telecommunication Transmission Systems*, McGraw-Hill, 1993.
4. *Global Deployment of SDH-Compliant Networks*, Special Issue, *IEEE Communications Magazine,* Vol. 28, Aug. 1990.
5. Lee, B. G., Kang, M., and Lee, J., *Broadband Telecommunications Technology*, Artech House, 1993.
6. Ballart, R., and Ching, Y., "SONET: Now It's the Standard Optical Network," *IEEE Communications Magazine,* Vol. 27, March 1989.
7. Bates, B., and Gregory, D., *Voice & Data Communications Handbook*, McGraw-Hill, 1996.
8. Molle, M., and Watson, G., "100Base-T / IEEE 802.12 / Packet Switching," *IEEE Communications Magazine,* Vol. 34, Aug. 1996, pp. 64–73.

CHAPTER 3

MATHEMATICAL TOOLS FOR DIGITAL TRANSMISSION ANALYSIS

3.1 INTRODUCTION

The study of digital wireless transmission is in large measure the study of (a) the conversion in a transmitter of a binary digital signal (often referred to as a *baseband* signal) to a modulated RF signal, (b) the transmission of this modulated signal from the transmitter through the atmosphere, (c) the corruption of this signal by noise, unwanted signals, and propagation anomalies, (d) the reception of this corrupted signal by a receiver, and (e) the recovery in the receiver, as best as possible, of the original baseband signal. In order to analyze such transmission, it is necessary to characterize mathematically, in the time, frequency, and probability domains, baseband signals, modulated RF signals, noise, propagation anomalies, and signals corrupted by noise, unwanted signals, and propagation anomalies. The purpose of this chapter is to review briefly the more prominent of those analytical tools used in such characterization—namely, spectral analysis and relevant statistical methods. Spectral analysis permits the characterization of signals in the frequency domain and provides the relationship between frequency domain and time domain characterizations. Noise and propagation anomalies are random processes leading to uncertainty in the integrity of a recovered signal. Thus no definitive determination of the recovered signal can be made. By employing statistical methods, however, computation of the fidelity of the recovered baseband signal is possible in terms of the probability that it's in error. The study of the basic principles of fixed wireless modulation, covered in Chapter 4, will apply several of the tools presented here. Those readers familiar with these tools may want to skip this chapter and proceed to Chapter 4.

3.2 SPECTRAL ANALYSIS OF NONPERIODIC FUNCTIONS

A nonperiodic function of time is a function that is nonrepetitive over time. A stream of binary data as typically transmitted by digital communication systems is a stream of non-periodic functions, each pulse having equal probability of being one or zero, independent of the value of other pulses in the stream. The analysis of the spectral properties of nonperiodic functions is thus an important component of the study of digital transmission.

3.2.1   The Fourier Transform

A nonperiodic waveform, $v(t)$ say, may be represented in terms of its frequency characteristics by the following relationship:

$$v(t) = \int_{-\infty}^{\infty} V(f)e^{j2\pi ft} df \qquad (3.1)$$

The factor $V(f)$ is the *amplitude spectral density* or the *Fourier transform*[1] of $v(t)$. It is given by

$$V(f) = \int_{-\infty}^{\infty} v(t)e^{-j2\pi ft} dt \qquad (3.2)$$

Because $V(f)$ extends from $-\infty$ to $+\infty$ (i.e., it exists on both sides of the zero frequency axis) it is referred to as a *two-sided* spectrum.

An example of the application of the Fourier transform that is useful in the study of digital communications is its use in determining the spectrum of a nonperiodic pulse. Consider a pulse $v(t)$ shown in Fig. 3.1(a), of amplitude V, and that extends from $t = -\tau/2$ to $t = \tau/2$. Its Fourier transform, $V(f)$, is given by

$$V(f) = \int_{-\tau/2}^{\tau/2} Ve^{-j2\pi ft} dt$$

$$= \frac{V}{-j2\pi f}\left[e^{-j2\pi f\tau/2} - e^{j2\pi f\tau/2}\right]$$

$$= V\tau \frac{\sin \pi f\tau}{\pi f\tau} \qquad (3.3)$$

The form $(\sin x)/x$ is well known and referred to as the *sampling function, Sa(x)*.[1] The plot of $V(f)$ is shown in Fig. 3.1(b). It will be observed that it is a continuous function. This is a common feature of the spectrum of all nonperiodic waveforms. We note also that it has zero crossings at $\pm 1/\tau, \pm 2/\tau, \ldots$.

The Fourier transform $V(f)$ of an impulse of unit strength is also a useful result. By definition an impulse $\delta(t)$ has zero value except at time $t = 0$, and an impulse of unit strength has the property

$$\int_{-\infty}^{\infty} \delta(t)\, dt = 1 \qquad (3.4)$$

Thus

$$V(f) = \int_{-\infty}^{\infty} \delta(t)e^{-2\pi jft} dt = 1 \qquad (3.5)$$

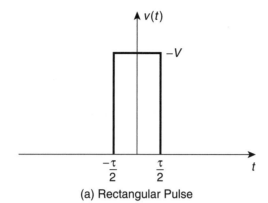

(a) Rectangular Pulse

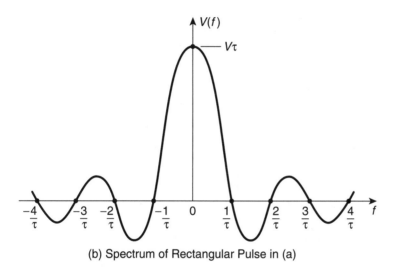

(b) Spectrum of Rectangular Pulse in (a)

FIGURE 3.1   Rectangular pulse and its spectrum.

Equation (3.5) indicates that the spectrum of an impulse $\delta(t)$ has a constant amplitude and phase and extends from $-\infty$ to $+\infty$.

A final example of the use of the Fourier transform is the analysis of what results in
the frequency domain when a signal $m(t)$, with Fourier transform $M(f)$, is multiplied by a
sinusoidal signal of frequency f_c. In the time domain the resulting signal is given by

$$v(t) = m(t)\cos 2\pi f_c t$$

$$= m(t)\left[\frac{e^{j2\pi f_c t} + e^{-j2\pi f_c t}}{2}\right] \tag{3.6}$$

and its Fourier transform is thus

$$V(f) = \frac{1}{2}\int_{-\infty}^{\infty} m(t)e^{-j2\pi(f+f_c)t}\,dt + \frac{1}{2}\int m(t)e^{-j2\pi(f-f_c)t}\,dt \tag{3.7}$$

Recognizing that

$$M(f) = \int_{-\infty}^{\infty} m(t)e^{-j2\pi ft} dt \qquad (3.8)$$

then

$$V(f) = \frac{1}{2}M(f+f_c) + \frac{1}{2}M(f-f_c) \qquad (3.9)$$

An amplitude spectrum $|M(f)|$, band limited to the range $-f_m$ to $+f_m$, is shown in Fig. 3.2(a). In Fig.3.2(b), the corresponding amplitude spectrum of $|V(f)|$ is shown.

3.2.2   Linear System Response

A linear system is one in which, in the frequency domain, the output amplitude at a given frequency bears a fixed ratio to the input amplitude at that frequency and the output phase at that frequency bears a fixed difference to the input phase at that frequency, irrespective

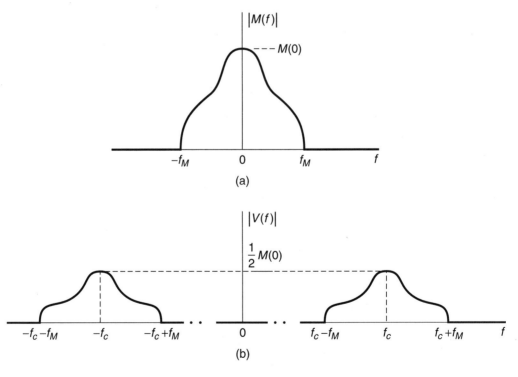

(a)

(b)

FIGURE 3.2   (a) The amplitude spectrum of a waveform with no special component beyond f_m. (b) The amplitude spectrum of the waveform in (a) multiplied by $\cos 2\pi f_c t$. (From Taub, H., and Schilling, D., *Principles of Communication Systems*, McGraw-Hill, 1971, and reproduced with the permission of the McGraw-Hill Companies.)

of the absolute value of the input signal. Such a system can be characterized by the complex transfer function, $H(f)$ say, given by

$$H(f) = |H(f)|e^{-j\theta(2\pi f)} \tag{3.10}$$

where $|H(f)|$ represents the absolute amplitude characteristic, and $\theta(2\pi f)$ the phase characteristic of $H(f)$.

Consider a linear system with complex transfer function $H(f)$, as shown in Fig.3.3, with an input signal $v_i(t)$, an output signal $v_o(t)$, and with corresponding spectral amplitude densities of $V_i(f)$, and $V_o(f)$. After transfer through the system, the spectral amplitude density of $V_i(f)$ will be changed to $V_i(f) H(f)$. Thus

$$V_o(f) = V_i(f)H(f) \tag{3.11}$$

and

$$v_o(t) = \int_{-\infty}^{\infty} V_i(f)H(f)e^{j2\pi ft}df \tag{3.12}$$

An informative situation is the one where the input to a linear system is an impulse function of unit strength. For this case, as per Eq. (3.5), $V_i(f) = 1$, and

$$V_o(f) = H(f) \tag{3.13}$$

Thus, the output response of a linear system to a unit strength impulse function is the transfer function of the system.

3.2.3   Energy and Power Analysis

In considering energy and power in communication systems, it is often convenient to assume that the energy is dissipated in a 1-ohm resistor, as with this assumption one need not keep track of the impact of the true resistance value, R say. When this assumption is made, we refer to the energy as the *normalized energy* and to the power as *normalized power*. It can be shown that the normalized energy E of a nonperiodic waveform $v(t)$, with a Fourier transform $V(f)$, is given by

$$E = \int_{-\infty}^{\infty} [v(t)]^2 dt = \int_{-\infty}^{\infty} |V(f)|^2 df \tag{3.14}$$

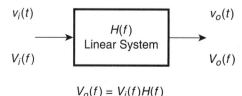

$$V_o(f) = V_i(f)H(f)$$

FIGURE 3.3   Signal transfer through a linear system.

The preceding relationship is called *Parseval's theorem.*[1] Should the actual energy be required, then it is simply E [as given in Eq. (3.14)] divided by R.

The *energy density, $D_e(f)$,* of a waveform is the factor $dE(f)/df$. Thus, by differentiating the right-hand side of Eq. (3.14), we have

$$D_e(f) = \frac{dE(f)}{df} = |V(f)|^2 \qquad (3.15)$$

For a nonperiodic function such as a single pulse, normalized energy is finite, but power, which is energy per unit time, approaches zero. Power is thus somewhat meaningless in this context. However, a train of binary nonperiodic adjacent pulses does have meaningful average normalized power. This power, P say, is equal to the normalized energy per pulse E, multiplied by f_s, the number of pulses per second; that is,

$$P = E f_s \qquad (3.16)$$

If the duration of each pulse is τ, then $f_s = 1/\tau$. Substituting this relationship and Eq. (3.14) into Eq. (3.16), we get

$$P = \frac{1}{\tau} \int_{-\infty}^{\infty} |V(f)|^2 \, df \qquad (3.17)$$

The *power spectral density, $G(f)$,* of a waveform is the factor $dP(f)/df$. Thus, by differentiating the right-hand side of Eq. (3.17), we have

$$G(f) = \frac{dP(f)}{df} = \frac{1}{\tau} |V(f)|^2 \qquad (3.18)$$

To determine the effect of a linear transfer function $H(f)$ on normalized power, we substitute Eq. (3.11) into Eq. (3.17). From this substitution we determine that the normalized power, P_o, at the output of a linear network, is given by

$$P_o = \frac{1}{\tau} \int_{-\infty}^{\infty} |H(f)|^2 |V_i(f)|^2 \, df \qquad (3.19)$$

Also, from Eq. (3.11), we have

$$\frac{|V_o(f)|^2}{\tau} = \frac{|V_i(f)|^2}{\tau} |H(f)|^2 \qquad (3.20)$$

Substituting Eq. (3.18) into Eq. (3.20), we determine that the power spectral density $G_o(f)$ at the output of a linear network is related to the power spectral density $G_i(f)$ at the input by the relationship

$$G_o(f) = G_i(f) |H(f)|^2 \qquad (3.21)$$

3.3 STATISTICAL METHODS

We now turn our attention away from the time and frequency domain and toward the probability domain where statistical methods of analysis are employed. As indicated in Section 3.1, such methods are required because of the uncertainty resulting from the introduction of noise and other factors during transmission.

3.3.1 The Cumulative Distribution Function and the Probability Density Function

A *random variable* $X^{1,2}$ is a function that associates a unique numerical value $X(\lambda_i)$ with every outcome λ_i of an event that produces random results. The value of a random variable will vary from event to event, and depending on the nature of the event will be either *discrete* or *continuous*. An example of a discrete random variable X_d is the number of heads that occur when a coin is tossed four times. As X_d can only have the values 0, 1, 2, 3, and 4, it is discrete. An example of a continuous random variable X_c is the distance of a shooter's bullet hole from the bull's eye. As this distance can take any value, X_c is continuous.

Two important functions of a random variable are the *cumulative distribution function* (*CDF*) and the *probability density function* (*PDF*).

The cumulative distribution function, $F(x)$, of a random variable X is given by

$$F(x) = P\left[X(\lambda) \leq x\right] \tag{3.22}$$

where $P[X(\lambda) \leq x]$ is the probability that the value $X(\lambda)$ taken by the random variable X is less than or equal to the quantity x.

The cumulative distribution function $F(x)$ has the following properties:

1. $0 \leq F(x) \leq 1$
2. $F(x_1) \leq F(x_2)$ if $x_1 \leq x_2$
3. $F(-\infty) = 0$
4. $F(+\infty) = 1$

The *probability density function* $f(x)$ of a random variable X is the derivative of $F(x)$ and thus is given by

$$f(x) = \frac{dF(x)}{dx} \tag{3.23}$$

The probability density function $f(x)$ has the following properties:

1. $f(x) \geq 0$ for all values of x
2. $\int_{-\infty}^{\infty} f(x)\, dx = 1$

Further, from Eqs. (3.22) and (3.23), we have

$$F(x) = \int_{-\infty}^{x} f(z)\, dz \tag{3.24}$$

The function within the integral is not shown as a function of x because, as per Eq. (3.22), x is defined here as a fixed quantity. It has been arbitrarily shown as a function of z, where z has the same dimension as x, $f(z)$ being the same PDF as $f(x)$). Some texts, however, show it equivalently as a function of x, with the understanding that x is used in the generalized sense.

The following example will help in clarifying the concepts behind the PDF, $f(x)$, and the CDF, $F(x)$. In Fig. 3.4(a) a four-level pulse amplitude modulated signal is shown. The amplitude of each pulse is random and equally likely to occupy any of the four levels. Thus, if a random variable X is defined as the signal level v, and $P(v = x)$ is the probability that $v = x$, then

$$P(v = -3) = P(v = -1) = P(v = +1) = P(v = +3) = 0.25 \qquad (3.25)$$

With this probability information we can determine the associated CDF, $F_{4L}(v)$. For example, for $v = -1$

$$F_{4L}(-1) = P(v \le -1) = P(v = -3) + P(v = -1) = 0.5 \qquad (3.26)$$

In a similar fashion, $F_{4L}(v)$ for other values of v may be determined. A plot of $F_{4L}(v)$ versus v is shown in Fig. 3.4(b).

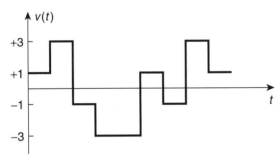

(a) 4-Level PAM Signal

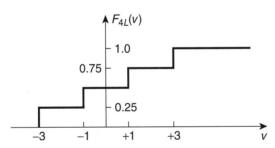

(b) Cumulative Distribution Function (CDF)
Associated with Levels of 4-Level PAM Signal

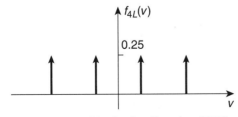

(c) Probability Distribution Function (PDF)
Associated with Levels of 4-Level PAM Signal

FIGURE 3.4　A four-level PAM signal and its associated CDF and PDF.

The PDF $f_{4L}(v)$ corresponding to $F_{4L}(v)$ can be found by differentiating $F_{4L}(v)$ with respect to v. The derivative of a step of amplitude V is a pulse of value V. Thus, since the steps of $F_{4L}(v)$ are of value 0.25,

$$f_{4L}(-3) = f_{4L}(-1) = f_{4L}(+1) = f_{4L}(+3) = 0.25 \tag{3.27}$$

A plot of $f_{4L}(v)$ versus v is shown in Fig. 3.4(c).

3.3.2   The Average Value, the Mean Squared Value, and the Variance of a Random Variable

The *average value* or *mean, m,* of a random variable X, also called the *expectation* of X, is also denoted either by $\overline{X}$ or $E(x)$. For a discrete random variable, X_d, where n is the total number of possible outcomes of values $x_1, x_2, \ldots, x_n$, and where the probabilities of the outcomes are $P(x_1), P(x_2), \ldots, P(x_n)$ it can be shown that

$$m = \overline{X_d} = E(X_d) = \sum_{i=1}^{n} x_i P(x_i) \tag{3.28}$$

For a continuous random variable X_c, with PDF $f_c(x)$, it can be shown that

$$m = \overline{X_c} = E(X_c) = \int_{-\infty}^{\infty} x \cdot f(x)\, dx \tag{3.29}$$

and that the *mean square value,* $\overline{X_c^2}$ or $E(X_c^2)$ is given by

$$\overline{X_c^2} = E(X_c^2) = \int_{-\infty}^{\infty} x^2 f(x)\, dx \tag{3.30}$$

Figure 3.5 shows an arbitrary PDF of a continuous random variable. A useful number to help in evaluating a continuous random variable is one that gives a measure of how widely spread its values are around its mean m. Such a number is the root mean square (rms) value of $(X - m)$ and is called the *standard deviation* σ of X.

The square of the standard deviation, σ^2, is called the variance of X and is given by

$$\sigma^2 = E\left[(X - m)^2\right] = \int_{-\infty}^{\infty} (x - m)^2 f(x) dx \tag{3.31}$$

The relationship between the variance σ^2 and the mean square value $E(X^2)$ is given by

$$\sigma^2 = E\left[(X - m)^2\right]$$
$$= E\left[X^2 - 2mX + m^2\right]$$
$$= E(X^2) - 2mE(X) + m^2 \tag{3.32}$$
$$= E(X^2) - m^2$$

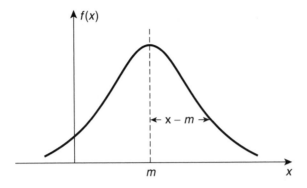

FIGURE 3.5   A Probability Distribution Function (PDF) of a continuous random variable.

We note that for the average value $m = 0$, the variance $\sigma^2 = E(X^2)$.

3.3.3   The Gaussian Probability Density Function

The *Gaussian* or, as it's sometimes called, the *normal* PDF[1,2] is very important to the study of wireless transmission and is the function most often used to describe *thermal noise*. Thermal noise is the result of thermal motions of electrons in the atmosphere, resistors, transistors, and so on and is thus unavoidable in communication systems. The Gaussian probability density function, $f(x)$, is given by

$$f(x) = \frac{1}{\sqrt{2\pi\sigma^2}} e^{-(x-m)^2/2\sigma^2} \tag{3.33}$$

where m is as defined in Eq. (3.28) and σ as defined in Eq. (3.31). When $m = 0$ and $\sigma = 1$ the *normalized Gaussian probability density function* is obtained. A graph of the Gaussian PDF is shown in Fig. 3.6(a).

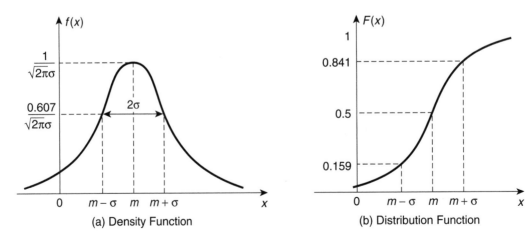

(a) Density Function

(b) Distribution Function

FIGURE 3.6   The Gaussian random variable.

The CDF corresponding to the Gaussian PDF is given by

$$F(x) = P\big[X(\lambda) \le x\big] = \int_{-\infty}^{x} \frac{e^{-(z-m)^2/2\sigma^2}}{\sqrt{2\pi\sigma^2}} dz \qquad (3.34)$$

When $m = 0$, the *normalized Gaussian cumulative distribution function* is obtained and is given by

$$F(x) = \int_{-\infty}^{x} \frac{e^{-z^2/2\sigma^2}}{\sqrt{2\pi\sigma^2}} dz \qquad (3.35)$$

A graph of the Gaussian cumulative distribution function is shown in Fig. 3.6(b). In practice, since the integral in Eq. (3.35) is not easily determined, it is normally evaluated by relating it to the well-known and numerically computed function, the *error function*. The error function of v is defined by

$$erf(v) = \frac{2}{\sqrt{\pi}} \int_{0}^{v} e^{-u^2} du \qquad (3.36)$$

and it can be shown that $erf(0) = 0$ and $erf(\infty) = 1$.

The function $[1 - erf(v)]$ is referred to as the *complementary error function, erfc(v)*. Noting that $\int_{0}^{v} = \int_{0}^{\infty} - \int_{v}^{\infty}$, we have

$$erfc(v) = 1 - erf(v)$$

$$= 1 - \left[\frac{2}{\sqrt{\pi}} \int_{0}^{\infty} e^{-u^2} du - \frac{2}{\sqrt{\pi}} \int_{v}^{\infty} e^{-u^2} du \right]$$

$$= 1 - \left[erf(\infty) - \frac{2}{\sqrt{\pi}} \int_{v}^{\infty} e^{-u^2} du \right]$$

$$= \frac{2}{\sqrt{\pi}} \int_{v}^{\infty} e^{-u^2} du \qquad (3.37)$$

Tabulated values of $erfc(v)$ are only available for positive values of v.

Using the substitution $u \equiv x/\sqrt{2}\sigma$, it can be shown[1] that the Gaussian CDF $F(x)$ of Eq. (3.35) may be expressed in terms of the complementary error function of Eq. (3.37) as follows:

$$F(x) = 1 - \frac{1}{2} erfc\left(\frac{x}{\sqrt{2}\sigma}\right) \qquad \text{for } x \ge 0 \qquad (3.38a)$$

$$= \frac{1}{2} erfc\left(\frac{|x|}{\sqrt{2}\sigma}\right) \qquad \text{for } x \ge 0 \qquad (3.38b)$$

3.3.4 The Rayleigh Probability Density Function

The propagation of wireless signals through the atmosphere is often subject to multipath fading. Such fading will be described in detail in Chapter 5. Multipath fading is best characterized by the *Rayleigh* PDF.[1] Other phenomena in wireless transmission are also characterized by the Rayleigh PDF, making it an important tool in wireless analysis. The Rayleigh probability density function $f(r)$ is defined by

$$f(r) = \frac{r}{\alpha^2} e^{-r^2/2\alpha^2}, \qquad 0 \le r \le \infty \tag{3.39a}$$

$$= 0, \qquad\qquad\qquad r < 0 \tag{3.39b}$$

and hence the corresponding CDF is given by

$$F(r) = P[R(\lambda) \le r] = 1 - e^{-r^2/2\alpha^2}, \qquad 0 \le r \le \infty \tag{3.40a}$$

$$= 0, \qquad\qquad\qquad r < 0 \tag{3.40b}$$

A graph of $f(r)$ as a function of r is shown in Fig. 3.7. It has a maximum value of $1/(\alpha\sqrt{e})$, which occurs at $r = \alpha$. It has a mean value $\overline{R} = \sqrt{\pi/2} \cdot \alpha$, a mean-square value $\overline{R^2} = 2\alpha^2$, and hence, by Eq. (3.32), a variance σ^2 given by

$$\sigma^2 = \left(2 - \frac{\pi}{2}\right)\alpha^2 \tag{3.41}$$

A graph of $F(r)$ versus $10 \log_{10}(r^2/2\alpha^2)$, which is from Feher,[3] is shown in Fig. 3.8. If the amplitude envelope variation of a radio signal is represented by the Rayleigh random variable R, then the envelope has a mean-square value of $\overline{R^2} = 2\alpha^2$, and hence the signal has an average power of $\overline{R^2}/2 = \alpha^2$. Thus, $10 \log_{10}(r^2/2\alpha^2)$, which equals $10 \log_{10}(r^2/2) - 10 \log_{10}(\alpha^2)$, represents the decibel difference between the signal power level when its amplitude is r and its average power. From Fig. 3.8 it will be noted that for signal power less than the average power by 10 dB or more, the distribution function $F(r)$ decreases by a factor of 10 for every 10-dB decrease in signal power. As a result, when fading radio signals exhibit this behavior, the fading is described as Rayleigh fading.

3.3.5 Thermal Noise

White noise[1] is defined as a random signal whose power spectral density is constant (i.e., independent of frequency). True white noise is not physically realizable since constant power spectral density over an infinite frequency range implies infinite power. However, thermal noise, which as indicated earlier has a Gaussian PDF, has a power spectral density that is relatively uniform up to frequencies of about 1000 GHz at room temperature (290K), and up to about 100 GHz at 29K.[4] Thus, for the purpose of practical communications

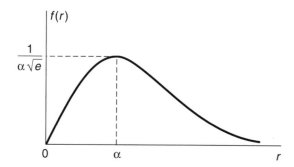

FIGURE 3.7 The Rayleigh probability density function. (From Taub, H., and Schilling, D., *Principles of Communication Systems,* McGraw-Hill, 1971, and reproduced with the permission of the McGraw-Hill Companies.)

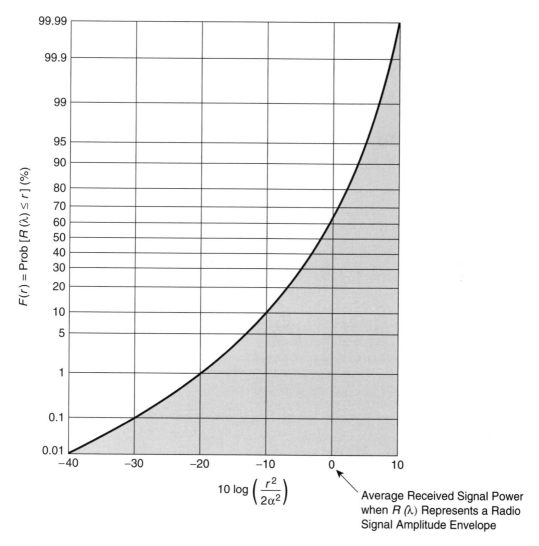

FIGURE 3.8 The Rayleigh cumulative distribution function. (By permission from Ref. 3.)

analysis, it is regarded as white. A simple model for thermal noise is one where the two-sided power spectral density $G_n(f)$ is given by

$$G_n(f) = \frac{N_0}{2} \text{ Watts / hertz} \tag{3.42}$$

where N_0 is a constant.

In a typical wireless communications receiver, the incoming signal and accompanying thermal noise is normally passed through a symmetrical bandpass filter centered at the carrier frequency f_c to minimize interference and noise. The width of the bandpass filter, W, is normally small compared to the carrier frequency. When this is the case the filtered noise can be characterized via its so-called *narrowband representation*.[1] In this representation, the filtered noise voltage, $n_{nb}(t)$, is given by

$$n_{nb}(t) = n_c(t)\cos 2\pi f_c t - n_s(t)\sin 2\pi f_c t \tag{3.43}$$

where $n_c(t)$ *and* $n_s(t)$ are Gaussian random processes of zero mean value, of equal variance and, further, independent of each other. Their power spectral densities, $G_{n_c}(f)$ and $G_{n_s}(f)$, extend only over the range $-W/2$ to $W/2$ and are related to $G_n(f)$ as follows:

$$G_{n_c}(f) = G_{n_s}(f) = 2G_n(f_c + f) \tag{3.44}$$

The relationship between these power spectral densities is shown in Fig. 3.9. This narrowband noise representation will be found to be very useful when we study carrier modulation methods.

3.3.6   Noise Filtering and Noise Bandwidth

In a receiver, a received signal contaminated with thermal noise is normally filtered to minimize the noise power relative to the signal power prior to demodulation. If, as shown in Fig. 3.10, the input two-sided noise spectral density is $N_0/2$, the transfer function of the real filter is $H_r(f)$, and the output noise spectral density is $G_{no}(f)$, then, by Eq. (3.21), we have

$$G_{no}(f) = \frac{N_0}{2}\left|H_r(f)\right|^2 \tag{3.45}$$

and thus the normalized noise power at the filter output, P_o, is given by

$$P_o = \int_{-\infty}^{\infty} G_{no}(f)\, df$$
$$= \frac{N_0}{2}\int_{-\infty}^{\infty}\left|H_r(f)\right|^2 df \tag{3.46}$$

A useful quantity to compare the amount of noise passed by one receiver filter versus another is the filter *noise bandwidth*.[1] The noise bandwidth of a filter is defined as the width of an ideal brick-wall (rectangular) filter that passes the same average power from a white noise source as does the real filter. In the case of a real low pass filter, it is assumed that

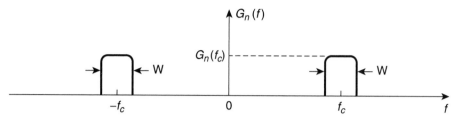

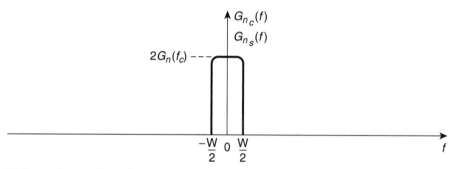

(a) Power Spectral Density of Bandpass Filtered Thermal Noise

(b) Power Spectral Density of Narrowband Noise Representation Components n_c and n_s

FIGURE 3.9 Spectral density relationships associated with narrowband representation of noise.

the absolute values of the transfer functions of both the real and brick-wall filters are normalized to one at zero frequency. In the case of a real bandpass filter, it is assumed that the brick-wall filter has the same center frequency as the real filter, f_c say, and that the absolute values of the transfer functions of both the real and brick-wall fitlers are normalized to one at f_c.

For an ideal brick-wall low pass filter of two-sided bandwidth B_n and $|H_{bw}(f)| = 1$ from $-B_n/2$ to $+B_n/2$

$$P_o = \frac{N_0}{2} B_n \tag{3.47}$$

Thus, from Eqs. (3.46) and (3.47) we determine that

$$B_n = \int_{-\infty}^{\infty} |H_r(f)|^2 df \tag{3.48}$$

White Noise

$G_n(f) = \dfrac{N_0}{2}$ → $H_r(f)$ → $G_{no}(f) = \dfrac{N_0}{2}|H_r(f)|^2$

FIGURE 3.10 Filtering of white noise.

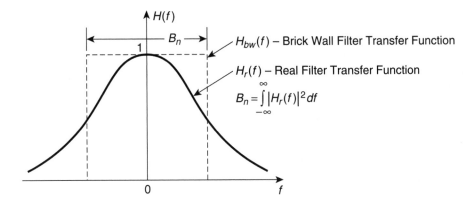

FIGURE 3.11   Low pass filter two-sided noise bandwidth, B_n.

Figure 3.11 shows the transfer function $H_{bw}(f)$ of a low pass brick-wall filter of two-sided noise bandwidth B_n superimposed on the two-sided transfer function $H_r(f)$ of a real filter.

REFERENCES

1. Taub, H., and Schilling, D., *Principles of Communication Systems*, McGraw-Hill, 1971.
2. Cooper, G. R., and McGillem, C. D., *Probabilistic Methods of Signal and System Analysis,* Holt, Rinehart and Winston, Inc., 1971.
3. Feher, K., editor, *Advanced Digital Communications,* Prentice Hall, 1987.
4. Members of the Technical Staff, Bell Telephone Laboratories, *Transmission Systems for Communications,* Revised Fourth Edition, Bell Telephone Laborities, Inc., 1971.

CHAPTER 4

FIXED WIRELESS DIGITAL MODULATION: THE BASIC PRINCIPLES

4.1 INTRODUCTION

Many methods of modulating the RF carrier have been applied to fixed wireless systems. However, over time, and as the state of the art has improved, certain methods have clearly become preferred due to their superior performance characteristics and/or ease of implementation. In particular, a generalized method called *quadrature amplitude modulation* (*QAM*) has proven to be the modulation of choice of many equipment providers. In this chapter these preferred methods will be reviewed along with some of the practical issues involved in their realization. First, however, the fundamentals of baseband transmission techniques will be reviewed, as these form the foundation on which the more complex techniques involving the modulation of a RF carrier are based.

4.2 PRINCIPLES OF BASEBAND DATA TRANSMISSION

4.2.1 Baseband Filtering

Data transported by communication systems is typically random or very close to random in nature. A stream of binary random full-length rectangular pulses (a full-length rectangular pulse is one of constant height for the entire pulse duration), where each pulse is equiprobable (probability of being positive is the same as being negative), is in fact a string of nonperiodic pulses. This is so because the value (polarity) of each pulse is entirely indedendent of all other pulses in the stream. For such a stream, of pulse duration τ and symbol rate $f_s = 1/\tau$ symbols per second, the two-sided amplitude spectral density is that of each of its nonperiodic pulses and thus as given in Eq. (3.3). The absolute value of its normalized positive (real) side is shown in Fig. 4.1. It occupies infinite bandwidth, with the first null at f_s. The spectrum within the first null is normally referred to as the main lobe. In communication systems bandwidth is normally at a premium. Thus the designer is motivated to filter the transmitted signal down to the minimum bandwidth possible without introducing errors into the transmission. Further, at the receiver, the incoming signal is normally filtered to minimize the negative effects of noise and interference. Filtering the signal results in changes to the shape of the original pulses, spreading their energy into adjacent pulses. This spreading effect is known as *dispersion* and can result in distortion of the pulse amplitude at the receiver sampling instant unless carefully controlled. In the *Nyquist criterion* on bandwidth transmission it is shown[1] that the minimum real channel bandwidth through which f_s independent symbols per second can be transmitted, without resulting in symbol amplitude distortion at the sampling instant, is the *Nyquist bandwidth* $f_n = f_s / 2$. Thus, for the rectangular pulse stream described previously, the minimum transmission bandwidth is half the width of the main lobe. Because the stream is binary, one transmitted symbol contains one information bit. Thus the bit rate f_b of the stream is the same as the symbol rate f_s and the minimum real transmission bandwidth f_n is $f_b / 2$.

Consider now the case of a 4-level PAM stream. Here two information bits are encoded into each symbol, the four symbol levels being encoded to represent 00, 01, 10, and 11. As a result, the symbol rate f_s is $f_b / 2$. In general, for *L*-level transmission systems, each transmitted symbol contains n information bits, where n is given by $n = \log_2 L$. Thus $f_s = f_b / n$, and the minimum transmission bandwidth $f_n = f_b / 2n$. For example, for an 8-level PAM signal, $n = 3$ and thus $f_n = f_b / 6$.

Let's initially look at the transmission of impulses, and then at the more practical case of pulse transmission. For impulse transmission, an ideal low-pass brick-wall filter, of bandwidth $f_s / 2$ and fixed time delay, will result in a nondistorted pulse amplitude at the sampling instant. Unfortunately, such a filter is unrealizable. Nyquist has shown, however, that if the amplitude characteristic of the brick-wall filter is modified to have a gradual roll off, with odd symmetry about the Nyquist bandwidth f_n, then a nondistorted pulse amplitude at the sampling instant is preserved. One class of such a filter that is most often used in digital communication systems is the so-called *raised cosine filter*. The amplitude characteristic of such a filter for impulse transmission, $H_{im}(f)$, consists of a flat portion followed by a roll-off portion that has a sinusoidal form. It is characterized in terms of its *roll-off factor* α, where α is defined

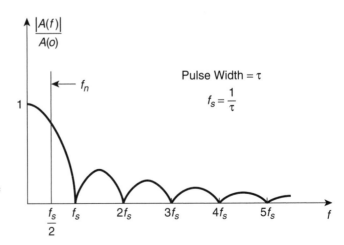

FIGURE 4.1   The absolute normalized amplitude spectral density of a train of rectangular pulses.

as $f_x / f_n, f_x$ being the amount of bandwidth used in excess of the Nyquist bandwidth. It is given mathematically by

$$H_{im}(f) = 1, \qquad\qquad\qquad\qquad 0 < f < f_n - f_x$$

$$= \frac{1}{2}\left[1 - \sin\frac{\pi}{2\alpha}\left(\frac{f}{f_n} - 1\right)\right], \qquad f_n - f_x < f < f_n + f_x \qquad (4.1)$$

$$= 0, \qquad\qquad\qquad\qquad f > f_n + f_x$$

where $\qquad \alpha = \dfrac{f_x}{f_n}$

Figure 4.2, which is from Feher,[1] is a graphical representation of $H_{im}(f)$.

The phase characteristic $\phi(f)$ of the raised cosine filter is linear over the frequency range where the amplitude response is greater than zero and is thus given by

$$\phi(f) = Kf \qquad\qquad 0 < f < f_n + f_x \qquad\qquad (4.2)$$

where K is a constant.

Because the input to the filter defined previously is a stream of impulses, then the amplitude spectral density at the filter input is of constant amplitude and, as per

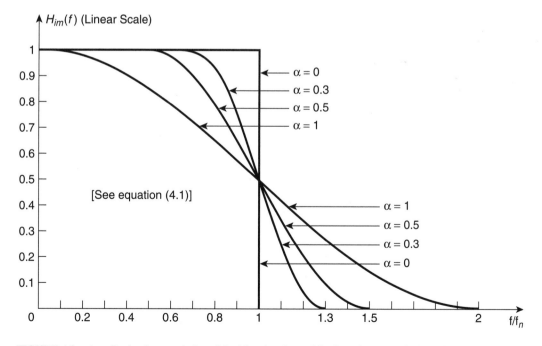

FIGURE 4.2   Amplitude characteristics of the Nyquist channel for impulse transmission. (By permission from Ref. 1.)

Eq. (3.13), the amplitude spectral density at the filter output $S_{rc}(f)$ has the identical characteristic of the filter transfer function $H_{im}(f)$, i.e.

$$S_{rc}(f) = H_{im}(f) \qquad (4.3)$$

A subtle but important point to note is that it is this output spectral density and hence the received pulse shape that results in nondistorted pulse amplitudes at the sampling instants, not the filter transfer function per se.

In practical systems, pulses of finite duration, not impulses, are used for digital transmission. A commonly used pulse is the full length rectangular pulse. We recall from Eq. (3.3) that the spectrum of such a pulse has a $(\sin x)/x$ form. For nondistorted pulse amplitudes at the sampling instances we desire that the spectral density, and hence the pulse shape, at the filter output of a transmission system conveying such pulses be the same as that for the impulse case discussed previously, namely $S_{rc}(f)$. To achieve this thus requires that the transfer function of the low-pass filter, $H_{rp}(f)$ say, be the filter transfer function for impulses modified by the factor $x/(\sin x)$. Thus $H_{rp}(f)$ is given by

$$H_{rp}(f) = \frac{\pi f / 2 f_n}{\sin(\pi f / 2 f_n)} H_{im}(f) \qquad (4.4)$$

Figure 4.3, which is also from Feher,[1] is a graphical representation of $H_{rp}(f)$.

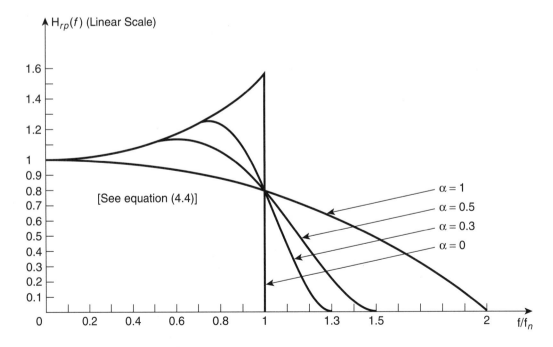

FIGURE 4.3   (a) Amplitude characteristics of the Nyquist channel for rectangular pulse transmission. (By permission from Ref. 1.)

The time response resulting from filtering to achieve a raised cosine spectral density $S_{rc}(f)$ as defined in Eq. (4.3) is given by[2]

$$h(t) = 2f_n \frac{\sin(2\pi f_n t)}{2\pi f_n t} \frac{\cos(2\pi f_n \alpha t)}{1 - (4 f_n \alpha t)^2} \tag{4.5}$$

and is shown in Fig. 4.4 for various values of α. We note that though $S_{rc}(0) = H_{im}(0) = 1$, the corresponding maximum pulse amplitude, $h(0)$, is given by

$$h(0) = 2f_n = f_s = 1/\tau \tag{4.6}$$

The shape of these time responses, irrespective of roll-off factor, explains why there is no pulse amplitude distortion at the sampling instant as a consequence of the spreading of adjacent pulses. Pulses are sampled at their maximum amplitude. However, with a raised cosine spectrum, spread pulses adjacent to a pulse being sampled are always at zero amplitude at the time of sampling.

The design of a filter with $\alpha = 0$, though desireable because it would result in a signal of minimum bandwidth, is not achievable. This is because its realization would require an infinite number of filter sections. In practice wireless communication systems use raised cosine filters with values of α as low as about 0.2 and as high as about 0.5.

The roll-off factor of a raised cosine filter is often expressed as the percentage of the bandwidth in excess of the Nyquist bandwidth. Thus a filter with $\alpha = 0.5$ is also referred to as a 50% roll-off or a 50% excess bandwidth filter.

The time responses shown in Fig. 4.4 are for perfect filters. In practice, raised cosine filters are never perfect. As a result, spread pulses adjacent to the pulse being sampled

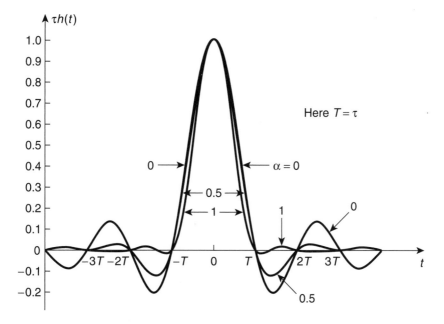

FIGURE 4.4   Time responses resulting from raised cosine amplitude spectral density. (By permission from Ref. 2.)

are not necessarily zero amplitude at the time of the sampling. Consequently, the sampled pulse is not always at its maximum nominal value but may be of a value that is more or less than expected. When this occurs it is referred to as *intersymbol interference (ISI)*.

In the laboratory intersymbol interference can be observed with the aid of an oscilloscope. On the vertical scale one displays the filtered response to the random pulse sequence under study, and the horizontal time base is set to the symbol duration. The resulting display is referred to as an *eye diagram* or an *eye pattern*. The inherent persistence of the oscilloscope's cathode-ray tube results in the display of superimposed pulse responses. Figure 4.5(a) shows what an eye pattern would look like that results from unlimited traces created by perfect raised cosine filtering of a binary stream of full-length rectangular pulses. Figure 4.5(b) shows what an eye pattern would look like that results from a few symbol traces created by imperfect filtering of a binary stream of full-length rectangular pulses. In Fig. 4.5(b), the difference d shown between the nominal peak amplitude and the minimum peak amplitudes is a measure of distortion caused by ISI, and the difference j between the nominal zero crossing time and the furthest removed zero crossing times is a measure of the *timing jitter*. The larger the ISI, the worse the error rate performance of the system. The larger the timing jitter, the larger the symbol sampling clock jitter, since the sampling clock is recovered via special circuitry from the incoming signal (Section 4.3.9.4). Jitter on the sampling clock can lead to degraded error rate performance.

The essential components of a baseband digital transmission system are shown in Fig. 4.6. The input signal can be a binary or multilevel (>2) pulse stream. The transmitter low-pass filter, with transfer function $T(f)$, is used to limit the transmitted spectrum. Noise and other interference are picked up by the transmission medium and fed into the

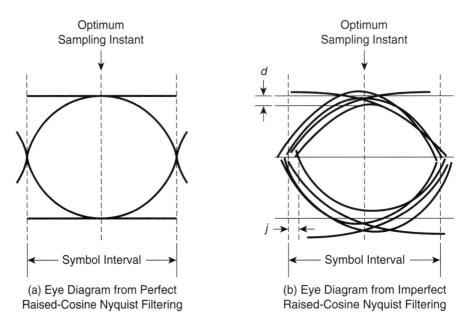

FIGURE 4.5   Eye diagrams from raised-cosine Nyquist filtering of a binary stream of full-length rectangular pulses.

receiver filter. The receiver filter, with transfer function $R(f)$, minimizes the noise and interference relative to the desired signal. The output of the receiver filter is fed to a decision threshold unit, which, for each pulse received, decides what was its most likely original level and outputs a pulse of this level. For a binary signal, it ouputs a pulse of amplitude $+V$, say, if the input pulse is above its decision threshold, which is 0 volts. If the input pulse is below 0 volts, it outputs a pulse of amplitude $-V$.

In designing the system, minimum ISI is desirable at the output of the receiver filter. This is normally achieved by employing filtering that results in a raised cosine amplitude spectral density S_{rc} at the receiver filter output. Thus, for full-length rectangular pulses, the combined transfer function of the transmitter filter and receiver filter (assuming that the transmission medium results in negligible impairment) should be as given in Eq. (4.4). There are an infinite number of ways of partitioning the total filtering transfer function between the transmitter filter and the receiver filter. Normally, the receiver filter is chosen to maximize the signal to noise ratio at its output as this optimizes the error rate performance in the presence of noise. It can be shown[2,3] that, for white Gaussian noise, the receiver filter transfer function $R(f)$ that accomplishes this is given by

$$R(f) = \left|S_{rc}(f)\right|^{\frac{1}{2}} \tag{4.7}$$

A filter with such a transfer function is referred to as a *root-raised cosine* filter.

The transmitter filter transfer function $T(f)$ is then chosen to maintain the desired composite characteristic; that is,

$$T(f) \cdot R(f) = H_{rp}(f) \tag{4.8}$$

Thus, by Eqs. (4.4) and (4.7),

$$T(f) \cdot \left|S_{rc}(f)\right|^{\frac{1}{2}} = \frac{\pi f / 2 f_n}{\sin\left(\pi f / 2 f_n\right)} S_{rc}(f)$$

and hence

$$T(f) = \frac{\pi f / 2 f_n}{\sin\left(\pi f / 2 f_n\right)} \left|S_{rc}(f)\right|^{\frac{1}{2}} \tag{4.9}$$

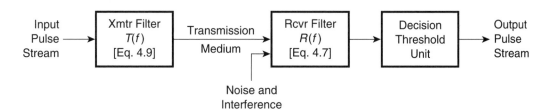

FIGURE 4.6   Basic baseband digital transmission system.

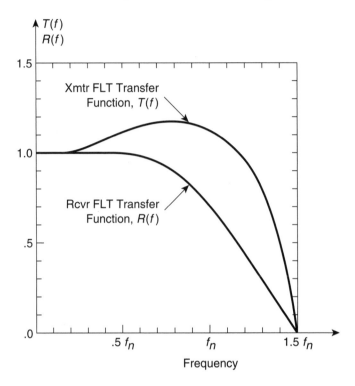

FIGURE 4.7   Amplitude transfer functions of transmitter and receiver filters when $\alpha = 0.5$.

Figure 4.7 shows plots of $R(f)$ and $T(f)$ for $\alpha = 0.5$.

4.2.2   Probability of Error of a Binary (2-level) PAM Signal in the Presence of White Gaussian Noise

Consider the case where the input pulse stream shown in Fig. 4.6 is a 2-level PAM signal of zero average amplitude and $R(f)$ and $T(f)$ are as defined in Eqs. (4.7) and (4.9). Then the output signal of the receiver filter, ignoring the corrupting effect of noise, will be as shown in Fig. 4.8. The decision threshold is set at the zero amplitude level. As a result, a positive pulse must be distorted at the sampling instant by a negative noise voltage n in excess of amplitude V for an error to occur. Similarly, a negative pulse must be distorted at the sampling instant by a positive noise voltage n in excess of amplitude V for an error to occur.

Let us assume that binary 1s are transmitted as positive pulses, and binary 0s are transmitted as negative pulses. If a 1 is transmitted but, as a result of noise, a 0 is decoded, the probability of such an occurrence is represented as

$$P_{e1} = P(0|1) \tag{4.10}$$

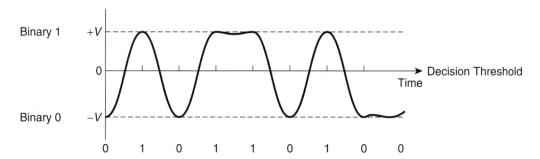

FIGURE 4.8 Polar binary pulse train with spectra filtered to be raised cosine.

and for the system described previously

$$P(0|1) = P[V - n < 0] = P[n > V] \tag{4.11}$$

Similarly, if a 0 is transmitted but a 1 is decoded, the probability of such an occurrence is represented by

$$P_{e0} = P(1|0) \tag{4.12}$$

and for the system described previously

$$P(1|0) = P[-V + n > 0] = P[n > V] \tag{4.13}$$

Thus

$$P(0|1) = P(1|0) = P[V > n] \tag{4.14}$$

If $P(1)$ represents the probability that a 1 is transmitted and $P(0)$ represents the probability that a 0 is transmitted, then the probability of bit error P_{be} of the binary transmission system can be represented by

$$P_{be} = P(1) \cdot P(0|1) + P(0) \cdot P(1|0) \tag{4.15}$$

For data transmission it is reasonable to assume that the binary digits 1 and 0 are equally probable; that is,

$$P(1) = P(0) = 1/2 \tag{4.16}$$

Thus, substituting Eqs. (4.14) and (4.16) into Eq. (4.15), we get

$$P_{be} = P[n > V] \tag{4.17}$$

Now

$$P[n > V] = 1 - P[n \le V] \tag{4.18}$$

and by Eq. (3.22)

$$P[n \le V] = F(V)$$

Thus

$$P_{be} = 1 - F(V) \qquad (4.19)$$

As the Gaussian noise is of zero mean ($m = 0$), its cumulative distribution function $F(V)$, which is the probability that the noise is less than $+V$, is given by Eq. (3.38a) and therefore

$$P_{be} = \frac{1}{2} erfc\left(\frac{V}{\sqrt{2}\sigma}\right) \qquad (4.20)$$

where σ^2 is the mean squared value of the noise voltage at the receiver filter output, which is in turn the normalized mean noise power at the receiver filter output P_{no}.

Equation (4.20) defines the probability of error in terms of the signal and noise at the receiver filter output. However, in communication systems we are more interested in the error performance as a function of the signal and noise at the receiver filter input.

The receiver filter of Fig. 4.6 is shown in Fig. 4.9 along with the frequency domain representation of its input signal and the time and frequency domain representations of its output signal. We recall from Eq. (4.6) that for $S_{rc}(0) = 1$, the maximum pulse amplitude $h(0) = 1 / \tau$. Thus, for a receiver filter output signal, $v_{ro}(t) = 1$, of maximum peak amplitude V, the spectral amplitude density at the receiver filter output $S_o(f)$ is given by

$$S_o(f) = \tau V \cdot S_{rc}(f) \qquad (4.21)$$

Now

$$S_o(f) = S_i(f) \cdot R(f) \qquad (4.22)$$

where $S_i(f)$ is the spectral amplitude density at the receiver filter input. Thus

$$S_i(f) = \frac{S_o(f)}{R(f)} \qquad (4.23)$$

For white Gaussian noise at its input and maximum signal to noise ratio at its output, $R(f)$, as indicated in Section 4.2.1, is given by Eq. (4.7). Substituting Eq. (4.21) and (4.7) into Eq. (4.23), we get

$$S_i(f) = \tau V \left| S_{rc}(f) \right|^{\frac{1}{2}} \qquad (4.24)$$

The power of the signal P_s at the receiver filter input is, by Eq. (3.17),

$$P_S = \frac{1}{\tau} \int_{-\infty}^{\infty} \left| S_i(f) \right|^2 df \qquad (4.25)$$

Substituting Eq. (4.24) into Eq. (4.25), we get

$$P_S = \tau V^2 \int_{-\infty}^{\infty} S_{rc}(f) \, df \qquad (4.26)$$

Now the raised cosine spectrum as characterized by Eq. (4.1) and as shown in Fig. 4.2 has unit value at $f = 0$ and odd symmetry about the Nyquist frequency $f_n = f_s / 2 = 1/2\tau$. Thus, for all values of the roll-off factor α

$$\int_0^\infty S_{rc}(f)\, df = \frac{1}{2\tau} \tag{4.27}$$

and therefore

$$\int_{-\infty}^\infty S_{rc}(f)\, df = \frac{1}{\tau} \tag{4.28}$$

Substituting Eq. (4.28) into Eq. (4.26), we get

$$P_S = V^2 \tag{4.29}$$

Having determined the relationship between the signal power at the receiver filter input and the filter output voltage, we now turn our attention to the relationship between the noise power at the receiver filter input, P_N, and the mean squared noise voltage at the filter output.

The normalized noise power P_{no} at the receiver filter output is, as per Eq. (3.46), given by

$$P_{no} = \frac{N_0}{2} \int_{-\infty}^\infty |R(f)|^2\, df \tag{4.30}$$

Substituting Eq. (4.7) into Eq. (4.30), we get

$$P_{no} = \frac{N_0}{2} \int_{-\infty}^\infty S_{rc}(f)\, df$$

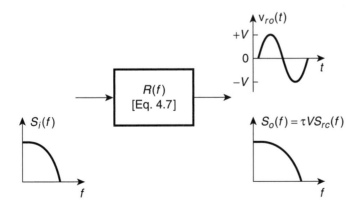

FIGURE 4.9 Binary PAM system receiver filter and its input and output signals.

and, by Eq. (4.28),

$$= \frac{N_0}{2\tau} \tag{4.31}$$

We note from Eq. (3.32) that, for Gaussian noise with zero mean, σ^2 is the mean squared value of the noise voltage which is in turn the normalized noise power. Thus

$$P_{no} = \sigma^2 \tag{4.32}$$

and therefore

$$\sigma^2 = \frac{N_0}{2\tau} \tag{4.33}$$

For the receive filter, the real (single-sided) Nyquist bandwidth $f_n = 1/2\tau$; thus the two-sided Nyquist bandwidth is equal to $1/\tau$. Hence, the noise power at the filter input in the two-sided Nyquist bandwidth, P_N, is given by

$$P_N = \frac{N_0}{2} \cdot \frac{1}{\tau} \tag{4.34}$$

Substituting Eq. (4.33) into Eq. (4.34), we get

$$P_N = \sigma^2 \tag{4.35}$$

In Eqs. (4.29) and (4.35) we now have the receiver filter input signal and the input noise in terms of the output signal and output noise. Specifically, we have

$$\frac{P_S}{2P_N} = \frac{V^2}{2\sigma^2} \tag{4.36}$$

Substituting Eq. (4.36) into Eq. (4.20), we get

$$P_{be} = \frac{1}{2} erfc \left[\left(\frac{P_S}{2P_N} \right)^{\frac{1}{2}} \right] \tag{4.37}$$

$$= Q \left[\left(\frac{P_S}{P_N} \right)^{\frac{1}{2}} \right] \tag{4.38}$$

where the Q function, which like the *erf* and the *erfc* is numerically computed, is given by

$$Q(z) = \frac{1}{2} erfc \left(\frac{z}{\sqrt{2}} \right) \tag{4.39}$$

We note that for this system the probability of bit error P_{be} is independent of the raised cosine spectrum excess bandwidth. A plot of the probability of error P_{be} versus the ratio of signal power to noise power in the two-sided Nyquist bandwidth for the two level ($L = 2$) polar system analyzed above is shown in Fig. 4.10.

4.2.3 Probability of Error of Multilevel PAM Signals in the Presence of White Gaussian Noise

Building on our analysis of 2-level PAM error performance, we now consider the generic case of even number multilevel PAM signals. We assume symbols transmitted at L possible levels, equally likely and equally spaced from each other. At the receiver output filter we assume that, with no corruption due ISI or noise, the symbol levels are of amplitude $\pm V, \pm 3V, \dots, \pm(L-1)V$. The decision thresholds are set at $0, \pm 2V, \dots, \pm(L-2)V$. Figure 4.11 shows the situation for an eight-level signal. An error occurs when the noise voltage at a sampling instant exceeds the value V, the distance between the uncorrupted level and the nearest decision threshold. As in the binary signal case, the two outermost signal levels, level 1 and level L, can be corrupted into an error zone by only one polarity of the noise voltage. For level 1 the noise polarity has to be positive and for level L the noise polarity has to negative. The probability of error of these levels is thus given by Eq. (4.20). All the inner signal levels, however, can be corrupted into an error zone by either polarity

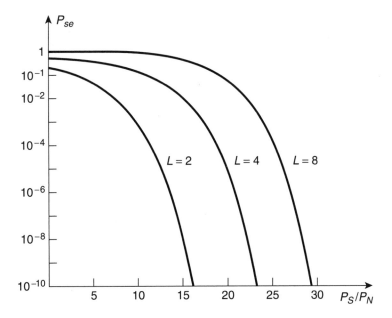

FIGURE 4.10   Probability of symbol error P_{se} versus P_s/P_N for L-level PAM, where P_s/P_N is the signal-to-noise power ratio in the Nyquist bandwidth.

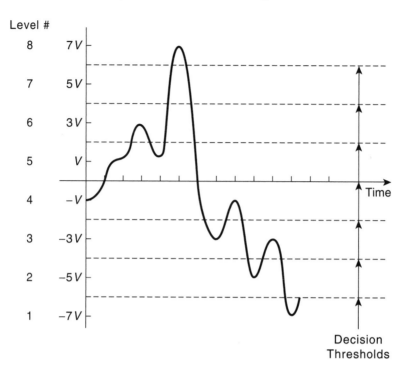

FIGURE 4.11   An 8-level PAM signal and associated decision thresholds.

of the noise. Thus the probability of error at these levels is twice that of the outer levels. Letting P_{ex} represent the probability of error of level X, we have

$$P_{e1} = P_{eL} = \frac{1}{2} erfc\left(\frac{V}{\sqrt{2}\sigma}\right) \tag{4.40}$$

and

$$P_{en}(n \neq 1, L) = 2P_{e1} \tag{4.41}$$

If $P(X)$ represents the probability that level X is transmitted, then P_{se}, the probability of symbol error of the L level transmission system, where all levels are assumed equally likely, is given by

$$P_{se} = P(1) \cdot P_{e1} + P(2) \cdot P_{e2} + ... P(L) \cdot P_{eL}$$

$$= \frac{1}{L}\left[2P_{e1} + (L-2)P_{en}\right]$$

$$= \frac{1}{L}\left[2P_{e1} + (L-2) \cdot 2P_{e1}\right]$$

$$= 2\left[1 - \frac{1}{L}\right]P_{e1} \tag{4.42}$$

and substituting Eq. (4.40) into Eq. (4.42), we get

$$P_{se} = \left[1 - \frac{1}{L}\right] erfc\left(\frac{V}{\sqrt{2}\sigma}\right)$$

(4.43)

We know from our earlier studies on the binary system that for a pulse amplitude level V at the receiver filter output the associated average symbol power at the filter input is V^2. Thus, for an L level system, where all levels are assumed equally likely, the average symbol power at the filter input P_s is given by

$$P_S = 2\left[P(\frac{L}{2}+1)\cdot V^2 + P(\frac{L}{2}+2)\cdot[3V]^2 + ...P(L)\cdot\left[(L-1)V\right]^2 \right]$$

$$= 2\cdot\frac{1}{L}\left[V^2 + [3V]^2 + ...[(L-1)V]^2\right]$$

$$= \frac{2V^2}{L}\left[1^2 + 3^2 + ...(L-1)^2\right]$$

(4.44)

Labeling the series $[1^2 + 3^2 + ...(L-1)^2]$ as S_{L-1}, we have

$$S_{L-1} = \left[1^2 + 2^2 + 3^2 + (L-1)^2\right] - \left[2^2 + 4^2 + ...(L-2)^2\right]$$

$$= \left[\sum_{x=1}^{x=L-1} x^2\right] - \left[2^2 \sum_{x=1}^{x=\frac{L-2}{2}} x^2\right]$$

(4.45)

But the series $1^2 + 2^2 + ...n^2$ is well known and given in closed form by

$$1^2 + 2^2 + ...n^2 = (n+1)(2n+1)/6$$

(4.46)

Substituting Eq. (4.46) into Eq. (4.45) leads to

$$S_{L-1} = \frac{L}{6}\left[L^2 - 1\right]$$

(4.47)

and substituting Eq. (4.47) into Eq. (4.44) gives

$$P_S = \frac{V^2}{3}\left[L^2 - 1\right]$$

(4.48)

Thus

$$V^2 = \frac{3P_S}{\left[L^2 - 1\right]}$$

(4.49)

The relationship between the noise power at the receiver filter input and output is the same as for the binary case. Thus, as per Eq. (4.35),

$$P_N = \sigma^2 \tag{4.50}$$

Substituting Eqs. (4.49) and (4.50) into Eq. (4.43), we get the following relationship for the probability of symbol error P_{se}:

$$P_{se} = \left[1 - \frac{1}{L}\right] erfc \left[\left(\frac{3}{2\left[L^2 - 1\right]} \cdot \frac{P_S}{P_N}\right)^{\frac{1}{2}}\right] \tag{4.51}$$

$$= 2\left[1 - \frac{1}{L}\right] Q \left[\left(\frac{3}{L^2 - 1} \cdot \frac{P_S}{P_N}\right)^{\frac{1}{2}}\right] \tag{4.52}$$

As expected, we note that for $L = 2$, P_{se} as per Eq. (4.52) equals P_{be} as per Eq. (4.39). Plots of P_{se} versus P_S / P_N for values of L equal to 4 and 8 are displayed in Fig. 4.10 in addition to the plot for $L = 2$ mentioned in Section 4.2.2.

4.3   LINEAR MODULATION SYSTEMS

Previously we discussed PAM baseband systems. Wireless communication systems, however, operate in assigned frequencies that are considerably higher than baseband frequencies. It is thus necessary to employ modulation techniques that shift the baseband data up to the operating frequency. In this section we consider linear modulation systems. These systems are so called because they exhibit a linear relationship between the baseband signal and the modulated RF carrier. As a result of this relationship, their performance in the presence of noise and other impairments can be deduced from their equivalent baseband forms (hence our earlier review of baseband systems). We will commence this study by reviewing so-called *double-sideband suppressed carrier (DSBSC)* modulation as this modulation forms the foundation on which many of the most widely used linear modulation methods are based.

4.3.1   Double-Sideband Suppressed Carrier (DSBSC) Modulation

A simplified DSBSC system for PAM signal transmission is shown in Fig. 4.12. First a polar PAM input signal, $a(t)$, with equiprobable symbols, is filtered with the low-pass filter, F_T, to limit its bandwidth to f_m, say, and the filtered signal $b(t)$ applied to a multiplier. Also feeding the multiplier is a sinusoidal signal at the desired carrier frequency f_c. As a result, the output signal of the multiplier, $c(t)$, is given by

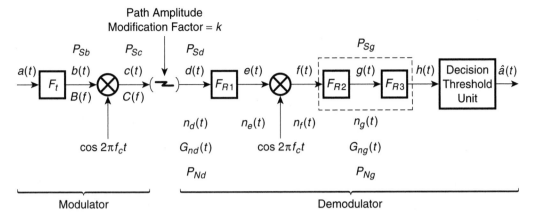

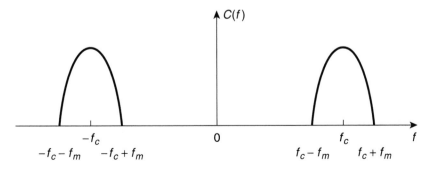

FIGURE 4.12 Simplified one-way DSBSC system for PAM transmission.

$$c(t) = b(t)\cos 2\pi f_c t \qquad (4.53)$$

If the amplitude spectral densities of $b(t)$ and $c(t)$ are represented as $B(f)$ and $C(f)$, respectively, then, by Eq. (3.9), $C(f)$ is given by

$$C(f) = \frac{1}{2}B(f + f_c) + \frac{1}{2}B(f - f_c) \qquad (4.54)$$

Thus, as shown in Fig. 4.13, $C(f)$ consists of two spectra. One is real and centered at f_c, the other imaginary and centered at $-f_c$, and each has a bandwidth $2f_m$ and an amplitude half that of $B(f)$. As these spectra are symmetrically disposed on either side of the carrier frequency, the signal is referred to as a *double-sideband (DSB)* signal. Further, as $b(t)$ is polar with equiprobable symbols, and thus has no fixed component, $c(t)$ contains no discrete carrier frequency component and is thus referred to as a *suppressed carrier* signal.

FIGURE 4.13 A DSBSC signal amplitude spectral density.

We assume that $c(t)$ travels over a linear transmission path and arrives at the demodulator input modified in amplitude by the factor k. Thus the input signal $d(t)$ to the receiver is given by

$$d(t) = k \cdot c(t) \tag{4.55}$$

The received signal $d(t)$ is passed through the bandpass filter F_{R1} to limit noise and interference. The bandwidth W of F_{R1} is normally greater than $2f_m$, the bandwidth of $d(t)$, so as to not impact the spectral density of $d(t)$. Assuming this to be the case, the output signal $e(t)$ of F_{R1} is given by

$$e(t) = d(t) \tag{4.56}$$

The signal $e(t)$ is fed to a multiplier. Also feeding the multiplier is the sinusoidal signal $\cos 2\pi f_c t$. As a result, the output of the multiplier $f(t)$ is given by

$$f(t) = e(t)\cos 2\pi f_c t \tag{4.57}$$

Substituting Eqs. (4.56), (4.55) and (4.53) into Eq. (4.57), we get

$$f(t) = k \cdot b(t)\cos^2\left(2\pi f_c t\right)$$

$$= \frac{k}{2}b(t)\left[1 + \cos\left(2 \cdot 2\pi f_c t\right)\right]$$

$$= \frac{k}{2}b(t) + \frac{k}{2}b(t)\cos(2 \cdot 2\pi f_c t) \tag{4.58}$$

Thus, by multiplying $e(t)$ by $\cos 2\pi f_c t$, a process referred to as *coherent detection*, we recover $b(t)$ and create a second signal with the same double-sided bandwidth as $b(t)$ but centered at $2f_c$. The signal $f(t)$ is fed into the low-pass filter F_{R2} that eliminates the component of the signal centered about $2f_c$ while leaving the baseband component undisturbed. Thus, the output of F_{R2}, $g(t)$, is given by

$$g(t) = \frac{k}{2}b(t) \tag{4.59}$$

The signal $g(t)$ is fed to F_{R3} for final pulse shaping prior to level detection in the decision threshold unit. In practice F_{R2} and F_{R3} are combined into one but are shown separately here to add clarity to the analysis. The output, $\hat{a}(t)$, of the decision threshold unit is a PAM signal that is the demodulator's best estimate of the modulator input signal, $a(t)$.

We know from our studies on baseband PAM signals that, for optimum probability of error performance in the presence of Gaussian noise, (a) the receiver baseband filter, F_{R3}, must be root raised cosine, and (b) the transmitter baseband shaping filter, F_T, must be such as to result in a raised cosine amplitude spectral density at the output of F_{R3}. With F_{R3} and F_T so chosen, then the probability of error P_e is given by Eq. (4.52), where P_S and P_N are defined at the input to F_{R3}. However, for the DSBSC system under study it is more appropriate to define error performance with respect to the signal to noise ratio at the input to the receiver. To do this we must, therefore, determine the relationship between the signal power to noise power ratio at the input to receiver filter F_{R1} and the equivalent ratio at the input to F_{R3}. This determination is somewhat tedious, and the impatient reader may want to skip directly to the result, which is given in Eq. (4.81). For those wanting to stay the course, we commence by determining the relationship between the power of the signal $d(t)$ at the input to F_{R1}, P_{Sd} say, and that of the signal $g(t)$ at the input to F_{R3}, P_{Sg} say.

Let the average normalized power of the signal $b(t)$ be P_{Sb} and the duration of each pulse of the PAM train be τ. Then, by Eq. (3.17), P_{Sb} is given by

$$P_{Sb} = \frac{1}{\tau} \int_{-\infty}^{\infty} |B(f)|^2 \, df \tag{4.60}$$

If the average normalized power of the signal $c(t)$ is P_{Sc}, then, also by Eq. (3.17), P_{Sc} is given by

$$P_{Sc} = \frac{1}{\tau} \int_{-\infty}^{\infty} |C(f)|^2 \, df \tag{4.61}$$

Substituting Eq. (4.54) into Eq. (4.61), we get

$$P_{Sc} = \frac{1}{\tau} \int_{-\infty}^{\infty} \frac{1}{4} |B(f + f_c)|^2 \, df + \frac{1}{\tau} \int_{-\infty}^{\infty} \frac{1}{4} |B(f - f_c)|^2 \, df \tag{4.62}$$

But $B(f + f_c)$ and $B(f - f_c)$ have the identical spectral content to $B(f)$, only they are shifted on the frequency axis. Thus

$$\int_{-\infty}^{\infty} |B(f + f_c)|^2 \, df = \int_{-\infty}^{\infty} |B(f - f_c)|^2 \, df = \int_{-\infty}^{\infty} |B(f)|^2 \, df \tag{4.63}$$

Substituting Eq. (4.63) into Eq. (4.62), we get

$$P_{Sc} = \frac{1}{2\tau} \int_{-\infty}^{\infty} |B(f)|^2 \, df$$

and by Eq. (4.60)

$$= \frac{1}{2} P_{Sb} \tag{4.64}$$

Since, as per Eq. (4.55), $d(t) = k \cdot c(t)$, then

$$P_{Sd} = k^2 P_{Sc}$$

$$= \frac{k^2}{2} P_{Sb} \tag{4.65}$$

From Eq. (4.59) we deduce that

$$P_{Sg} = \frac{k^2}{4} P_{Sb} \tag{4.66}$$

Thus, from Eqs. (4.65) and (4.66) we determine that

$$P_{Sg} = \frac{1}{2} P_{Sd} \tag{4.67}$$

Equation (4.67) indicates that, as a result of the coherent detection of the incoming signal, $d(t)$, the average power of the resulting baseband signal $g(t)$ is half that of $d(t)$.

We now determine the relationship between the noise power P_{Nd} at the demodulator input and the noise power P_{Ng} at the input to the baseband-shaping filter F_{R3}. To do this, we must first, however, determine the relationship between the noise power spectral density $G_{nd}(f)$ at the demodulator input and the noise power spectral density $G_{ng}(f)$ at the input to F_{R3}.

At the demodulator input we assume that the received noise $n_d(t)$ is white Gaussian of two-sided power spectral density $G_{nd}(f)$ given by

$$G_{nd}(f) = N_0 / 2 \tag{4.68}$$

The received noise, like the desired signal, is filtered by the bandpass filter F_{R1} of bandwidth W. As a result, the noise signal $n_e(t)$ at the output of F_{R1} is no longer white. However, within the filter passband frequencies, its power spectral density $G_{ne}(f)$ is still $N_0 / 2$. Because the noise has been filtered to occupy a bandwidth that is small compared to f_c we can represent $n_e(t)$ by its narrow-band representation. Thus, by Eq. (3.43), we have

$$n_e(t) = n_{ec}(t)\cos 2\pi f_c t - n_{es}(t)\sin 2\pi f_c t \tag{4.69}$$

Like the desired signal $e(t)$, $n_e(t)$ is multiplied by the sinusoidal signal $\cos 2\pi f_c t$. As a result, the noise component of the signal at the multiplier output, $n_f(t)$, is given by

$$n_f(t) = n_e(t)\cos 2\pi f_c t \tag{4.70}$$

Substituting Eq. (4.69) into Eq. (4.70), we get

$$n_f(t) = n_{ec}(t)\cos^2\left(2\pi f_c t\right) - n_{es}(t)\sin 2\pi f_c t \cdot \cos 2\pi f_c t$$

$$= \frac{1}{2}n_{ec}(t) + \frac{1}{2}n_{ec}(t)\cos 4\pi f_c t - \frac{1}{2}n_{es}(t)\sin 4\pi f_c t \tag{4.71}$$

Filter F_{R2} removes the components of $n_f(t)$ centered about $2f_c$. Thus, the noise signal $n_g(t)$ at the output of F_{R2} is given by

$$n_g(t) = \frac{1}{2}n_{ec}(t) \tag{4.72}$$

From Eq. (4.72) we deduce that the power spectral density $G_{ng}(f)$ of $n_g(t)$ is related to the power spectral density $G_{nec}(f)$ of $n_{ec}(t)$ as follows:

$$G_{ng}(f) = \frac{1}{4} G_{nec}(f) \qquad (4.73)$$

By Eq. (3.44) we have

$$G_{nec}(f) = 2 G_{ne}(f_c + f) \qquad (4.74)$$

Thus, by substituting Eq. (4.74) into Eq. (4.73) we have

$$G_{ng}(f) = \frac{1}{2} G_{ne}(f_c + f) \qquad (4.75)$$

But

$$G_{ne}(f_c + f) = \frac{N_0}{2}, \qquad |f| < \frac{W}{2} \qquad (4.76)$$

Therefore, substituting Eq. (4.76) into Eq. (4.75), we get

$$G_{ng}(f) = \frac{N_0}{4}, \qquad |f| < \frac{W}{2} \qquad (4.77)$$

Since the two-sided noise spectral density G_{nd} at the demodulator input is $N_0/2$, Eq. (4.77) indicates that the two-sided noise spectral density $G_{ng}(f)$ at the input to F_{R3}, over the bandwidth $-W/2$ to $+W/2$, is one-half that at the input to the demodulator. Thus the noise power P_{Ng} in the two-sided Nyquist bandwidth at the input to F_{R3} is given by

$$P_{Ng} = \frac{1}{2} \cdot \frac{N_0}{2} W \qquad (4.78)$$

and P_{Nd}, the noise power in the two-sided Nyquist bandwidth at the input to the demodulator, is given by

$$P_{Nd} = \frac{N_0}{2} \cdot W \cdot 2 = N_0 W \qquad (4.79)$$

From Eqs. (4.78) and (4.79), we have

$$P_{Ng} = \frac{P_{Nd}}{4} \qquad (4.80)$$

and finally, from Eqs. (4.67) and (4.80) we have the relationship we set out to determine, i.e.,

$$\frac{P_{Sg}}{P_{Ng}} = 2 \frac{P_{Sd}}{P_{Nd}} \qquad (4.81)$$

Equation (4.81) indicates that the process of double-sideband coherent detection results in a doubling of the signal to noise ratio from predetection to postdetection. This improvement can be explained by recognizing that at the detector output sideband voltages of the signal add coherently whereas sideband voltages of the noise add incoherently.

Substituting Eq. (4.81) into Eq. (4.52), we get, for PAM modulated DSBSC systems, a symbol P_{se} given by

$$P_{se} = 2\left[1 - \frac{1}{L}\right]Q\left[\left(\frac{6}{L^2-1}\cdot\frac{P_S}{P_N}\right)^{\frac{1}{2}}\right] \tag{4.82}$$

where P_S is the average signal power at the demodulator input and P_N is the noise power in the two-sided Nyquist bandwidth at the demodulator input.

4.3.2   Binary Phase Shift Keying (BPSK)

A special case of PAM transmission via a DSBSC system is when the PAM signal $a(t)$ in Fig. 4.12 has a binary, polar, NRZ format. In this situation, if the filtered signal $b(t)$ has maximum peak amplitude of $\pm b$ volts say, then the modulated signal $c(t)$ varies between $c_1(t)$ and $c_0(t)$ as $b(t)$ varies between $+b$ and $-b$, where

$$c_1(t) = b\cos 2\pi f_c t$$
$$c_0(t) = -b\cos 2\pi f_c(t) \tag{4.83}$$

$$= b\cos(2\pi f_c t + \pi) \tag{4.84}$$

When $b(t)$ is positive, the phase of $c(t)$ is 0°. When $b(t)$ is negative, the phase of $c(t)$ is π radians or $-180°$. Thus the phase has only two states. This modulation is referred to as *binary phase shift keying* (*BPSK*) and represents the simplest linear modulation scheme. Fig. 4.14 shows typical examples of signals $a(t)$, $b(t)$ and $c(t)$. Fig. 4.15 shows the *signal space* or *vector* or *constellation* diagram of $c(t)$. This diagram portrays both the amplitude and phase of $c(t)$ at the instances when the modulating signal $b(t)$ is at its peak.

Since, for BPSK, individual input bits are converted to individual symbols, it follows that the probability of bit error $P_{be(BPSK)}$ is the same as the probability of symbol error $P_{se(BPSK)}$. As the number of PAM levels L is 2, $P_{se(BPSK)}$ and hence $P_{be(BPSK)}$ is given by Eq. (4.82) to be

$$P_{se(BPSK)} = P_{be(BPSK)} = Q\left[\left(\frac{2P_S}{P_N}\right)^{\frac{1}{2}}\right] \tag{4.85}$$

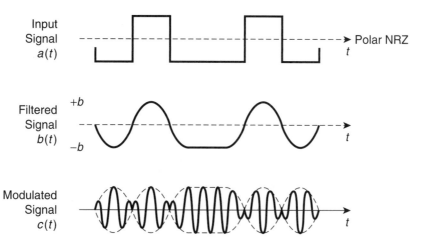

FIGURE 4.14   Typical BPSK signals.

where P_S = the average signal power at the demodulator input and P_N = the noise power in the two-sided Nyquist bandwidth at the demodulator input.

So far probability of error has been defined in terms of a signal to noise ratio. In digital communication systems, however, it is just as common to define P_e in terms of the ratio of the energy per bit E_b in the received signal to the noise power density N_0 at the receiver input. Defining P_{be} in terms of E_b / N_0 makes it easy to compare the error performance of different modulation systems for the same bit rate. Given that, for a BPSK system with bit rate f_b, and hence a single-sided, double-sideband Nyquist band-width also of f_b,

$$P_S = E_b \cdot f_b \tag{4.86}$$

and

$$P_N = \frac{N_0}{2} \cdot f_b \cdot 2 = N_0 f_b \tag{4.87}$$

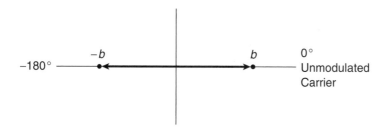

FIGURE 4.15   Signal space diagram of BPSK modulated signal.

Thus

$$\frac{P_S}{P_N} = \frac{E_b}{N_0} \tag{4.88}$$

and hence

$$P_{be(BPSK)} = Q\left[\left(2\frac{E_b}{N_0}\right)^{\frac{1}{2}}\right] \tag{4.89}$$

Figure 4.16 shows the power spectral density of BPSK when the modulating signal is unfiltered. This power spectral density has the same $[\sin x\,/\,x]^2$ form as that of the two-sided baseband signal, except that it is shifted in frequency by f_c. It can be shown[4] to be given by

$$G_{BPSK}(f) = P_S\tau_b\left[\frac{\sin \pi(f - f_c)\tau_b}{\pi\left(f - f_c\right)\tau_b}\right]^2 \tag{4.90}$$

where τ_b, which is equal to $1\,/\,f_b$, is the bit duration of the baseband signal.

Also shown in Fig. 4.16 is the single-sided, double-sideband Nyquist bandwidth of the system, which is equal to f_b. Thus, at its theoretical best, BPSK is capable of transmitting 1 bit per second in each Hertz of transmission bandwidth. The system is therefore said to have a maximum *spectral efficiency* of 1 bit/s/Hz. Because, as indicated earlier, fil-

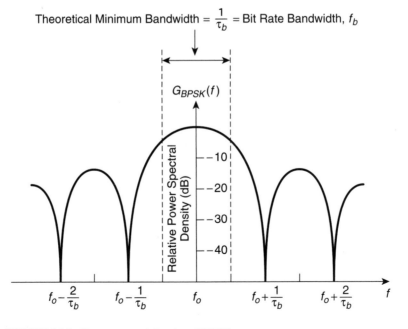

FIGURE 4.16   Power spectral density of BPSK.

tering to 0% excess bandwidth to achieve the Nyquist bandwidth is not practical, real BPSK systems have spectral efficiencies less than 1 bit/s/Hz. For an excess bandwidth of 25% say, then data at a rate of 1 bit/s requires 1.25 Hertz of bandwidth, leading to spectral efficiency of 0.8 bits/s/Hz.

As we shall see in succeeding sections, spectral efficiencies much greater than that afforded by BPSK are easily realizable. As a result, BPSK is rarely used in wireless communication networks, where, as a rule, available spectrum is limited and thus highly valued. Nonetheless, an understanding of its operating principles is very valuable in analyzing *quadrature phase shift keying (QPSK)*, a popular modulation technique.

4.3.3 Quadrature Amplitude Modulation (QAM)

The PAM-modulated DSBSC systems described previously are only capable of amplitude modulation accompanied by 0° or 180° phase shifts. However, by adding a *quadrature branch* as shown if Fig. 4.17, it becomes possible to generate signals with any desired amplitude and phase. In the quadrature branch a second PAM baseband signal is multiplied with a sinusoidal carrier of frequency f_c, identical to that of the *in-phase* carrier but delayed in phase by 90°. The outputs of the two multipliers are then added together to form a *quadrature amplitude modulated (QAM)* signal.

Labeling the in-phase PAM filtered signal $b_i(t)$ and the quadrature PAM filtered signal $b_q(t)$, then the summed output of the modulator, $c(t)$, is given by

$$c(t) = b_i(t)\cos 2\pi f_c t + b_q(t)\cos(2\pi f_c t - \pi/2)$$

$$= b_i(t)\cos 2\pi f_c t + b_q(t)\sin 2\pi f_c t \tag{4.91}$$

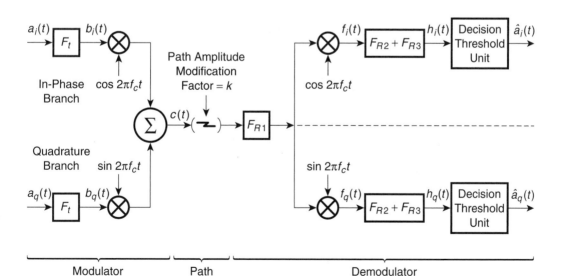

FIGURE 4.17   Simplified one-way quadrature amplitude modulated system.

At the QAM demodulator the incoming signal is passed through the bandpass filter F_{R1} to limit noise and interference. It is then divided into two and each branch inputted to a multiplier, one multiplier being fed also with an in-phase carrier, $\cos 2\pi f_c t$, the other with a quadrature carrier, $\sin 2\pi f_c t$.

Ignoring the effects of noise and interference, the output signal $f_i(t)$ of the in-phase multiplier is given by

$$f_i(t) = k \cdot c(t) \cdot \cos 2\pi f_c t$$

$$= k \cdot b_i(t) \cos^2 2\pi f_c t - k \cdot b_q(t) \sin 2\pi f_c(t) \cdot \cos 2\pi f_c t$$

$$= \frac{k}{2} b_i(t) + \frac{k}{2} b_i(t) \cos(2 \cdot 2\pi f_c t) - \frac{k}{2} b_q(t) \sin(2 \cdot 2\pi f_c t) \qquad (4.92)$$

The only difference between $f_i(t)$ and $f(t)$ of Eq. (4.58) for an in-phase only modulated system is the final component of Eq. (4.92). However, this component, like the second component in the Eq. (4.92), is spectrally centered at $2f_c$ and is filtered prior to decision threshold detection, leaving only the original quadrature modulating signal $b_i(t)$.

Again ignoring the effects of noise and interference, the output signal $f_q(t)$ of the quadrature multiplier is given by

$$f_q(t) = k \cdot c(t) \cdot \sin 2\pi f_c t$$

$$= k \cdot b_q(t) \sin^2 2\pi f_c t + k \cdot b_i(t) \cos 2\pi f_c t \cdot \sin 2\pi f_c t$$

$$= \frac{k}{2} b_q(t) - \frac{k}{2} b_q(t) \cos(2 \cdot 2\pi f_c t) + \frac{k}{2} b_i(t) \sin(2 \cdot 2\pi f_c t) \qquad (4.93)$$

As with the output $f_i(t)$ from the in-phase multiplier, $f_q(t)$ consists of the original in-phase modulating signal $b_q(t)$ as well as two components centered spectrally at $2f_c$ that are filtered prior to decision threshold unit. Thus, by quadrature modulation, it is possible to transmit two independent bit streams on the same carrier with no interference of one stream with the other, given ideal conditions.

4.3.4   Quaternary Phase Shift Keying (QPSK)

Quaternary (or quadrature) phase shift keying (QPSK) is one of the simplest implementations of quadrature amplitude modulation and is sometimes referred to as *4-QAM*. In it, the modulated signal has four distinct states. A block diagram of a conventional, simplified, QPSK system is shown in Fig. 4.18(a). The binary NRZ input data stream $a_{in}(t)$ of bit rate f_b and bit duration τ_b, is fed to the modulator, where it is converted by a serial to parallel converter into two NRZ streams, an *I* stream labeled $a_i(t)$ and a *Q* stream labeled $a_q(t)$, each of symbol rate f_B, half that of f_b, and symbol duration τ_B, twice that of τ_b. The relationship between the data streams $a_{in}(t)$, $a_i(t)$, and $a_q(t)$ is shown in Fig. 4.18(b). The *I* and *Q* streams undergo standard QAM processing as described in Section 4.3.3. The in-phase multiplier if fed by the carrier signal $\cos 2\pi f_c t$. The quadrature multiplier is fed by the carrier signal delayed by $90°$ to create the signal $\sin 2\pi f_c t$. The output of each multiplier is a BPSK signal. The BPSK output signal of the in-phase carrier driven multiplier has phase values of $0°$ and $180°$ relative to the in-phase carrier, and the BPSK output signal of the

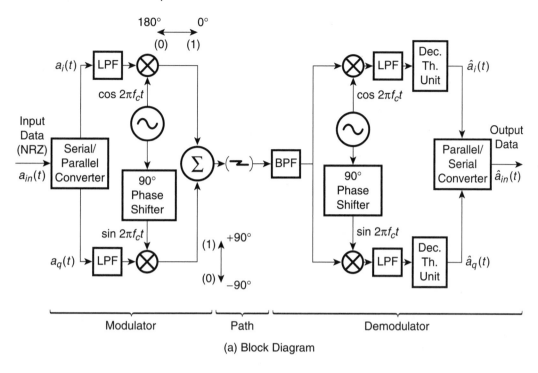

(a) Block Diagram

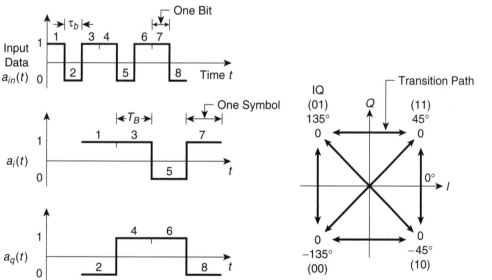

Note: Data in bit number 'n' is converted
to I/Q baud number 'n'.

Bit Rate = $\frac{1}{\tau_b}$ Baud or Symbol Rate = $\frac{1}{\tau_B}$

(b) Modulator Data Streams

(c) Signal Space Diagram

FIGURE 4.18   QPSK system representation.

quadrature carrier driven multiplier has phase values of 90° and 270° relative to the in-phase carrier. The multiplier outputs are summed to give a four-phase signal. Thus, QPSK can be regarded as two associated BPSK (ABPSK) systems operating in quadrature.

The four possible output signal states of the modulator, their *IQ* digit combinations, and their possible transitions from one state to another are shown in Fig. 4.18(c). We note that either 90° or 180° phase transitions are possible. For example, a 90° phase transition occurs when the *IQ* combination changes from 00 to 10, and a 180° phase transition occurs when the *IQ* combination changes from 00 to 11. For a system where $a_i(t)$ and $a_q(t)$ are unfiltered prior to application to the multipliers, phase transitions occur instantaneously and thus the signal has a constant amplitude. However, for systems where these signals are filtered, as is normally the case to limit the radiated spectrum, phase transitions occur over time and the modulated signal has an amplitude envelope that varies with time. In particular, a 180° phase change results in a change over time in amplitude envelope value from maximum to zero and back to maximum.

In the demodulator, as a result of quadrature demodulation, signals $\hat{a}_i(t)$ and $\hat{a}_q(t)$, estimates of the original modulating signals, are produced. These signals are then recombined in a parallel to serial converter to form $\hat{a}_{in}(t)$, an estimate of the original input signal to the modulator.

As indicated previously, QPSK can be regarded as two associated BPSK systems operating in quadrature. From a spectral point of view at the modulator output, two BPSK signal spectra are superimposed on each other. The BPSK symbol duration is τ_B. But $\tau_B = 2\tau_b$. Thus the spectral density of each BPSK signal, and hence of the QPSK signal, is given by Eq. (4.90), but with τ_b replaced by $2\tau_b$. Making this replacement, we get

$$G_{QPSK}(f) = 2P_S\tau_b \left[\frac{\sin 2\pi(f - f_c)\tau_b}{2\pi(f - f_c)\tau_b} \right]^2 \tag{4.94}$$

A graph of $G_{QPSK}(f)$ is shown in Fig. 4.19. We note that the widths of the main lobe and side lobes are half that for BPSK given the same bit rate for each system. As a result, the maximum spectral efficiency of QPSK is twice that of BPSK (i.e., 2 bits/s/Hz).

Because QPSK can be treated as two *associated BPSK* (*ABPSK*) systems operating in quadrature, with the QPSK bit stream being in fact the combined ABPSK bit streams, the probability of bit error of the QPSK system, $P_{be(QPSK)}$, is the same as the probability of bit error of the ABPSK systems, $P_{be(ABPSK)}$. That is,

$$P_{be(QPSK)} = P_{be(ABPSK)} \tag{4.95}$$

By Eq. (4.85), $P_{be(ABPSK)}$ is given by

$$P_{be(ABPSK)} = Q\left[\left(\frac{2P_{S(ABPSK)}}{P_{N(ABPSK)}} \right)^{\frac{1}{2}} \right] \tag{4.96}$$

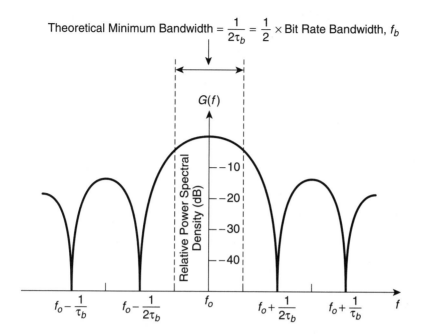

FIGURE 4.19 Power spectral density fo QPSK and offset QPSK.

For each ABPSK system, the average signal power $P_{S(ABPSK)}$ at the demodulator input is half that of $P_{S(QPSK)}$, that is,

$$P_{S(ABPSK)} = \frac{1}{2} P_{S(QPSK)} \qquad (4.97)$$

Also,

$$P_{S(QPSK)} = E_b \cdot \frac{1}{\tau_b} \qquad (4.98)$$

Thus, by Eqs. (4.97) and (4.98), we have

$$P_{S(ABPSK)} = \frac{E_b}{2\tau_b} \qquad (4.99)$$

The Nyquist bandwidth of the ABPSK signal is equal to $1/\tau_b$. Thus $P_{N(ABPSK)}$, the noise in the ABPSK two-sided Nyquist bandwidth at the receiver input, is given by

$$P_{N(ABPSK)} = \frac{N_0}{2} \cdot 2 \cdot \frac{1}{\tau_B}$$

$$= \frac{N_0}{2\tau_b} \qquad (4.100)$$

Substituting Eqs. (4.100) and (4.99) into Eq. (4.96), and then Eq. (4.96) into Eq. (4.95), we get

$$P_{be(QPSK)} = Q\left[\left(\frac{2E_b}{N_0}\right)^{\frac{1}{2}}\right]$$

(4.101)

We note that this relationship is identical to that for the probability of bit error versus E_b / N_0 for BPSK. Graphs of P_{be} versus E_b / N_0 for QPSK and other linear modulation methods covered in this chapter are shown in Fig. 4.20.

In summary, for the same bit rate, the spectral efficiency of QPSK is twice that of BPSK with no loss in the probability of bit error performance. The QPSK hardware is, however, somewhat more complex than that required for BPSK.

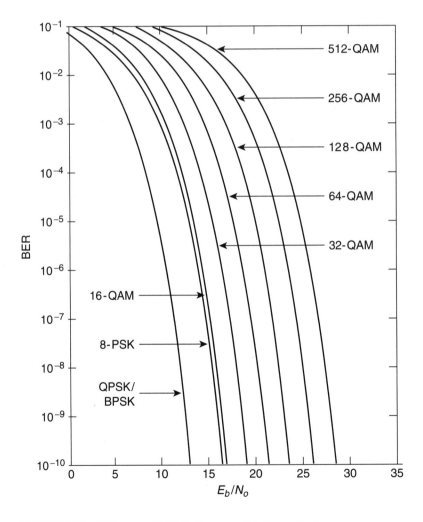

FIGURE 4.20   BER versus E_b/N_0 for linear modulation methods.

4.3.5   Offset Quaternary Phase Shift Keying (OQPSK)

Offset quaternary phase shift keying[5] (*OQPSK*), sometimes referred to as *staggered QPSK,* is very similar to conventional QPSK. A block diagram of a simplified OQPSK system is shown in Fig. 4.21(a). The difference between this system and the QPSK system discussed previously lies in the timing relationship between the streams $a_i(t)$ and $a_q(t)$ as they enter the multipliers. As with QPSK, the incoming stream $a_{in}(t)$ is applied to a serial to parallel converter. However, with OQPSK, one of the converter output streams, $a_q(t)$ in the case shown in Fig. 4.21(a) is "offset" or "staggered" with respect to the other by delaying it in time by an amount equal to the incoming signal bit duration τ_b. This results in the relationship between the data streams $a_{in}(t)$, $a_i(t)$, and $a_q(t)$ as shown in Fig. 4.21(b). The four possible output signal states of the modulator, their *IQ* combinations, and their possible transitions from one state to another are shown in Fig. 4.21(c). As can be seen, the resulting phase states at the modulator output are the same as for QPSK. However, because both data streams applied to the multipliers can never be in transition simultaneously, only one of the vectors that comprise the modulator output can change at any one time. The result is that only 90° phase transitions can occur in the modulator output signal. Like QPSK, if the *I* and *Q* data steams are unfiltered prior to application to the multipliers, phase transitions occur instantaneously and the resulting OQPSK signal has a constant amplitude envelope. Also, like QPSK, when these data streams are filtered, phase transitions occur over time and the resulting OQPSK signal has an amplitude envelope that varies with time. The significant difference with QPSK, however, is that its amplitude envelope varies only between its maximum value and a value $1/\sqrt{2}$ of maximum. This lower-amplitude envelope variation versus that of QPSK imparts certain advantages to OQPSK as compared to QPSK in systems including nonlinear elements.[6] For example, when a filtered OQPSK signal is amplified prior to transmission by an amplitude limiting device, there is only partial regeneration of the spectrum amplitude back to the unfiltered level. For QPSK under the same circumstances, however, there is complete regeneration to the unfiltered level.

The demodulator shown in Fig. 4.21(a) is identical to that shown in Fig. 4.18(a) for QPSK with the exception that the regenerated data stream $\hat{a}_i(t)$ is delayed by τ_b so that when combined with $\hat{a}_q(t)$, an estimate $\hat{a}_{in}(t)$ of the original input data stream is created.

Not surprisingly, for a random and equiprobable binary input data stream, the spectral density $Q_{OQPSK}(f)$ of an OQPSK modulated carrier is identical to that for QPSK for identical filtering and a linear transmission environment. Also, the probability of bit error performance of OQPSK is identical to that of QPSK for identical filtering and a linear system environment. In a highly filtered, nonlinear environment, the probability of error performance differs slightly.[6]

4.3.6   High-Order 2^{2n}-QAM

Though relatively easy to implement and robust in performance, linear four-phase systems such as QPSK and OQPSK do not often afford the desired spectral efficiency in commercial fixed-wireless systems. Higher-order QAM systems, however, do permit higher spectral efficiencies and have become very popular. A common class of QAM systems allowing high spectral density is one where the number of states is 2^{2n}, where n equals $2, 3, 4, \ldots$ A generalized and simplified block diagram of a 2^{2n}-QAM system is shown in

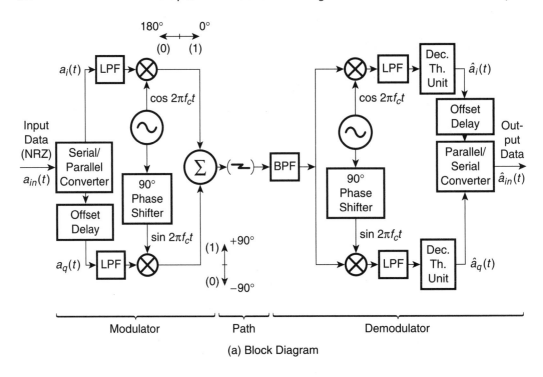

(a) Block Diagram

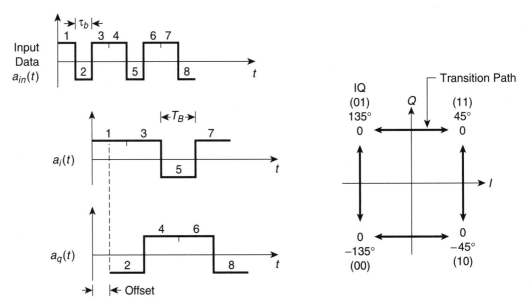

Note: Data in bit number '*n*' is converted
to I/Q baud number '*n*'.

(b) Modulator Data Streams

Note: No diagonal transition paths
as with QPSK

(c) Signal Space Diagram

FIGURE 4.21   Offset QPSK system representation.

Fig. 4.22. The difference between this generalized system and the QPSK system shown in Fig. 4.18 is that (a) in the generalized modulator, the I and Q signals $a_i(t)$ and $a_q(t)$ are each fed to a 2 to 2^n level converter prior to filtering and multiplication with the carrier, and (b) in the generalized demodulator, the outputs of the decision threshold units are each fed to a 2^n to 2 level converter prior to being combined in a parallel to serial converter. 2^n -QAM systems have been deployed commercially for values of n from 2 to 4.

For n equal to 2, a 16-QAM system is derived. In such a system incoming symbols to each modulator level converter are paired, and output symbols, in the form of signals at one of four possible amplitude levels, are generated in accordance with the coding table shown in Fig. 4.23(a). The duration of these output symbols, τ_{B4L} say, is twice that of τ_B, the duration of the input symbols. As a result of the application of the four-level signals to the multipliers, the output of each multiplier is a 4-level amplitude modulated DSBSC signal, and the combined signal at the modulator output is a QAM signal with sixteen states. Thus, 16-QAM can be treated as two 4-level PAM DSBSC systems operating in quadrature. The constellation diagram of a16-QAM signal is shown in Fig. 4.23(b). From this figure it is clear that 16-QAM has an amplitude envelope that varies considerably over time, irrespective of whether the signal has been filtered or not, and thus must be transmitted over a highly linear system if it is to preserve its spectral properties.

As τ_B, the duration of symbols from the serial to parallel converter, is equal to $2\tau_b$, where τ_b is the duration of incoming bits to the modulator, it follows that

$$\tau_{B4L} = 4\tau_b \tag{4.102}$$

Using Eq. (4.102) and the same logic used to determine the spectral density of QPSK, it can be shown that $G_{16\text{-}QAM}(f)$, the spectral density of 16-QAM, is given by

$$G_{16\text{-}QAM}(f) = 4P_s\tau_b \left[\frac{\sin 4\pi(f - f_c)\tau_b}{4\pi(f - f_c)\tau_b} \right]^2 \tag{4.103}$$

$G_{16\text{-}QAM}(f)$ is such that its main lobes and side lobes are one-fourth as wide as those of BPSK. As a result the maximum spectral efficiency of 16-QAM is 4 bits/s/Hz, twice that of four-phase systems such as QPSK and OQPSK.

Applying yet again the concepts used in the QPSK analysis, we recognize that the probability of bit error of a 2^{2n}-QAM system, $P_{be(QAM)}$, is the same as the probability of bit error of each 2^n-level PAM *associated DSBSC (ADSB)* system, $P_{be(ADSB)}$ say; that is,

$$P_{be(QAM)} = P_{be(ADSB)} \tag{4.104}$$

In order to determine the bit error probability $P_{be(ADSB)}$ of a 16-QAM system we first determine the 4-level symbol error probability $P_{be(ADSB)}$. Substituting $L = 4$ into Eq. (4.82), we have

$$P_{se(ADSB)} = \frac{3}{2} Q \left[\left(\frac{2P_{S(ADSB)}}{5P_{N(ADSB)}} \right)^{\frac{1}{2}} \right] \tag{4.105}$$

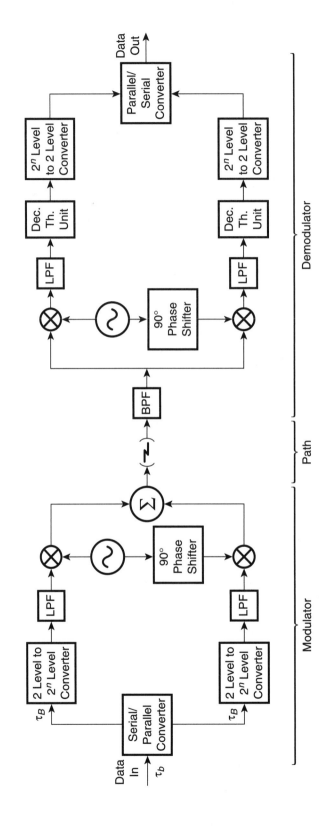

FIGURE 4.22   Generalized block diagram of a 2^{2n}-QAM system.

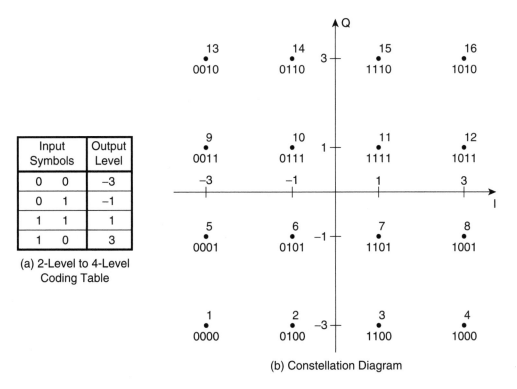

Input Symbols		Output Level
0	0	−3
0	1	−1
1	1	1
1	0	3

(a) 2-Level to 4-Level Coding Table

(b) Constellation Diagram

FIGURE 4.23   16-QAM level converter coding table and constellation diagram.

For each ADSB system, the average signal power $P_{S(ADSB)}$ at the receiver input is half that of the 16-QAM system; that is,

$$P_{S(ADSB)} = \frac{1}{2} P_{S(16-QAM)}$$

(4.106)

Also,

$$P_{S(16-QAM)} = E_b \cdot \frac{1}{\tau_b}$$

(4.107)

Thus, by Eqs. (4.106) and (4.107), we have

$$P_{S(ADSB)} = \frac{E_b}{2\tau_b}$$

(4.108)

The Nyquist bandwidth of the ADSB system is equal to $1/\tau_{B4I}$. Thus $P_{N(ADSB)}$, the noise in the two-sided ADSB Nyquist bandwidth at the receiver input, is given by

$$P_{N(ADSB)} = \frac{N_0}{2} \cdot 2 \cdot \frac{1}{\tau_{B4l}}$$

$$= \frac{N_0}{4\tau_b} \tag{4.109}$$

Substituting Eqs. (4.109) and (4.108) into Eq. (4.105), we get

$$P_{se(ABSB)} = \frac{3}{2} Q \left[\left(\frac{4}{5} \frac{E_b}{N_0} \right)^{\frac{1}{2}} \right] \tag{4.110}$$

In a 2-level to L-level converter, an output symbol, which can have one of L levels, is created from $\log_2 L$ bits. The 2-level to 4-level coding shown in Fig. 4.23 (a) is an example of *Gray coding*. In Gray coding, the bits that create any pair of adjacent levels differ by only one bit. Thus, in a 2-level to L-level converter with Gray coding, and where only adjacent level errors are significant, only one bit out of $\log_2 L$ bits will be in error for every symbol error. As a result, the probability of bit error $P_{be(ADSB)}$ is related to the probability of symbol error by the following relationship:

$$P_{be(ADSB)} = \frac{P_{se(ADSB)}}{\log_2 L} \tag{4.111}$$

For a 16-QAM system, where $L = 4$ and hence $\log_2 L = 2$, $P_{be(ADSB)}$ and hence, by Eq. (4.104), $P_{be(16\text{-}QAM)}$ is given by

$$P_{be(16-QAM)} = \frac{3}{4} Q \left[\left(\frac{4}{5} \frac{E_b}{N_0} \right)^{\frac{1}{2}} \right] \tag{4.112}$$

A graph of $P_{be(16\text{-}QAM)}$ versus E_b / N_0 is shown in Fig. 4.20. It will be observed that for a probability of bit error of 10^{-3} the E_b / N_0 required for 16-QAM is 3.8 dB greater than that required for QPSK. Thus the doubling of the spectral efficiency achieved by 16-QAM relative to QPSK comes at the expense of probability of bit error performance.

As mentioned previously, the 2-level to 4-level coding shown in Fig. 4.23(a) is an example of Gray coding. It is interesting to note that, as a result of Gray coding in the I and Q channels, the 16 states shown in the signal space diagram in Fig. 4.23(b) are also Gray coded. Thus an error resulting from one of these states being decoded as one of its closest adjacent states will result in only one bit being in error.

For n equal to 3 a 64-QAM system is derived, two 8-level PAM DSBSC signals being combined in quadrature. The 8-level PAM signals are created by grouping the incoming symbols to the level converter into sets of three and using these three-digit codewords to derive the 8 output levels.

For n equal to 4 a 256-QAM system is derived, two 16-level PAM DSBSC signals being combined in quadrature. Here the 16-level PAM signals are created by grouping the incoming symbols to the level converter into sets of four and using these four-digit code words to derive the 16 output levels.

Using the same logic as applied to the analysis of 16-QAM, but with generalized equations, it can be shown that the maximum spectral efficiency of 2^{2n}-QAM is $2n$ bits/s/Hz. Thus that of 64-QAM is 6 bits/s/Hz and that of 256-QAM is 8 bits/s/Hz.

Also, again using the same reasoning as applied to the 16-QAM analysis, it can be shown that for Gray coded 2^{2n}-QAM the generalized equation for probability of bit error P_e versus E_b / N_0 is

$$P_{be(2^{2n}-QAM)} = \frac{2}{\log_2 L}\left[1 - \frac{1}{L}\right]Q\left[\left(\frac{6\log_2 L}{L^2 - 1}\frac{E_b}{N_0}\right)^{\frac{1}{2}}\right] \tag{4.113}$$

Thus, for 64-QAM we have

$$P_{be(64-QAM)} = \frac{7}{12}Q\left[\left(\frac{2}{7}\frac{E_b}{N_0}\right)^{\frac{1}{2}}\right] \tag{4.114}$$

and for 256-QAM we have

$$P_{be(256-QAM)} = \frac{15}{32}Q\left[\left(\frac{8}{85}\frac{E_b}{N_0}\right)^{\frac{1}{2}}\right] \tag{4.115}$$

The probability of error relationships for 64- and 256-QAM are also shown in Fig. 4.20. We observe that as the number of QAM states increases, the P_e performance decreases.

In the probability of error analysis outlined previously, the QAM system probability of bit error P_{be} was deduced by determining first the symbol probability of error $P_{se(ADSB)}$, and from that the bit probability of error of the associated PAM DSBSC system. It should be noted, however, that the same results can be deduced by using $P_{se(ADSB)}$ to determine the probability of symbol error of the QAM system, $P_{se(QAM)}$, and from this determining $P_{be(QAM)}$. Using this latter approach, we commence by defining the probability of error of the associated DSBSC systems in terms of the probability of error of its two outermost symbols, P_{e1} say. For an L-level DSBSC system, the average probability of symbol error is the sum of the individual probabilities of symbol error divided by the number of symbols. From earlier analysis we know that the probability of error of the inner symbols is $2P_{e1}$. There are $(L - 2)$ such symbols. Based on the previous data it is easy to compute the average probability of symbol error. In fact, the reader may recall that this computation has already been done in Section 4.2.3, where PAM systems are analyzed. Thus, by Eq. (4.42) we have

$$P_{se(ADSB)} = 2\left[1 - \frac{1}{L}\right]P_{e1} \tag{4.116}$$

Turning now to the probability of symbol error of the QAM system, we note that any symbol in the QAM array is subject to incorrect detection due to incorrect decisions in both the I and Q planes. Thus the probability of *correct* QAM detection of an *individual* symbol, $P_{isc(QAM)}$, is given by

$$P_{isc(QAM)} = \left(1 - P_{ise(ADSBI)}\right)\left(1 - P_{ise(ADSBQ)}\right) \tag{4.117}$$

where $P_{ise(ABSCI)}$ is the probability of in-phase error of the individual symbol and $P_{ise(ABSCQ)}$ is the probability of quadrature error of the individual symbol.

From Eq. (4.117) it follows that the probability of *incorrect* QAM detection of an individual symbol—that is, the probability of error of an individual QAM symbol, $P_{ise(QAM)}$,— is given by

$$P_{ise(QAM)} = 1 - P_{isc(QAM)}$$

$$= 1 - \left(1 - P_{sei(ADSBI)}\right)\left(1 - P_{sei(ADSBQ)}\right)$$

$$\simeq P_{sei(ADBSI)} + P_{sei(ADSBQ)} \tag{4.118}$$

for very low values of $P_{ise(ABSCI)}$ and $P_{ise(ABSCQ)}$, which is normally the case.

As was done for the associated DSBSC systems, we calculate the $L \times L$ QAM average probability of symbol error by summing the individual probabilities of symbol error and dividing this sum by the number of symbols. Starting with the four corner symbols [symbols 1, 4, 13, and 16 in Fig. 4.23(b) for the 16-QAM case], the probability of error of an individual corner symbol in the I plane or Q plane is P_{e1} as defined for the ADSB systems. Thus, by Eq. (4.118), the probability of such a symbol in the two-dimensional QAM plane is $2P_{e1}$.

For the symbols on the outer periphery, excluding the corner symbols [symbols 2, 3, 8, 12, 5, 9, 14, and 15 in Fig. 4.23(b) for the 16-QAM case], the probability of error is $3P_{e1}$. This is easily demonstrated by considering one such symbol in Fig. 4.23(b). Considering symbol 2, for example, we note that in the I plane its probability of error is $2P_{e1}$, whereas in the Q plane its probability of error is $1P_{e1}$. Hence, by Eq. (4.118) its probability of error in the QAM plane is $3P_{e1}$. There are $4(L - 2)$ such symbols.

The remaining symbols are the inner symbols [symbols 6, 7, 10, and 11 in Fig. 4.23(b) for the 16-QAM case]. By analyzing the inner symbols in Fig. 4.23(b), it becomes clear that they have a probability of error of $4P_{e1}$. There are $(L - 2)^2$ such symbols.

Using the preceding data, it is easily shown that the average symbol probability of error of a $L \times L$ QAM array, $P_{se(QAM)}$, is given by

$$P_{se(QAM)} = 4\left[1 - \frac{1}{L}\right]P_{e1} \tag{4.119}$$

Comparing Eq. (4.119) with Eq. (4.116), we get the relationship we seek; that is,

$$P_{se(QAM)} = 2P_{se(ADSB)} \tag{4.120}$$

Now for an M array QAM system with Gray coding applied to the I and Q level converters, the M modulated signal states are also Gray coded. For this situation it can be shown, using the same logic as applied above to 2-level to L-level converters, that the probability of bit error $P_{be(QAM)}$ is given by

$$P_{be(QAM)} = \frac{1}{\log_2 M} P_{se(QAM)} \tag{4.121}$$

For the rectangular array QAM systems under consideration, $M = L^2$. Hence

$$P_{be(QAM)} = \frac{1}{2\log_2 L} P_{se(QAM)}$$

$$= \frac{1}{\log_2 L} P_{se(ADSB)} \tag{4.122}$$

It will be observed that Eq. (4.122) gives the same relationship between $P_{be(QAM)}$ and $P_{se(ADSB)}$ as provided when Eq. (4.111) is substituted into Eq. (4.104).

4.3.7   High-Order 2^{2n+1}-QAM

For the 2^{2n}-QAM systems described previously, the signal space consists of a full rectangular array of states with each state coded by an even number of bits. When each state is coded by an odd number of bits, however, it's not possible to create a full rectangular array. Such systems have 2^{2n+1} states with each state coded with $2n + 1$ bits. For n equals 1 we get an 8-QAM system. Many constellations have been proposed for 8-QAM.[6] However, as there appears to be no obvious "standard" constellation and as 8-QAM is rarely used in fixed wireless communications, it won't be reviewed here other than to point out that it affords a spectral efficiency of 3 bits/s/Hz. For $n = 2$, 3, and 4, we get 32-QAM, 128-QAM, and 512-QAM, respectively. The 32- and 128-QAM systems are very popular and 512-QAM, though challenging, has been offered commercially.

With 2^{2n+1}-QAM modulators, because an odd number of bits is used to code each signal state, it is not possible to divide the incoming bit stream into two and then use 2-level to L-level PAM modulation in order to create the QAM signal. Instead, the incoming signal of bit rate f_b and duration τ_b is fed to a logic circuit that uses a lookup table to create in-phase and quadrature L level symbols of duration τ_B, where

$$\tau_B = (2n+1)\tau_b \tag{4.123}$$

In the demodulator, the L level outputs of the I and Q decision threshold units are fed to a logic circuit, which, via its lookup table, re-creates the original bits. A simplified block diagram of a 2^{2n+1}-QAM system is shown in Fig. 4.24.

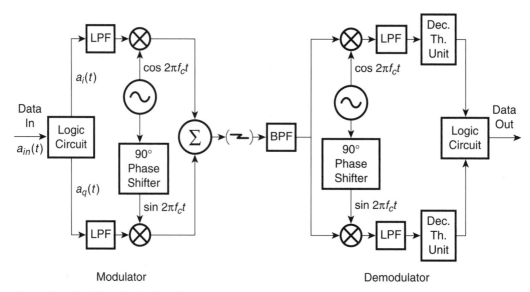

Modulator   Demodulator

Note: The Modulator Logic Circuit uses a look up table to convert odd number input sequences to
      I and Q symbol streams. The Demodulator Logic Circuit uses a look up table to convert
      I and Q symbol streams to odd number output sequences.

FIGURE 4.24   Simplified block diagram of 2^{2n+1}-QAM system.

Because of the freedom afforded by the lookup table design, many 2^{2n+1} signal state constellations are possible. A very common class is the *cross-constellation* class, so called because of its cross appearance. Constellations in this class are essentially full rectangular $L \times L$ arrays, where $L = 3 \times 2^{n-1}$, but with corner states removed.

Fig. 4.25(a) shows a 32-QAM cross constellation, and Fig.4.25(b) shows typical 32-QAM modulator logic circuit signals. The constellation is a 6×6 array with the four corner states removed. From the figure it is clear that this constellation can be created by summing two six-level PAM DSBSC signals in quadrature. Each state in the 32-QAM constellation is equally likely. However, unlike 2^{2n}-QAM, each level in each of the 6-level DSBSC signals is not equally likely. In Fig. 4.25(a) the probability of level X occurrence, $P(X)$, in the in-phase channel is shown. Also shown in Fig. 4.25(a) are the six PAM levels that drive the in-phase multiplier. The same probabilities and PAM levels apply to the equivalent levels in the quadrature channel.

It is not possible to determine the probability of bit error of 32-QAM, $P_{be(32\text{-}QAM)}$, by determining the probability of bit error of the associated 6-level PAM DSBSC systems as the bit groupings driving these systems are not independent of each other. It is possible to be determined, however, by the following method developed by the author. First, the probability of symbol error of the associated DSBSC systems, $P_{se(ADSB)}$, is determined both in terms of E_b / N_0 and P_{e1}, the probability of error of the two outermost DSBSC symbols. Next, this information is used to deduce the probability of symbol error of the 32-QAM system, $P_{se(32\text{-}QAM)}$. Finally, $P_{be(32\text{-}QAM)}$ is determined from $P_{se(32\text{-}QAM)}$ and the specifics of the constellation coding.

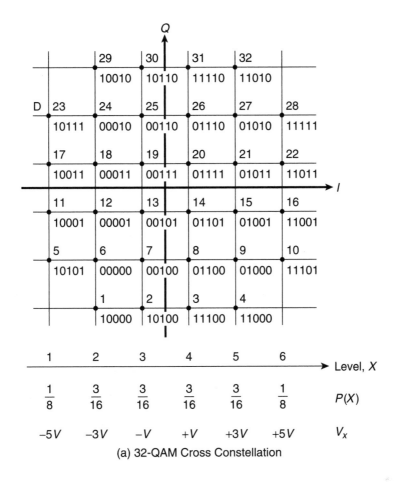

(a) 32-QAM Cross Constellation

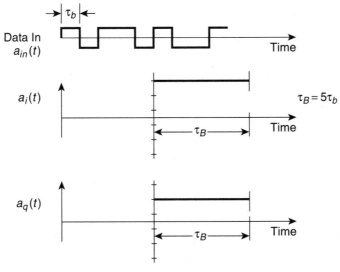

(b) Modular Logic Circuit Signals

FIGURE 4.25   32-QAM constellation and logic circuits signals.

To determine $P_{se(ADSB)}$ in terms of E_b / N_0 we must first determine the probability of error of the associated 6-level PAM signals, $P_{se(PAM)}$. Using logic similar to that applied in Section 4.2.3, but taking into account the varying probabilities of the symbol levels, it can be shown that if P_{e1} is the probability of symbol error of the two outer levels, levels 1 and 6, then the probability of symbol error of the four inner levels, levels 2, 3, 4, and 5, is $2P_{e1}$. Thus the average probability of symbol error P_e is given by

$$P_{se(PAM)} = P(1) \cdot P_{e1} + P(2) \cdot P_{e2} + \ldots P(6) \cdot P_{e6}$$

$$= 2 \cdot \frac{1}{8} \cdot P_{e1} + 4 \cdot \frac{3}{16} \cdot 2 P_{e1}$$

$$= \frac{7}{4} P_{e1} \qquad\qquad (4.124a)$$

$$= \frac{7}{8} erfc\left(\frac{V}{\sqrt{2}\sigma}\right) \qquad\qquad (4.124b)$$

Defining the voltage of level X as V_x, the average symbol power $P_{S(PAM)}$ is given by

$$P_{S(PAM)} = P(1) \cdot V_1^2 + P(2) \cdot V_2^2 + \ldots P(6) \cdot V_6^2$$

$$= 2 \cdot \frac{1}{8} \cdot (5V)^2 + 2 \cdot \frac{3}{16} \cdot (3V)^2 + 2 \cdot \frac{3}{16} \cdot V^2$$

$$= 10V^2 \qquad\qquad (4.125)$$

Substituting Eqs. (4.125) and (4.50) into Eq. (4.124b), we get

$$P_{se(PAM)} = \frac{7}{8} erfc\left[\left(\frac{1}{20}\frac{P_{S(PAM)}}{P_{N(PAM)}}\right)^{\frac{1}{2}}\right] \qquad\qquad (4.126a)$$

$$= \frac{7}{4} Q\left[\left(\frac{1}{10}\frac{P_{S(PAM)}}{P_{N(PAM)}}\right)^{\frac{1}{2}}\right] \qquad\qquad (4.126b)$$

Applying now Eq. (4.81) to Eq. (4.126b), we determine that $P_{se(ADSB)}$ is given by

$$P_{se(ADSB)} = \frac{7}{4} Q\left[\left(\frac{1}{5}\frac{P_{S(ADSB)}}{P_{N(ADSB)}}\right)^{\frac{1}{2}}\right] \qquad\qquad (4.127)$$

For each associated DSB system, the average signal power $P_{S(ADSB)}$ at the demodulator input is half that of the 32-QAM system; that is,

$$P_{S(ADSB)} = \frac{1}{2}P_{S(32\text{-}QAM)} \tag{4.128}$$

Also,

$$P_{S(32\text{-}QAM)} = E_b \cdot \frac{1}{\tau_b} \tag{4.129}$$

Thus, by Eqs. (4.128) and (4.129) we have

$$P_{S(ADSB)} = \frac{E_b}{2\tau_b} \tag{4.130}$$

The Nyquist bandwidth of the ADSB signal is equal to $1/\tau_{B6l}$, where $\tau_{B6l} = 5\tau_b$. Thus $P_{N(ADSB)}$, the noise in the two-sided ADSB Nyquist bandwidth at the demodulator input, is given by

$$P_{N(ADSB)} = \frac{N_0}{2} \cdot 2 \cdot \frac{1}{\tau_{B6l}}$$

$$= \frac{N_0}{5\tau_b} \tag{4.131}$$

Substituting Eqs. (4.130) and (4.131) into Eq. (4.127), we get

$$P_{se(ABSB)} = \frac{7}{4}Q\left[\left(\frac{1}{2}\frac{E_b}{N_0}\right)^{\frac{1}{2}}\right] \tag{4.132}$$

To determine $P_{se(ADSB)}$ in terms of the probability of error of the two outermost symbols of the associated DSBSC system, P_{e1}, the procedure is the same as that for $P_{se(PAM)}$ outlined in Eq. (4.124a). Thus

$$P_{se(ADSB)} = \frac{7}{4}P_{e1} \tag{4.133}$$

Turning now to the probability of symbol error of 32-QAM, $P_{se(32\text{-}QAM)}$, we proceed along the same lines as was used for determining $P_{se(16\text{-}QAM)}$ in Section 4.3.6. Starting with the 16 edge symbols, we recognize that the probability of error of symbols in these locations is $3P_{e1}$. For the remaining 16 inner symbols, we recognize that the probability of error of such symbols is $4P_{e1}$. Utilizing this data and the fact that all symbols are equally likely, we derive the following relationship for $P_{se(32\text{-}QAM)}$:

$$P_{se(32-QAM)} = \frac{1}{32}\left[16 \cdot 3P_{e1} + 16 \cdot 4P_{e1}\right]$$

$$= \frac{7}{2}P_{e1} \tag{4.134}$$

Comparing Eq. (4.134) with Eq. (4.133), we get

$$P_{se(32-QAM)} = 2P_{se(ADSB)} \tag{4.135}$$

Substituting Eq. (4.135) into Eq. (4.132), we get

$$P_{se(32-QAM)} = \frac{7}{2}Q\left[\left(\frac{1}{2}\frac{E_b}{N_0}\right)^{\frac{1}{2}}\right] \tag{4.136}$$

Finally, to determine the probability of *bit* error we need to know not only $P_{se(32-QAM)}$ but also the symbol coding applied. Gray coding leads to the best coding gain but, unfortunately, it is not possible to Gray code 32-QAM.[7] Were it possible, then the bit error rate would 1/5 of the symbol error rate. Since codes are available that lead to some gain, then the bounds on $P_{be(32-QAM)}$ are given by

$$P_{be(32-QAM)} > \frac{7}{10}Q\left[\left(\frac{1}{2}\frac{E_b}{N_0}\right)^{\frac{1}{2}}\right]$$

$$< \frac{7}{2}Q\left[\left(\frac{1}{2}\frac{E_b}{N_0}\right)^{\frac{1}{2}}\right] \tag{4.137}$$

The coding shown in Fig. 4.25 is an example of quasi-Gray coding. Careful examination will reveal that all symbols differ from adjacent symbols by only one bit with the exception of symbols 5, 10, 23 and 28, which differ from their adjacent horizontal symbols by three bits. Such coding should lead to a bit error rate that's very close to 1/5 of the symbol error rate.

As indicated previously, the symbol duration τ_{B6l} of the 6 level PAM streams generated by the modulator logic circuit is equal to $5\tau_b$. Using this relationship and the same logic used to determine the spectral density of QPSK, it can be shown that $G_{32-QAM}(f)$, the spectral density of 32-QAM, is given by

$$G_{32-QAM}(f) = 5P_s\tau_b\left[\frac{\sin 5\pi(f - f_c)\tau_b}{5\pi(f - f_c)\tau_b}\right]^2 \tag{4.138}$$

$P_{32-QAM}(f)$ is such that its main lobes and side lobes are one fifth as wide as those of BPSK. As a result, the maximum spectral efficiency off 32-QAM is 5 bits/s/Hz.

The 128-QAM cross constellation is a 12×12 array with four states removed from each corner. The symbol duration of the 12-level PAM signals is 7 times that of the incoming signal. As a result, the spectral efficiency of 128-QAM is 7 bits/s/Hz. Using the same methodology as that used for 32-QAM, it can be shown that the probability of bit error of 128-QAM is given by

$$
P_{be(128-QAM)} > \frac{15}{28} Q \left[\left(\frac{7}{41} \frac{E_b}{N_0} \right)^{\frac{1}{2}} \right]
$$

$$
< \frac{15}{4} Q \left[\left(\frac{7}{41} \frac{E_b}{N_0} \right)^{\frac{1}{2}} \right]
$$

(4.139)

The 512-QAM cross constellation is a 24×24 mega-array with 16 states removed from each corner. The symbol duration of the 24-level PAM signals is nine times that of the incoming signal. As a result, the spectral efficiency of 512-QAM is a highly impressive 9 bits/s/Hz. It can be shown, again with the same methodology as that used for 32-QAM, that the probability of bit error of 512-QAM is given by

$$
P_{be(512-QAM)} > \frac{31}{72} Q \left[\left(\frac{9}{165} \frac{E_b}{N_0} \right)^{\frac{1}{2}} \right]
$$

$$
< \frac{31}{8} Q \left[\left(\frac{9}{165} \frac{E_b}{N_0} \right)^{\frac{1}{2}} \right]
$$

(4.140)

Graphs of the lower bound of $P_{be(32-QAM)}$, $P_{be(128-QAM)}$, and $P_{be(512-QAM)}$ versus E_b / N_0 are shown in Fig. 4.20.

4.3.8 8-PSK

By the clever manipulation of the input data stream, an 8-PSK signal can, like the QAM signals described previously, be created as the sum of in-phase and quadrature DSBSC signals.[2] The Gray coded constellation diagram of 8-PSK is shown in Fig. 4.26.

Unfiltered 8-PSK, like QPSK, has a constant amplitude envelope. However, when filtered to limit the spectrum, as it almost always is, the envelope varies between its peak level and zero. It has a spectral efficiency of 3 bits/s/Hz and its probability of bit error $P_{be(8-PSK)}$ versus E_b / N_0 performance, assuming Gray coding, is given by[8]

$$
P_{be(8-PSK)} = \frac{2}{3} Q \left[\left(0.878 \frac{E_b}{N_0} \right)^{\frac{1}{2}} \right]
$$

(4.141)

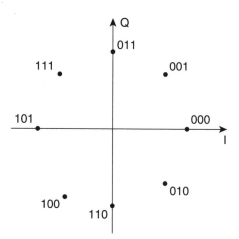

FIGURE 4.26   Gray-coded
constellation diagram of 8-PSK.

A graph of $P_{be(8-PSK)}$ versus E_b / N_0 is shown in Fig. 4.20.

8-PSK found several applications in the early days of digital fixed wireless as a way to improve on the spectral efficiency of four-state systems such as QPSK. However, it was soon supplanted by 16-QAM and higher-level QAM techniques as its error performance, as can be seen in Fig. 4.20, is only marginally better than that of 16-QAM, yet 16-QAM affords a higher spectral efficiency.

4.3.9   Modem Realization Techniques

The modems described previously provide a theoretical understanding of their operation and performance. However, to realize such modems in practice, several techniques must be applied. For example, in the demodulator, the identical carrier as available in the modulator has been assumed, but how in practice is this made available? In this section, a number of key implementation techniques, required even for performance in a linear transmission environment, are reviewed. Fig. 4.27(a) shows a quadrature type modulator that indicates the placement of the techniques covered. Similarly, Fig. 4.27(b) indicates the placement in the associated demodulator.

4.3.9.1   Line Code to NRZ/NRZ to Line Code Conversion

Data presented to a digital radio transmitter is normally in the form of one of the many line codes addressed in Section 2.2.5, the line code being determined by the data rate. However, digital modems prefer to process data in the bipolar NRZ format. Thus, the first modulator circuitry that the incoming data signal is fed to is normally a data rate specific line code to NRZ converter. For example, if the incoming signal is a DS3 one, then, from Table 2.1, it is seen that its line code is B3ZS and thus a B3ZS to NRZ conversion is required. In the demodulator, the last circuitry is normally a NRZ to line code converter in order to return the data to its original input format.

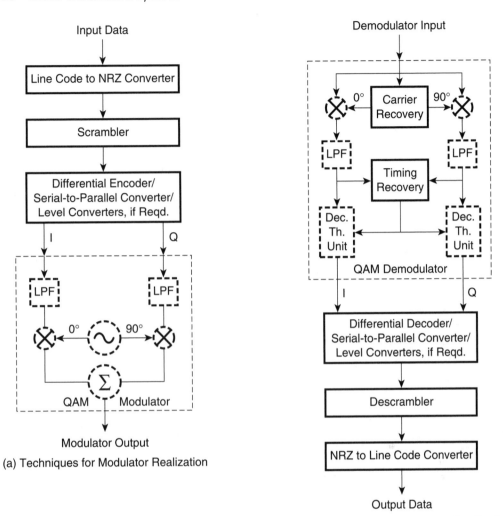

FIGURE 4.27   Modem realization techniques.

4.3.9.2   Scrambling/Descrambling

Following the modulator line code converter is normally a scrambler.[9] Its function is to eliminate from the incoming data stream (a) any periodic data pattern, (b) long sequences of ones or zeros, and (c) any direct current (DC) component that may occur as a result of these long sequences. A scrambler achieves this by *scrambling* the data stream, using a shift register to generate a repetitive but long *pseudorandom* (almost random) bit sequence and logically combining this sequence with the incoming data. The scrambled output data assumes properties similar to that of the pseudorandom sequence, irrespective of the input data properties.

The removal of any DC component by the scrambler allows the use of alternating current (AC) coupled circuitry which is normally easier to implement. By eliminating periodic patterns in the modulating data, scrambling guarantees that the radiated spectrum is essentially uniformly distributed and free of spectral lines. Such lines, if significant, can cause unwanted interference in adjacent RF channels. In the demodulator, the frequent transitions in the modulating signal, resulting from the elimination of long sequences of ones or zeros, are required for accurate timing recovery.

The output of the demodulator parallel to serial converter is fed to a *descrambler* in order to generate the estimate of the original signal. Because the descrambler has to know the specifics of the pseudorandom generator in the scrambler in order to function, the scrambler/descrambler circuitry serves as a form of encryption. In fact, some manufacturers provide user programmable scrambler/descrambler circuits so as to give the user direct control over the scrambling sequence.

4.3.9.3 Carrier Recovery

In the theoretical coherent demodulation analyses presented previously, it has simply been assumed that a replica of the original carrier signal in the modulator, identical in frequency and phase, is available in the demodulator without indicating how such a carrier is obtained. Not surprisingly, the carrier is extracted from the received signal. It is less than intuitive, however, how such extraction is achieved. This is because, in DSBSC systems modulated with random equiprobable data, the resulting transmitted spectrum is continuous, containing no discrete carrier component. Further, the received spectrum is contaminated with noise and possibly unwanted interference. The process of extracting a carrier from such a spectrum is referred to as *carrier recovery,* and many ingenious ways have been devised to do this. In this section we shall briefly review some of these methods.

There are two conceptually different approaches to the problem of carrier recovery. One is simply to change the rules of the game by adding the carrier as a discrete component to the modulated signal and in the demodulator filtering it out and locking onto it with a phase lock loop. This approach, though straightforward, is at the expense of the energy per bit transmitted and is not often employed. The other approach is to extract the carrier from the continuous received spectrum despite its obscurity. Three successful methods using this approach are (1) the multiply-filter-divide method, (2) the Costas loop method, and (3) the decision directed method.

The Multiply-Filter-Divide Method. The *multiply-filter-divide method* is the simplest of the three to implement. A block diagram of a carrier recovery system employing this method is shown in Fig. 4.28. The incoming signal is fed to bandpass filter to limit noise and the output of the filter frequency multiplied. The latter can be accomplished via a nonlinear device such as a saturating amplifier or diode. The desired harmonic of the multiplier is then "filtered" with a *phase lock loop* (*PLL*)[10] tuned to that harmonic. The harmonic that is filtered depends on the modulated signal. Finally, the output of the *voltage-controlled oscillator* (*VCO*) of the PLL is frequency divided to obtain a signal at the desired carrier frequency, which may, depending on the multiplication, contain a fixed phase shift relative the received carrier. If this is the case, then a phase shifter is added to bring the phase back to that of the received carrier.

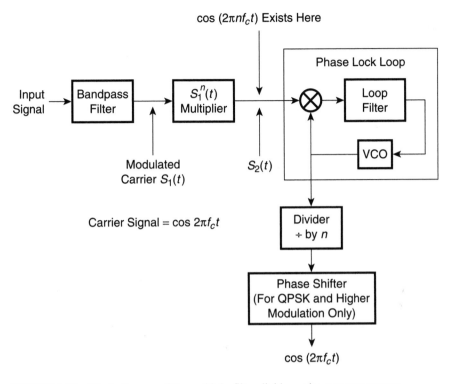

FIGURE 4.28   Block diagram of the multiply-filter-divide carrier recovery system.

For BPSK, the simplest DSBSC system, the desired frequency multiplication is 2. This is accomplished by raising the signal to the power of 2 (i.e., squaring it). To understand why this gives the desired result, let's follow mathematically the signal through the system. To simplify this analysis noise accompanying the signal is ignored. Let $s_1(t)$, the BPSK filtered input signal to the multiplier, be given by

$$s_1(t) = \cos\left[2\pi f_c t + \pi(i-1)\right],\ i = 1,\ 2 \tag{4.142}$$

where $\cos 2\pi f_c t$ is the carrier signal and $\pi(i-1)$ represents the modulation of the carrier phase ($0°$ and $180°$). Then $s_2(t)$, the output of the multiplier, is given by

$$s_2(t) = s_1^{\,2}(t)$$

$$= \frac{1}{2}\left[1 + \cos\left(2\pi \cdot 2 f_c t + 2\pi(i-1)\right)\right]$$

$$= \frac{1}{2}\left[1 + \cos\left(2\pi \cdot 2 f_c t\right)\right] \tag{4.143}$$

Equation (4.143) indicates that the process of squaring the input signal to the multiplier results in a DC component and a signal of frequency $2f_c$ but, importantly, devoid of the modulation component. The PLL locks onto the $2f_c$ component and minimizes the variations resulting from noise. The output of the PLL is divided by 2 to give a received carrier $\cos(2\pi f_c t)$. Unfortunately, this division results in a 180° phase ambiguity in the recovered carrier. To understand this we note that it is equally correct to write the PLL output signal as $\cos(2\pi \cdot 2f_c t)$ and $\cos(2\pi \cdot 2f_c t + 2\pi)$. However, dividing the former by 2 yields $\cos(2\pi f_c t)$ whereas dividing the latter by 2 yields $\cos(2\pi f_c t + \pi)$. This phase ambiguity must be resolved prior to final decoding since, if not, it could result in a 100% bit error rate! Resolving this class of ambiguity can be accomplished by *differential encoding,* which will be addressed in the next section.

For QPSK the input signal is raised to the fourth power. Let $s_1(t)$, the QPSK filtered input signal to the multiplier, be given by

$$s_1(t) = \cos\left[2\pi f_c t + \frac{\pi}{2}\left(i - \frac{1}{2}\right)\right], \ i = 1, 2, 3, 4 \tag{4.144}$$

where $\cos 2\pi f_c t$ is the carrier signal and $(\pi/2) \times (i - 1/2)$ represents the modulation of the carrier phase (45°, 135°, 225° and 315°). Then $s_2(t)$, the output of the multiplier, is given by

$$s_2(t) = s_1{}^4(t)$$

$$= \frac{3}{8} + \frac{1}{2}\cos\left(2\pi \cdot 2f_c t + \pi\left(i - \frac{1}{2}\right)\right) + \frac{1}{8}\cos\left(2\pi \cdot 4f_c t + 2\pi\left(i - \frac{1}{2}\right)\right)$$

$$= \frac{3}{8} + \frac{1}{2}\cos\left(2\pi \cdot 2f_c t + \pi\left(i - \frac{1}{2}\right)\right) + \frac{1}{8}\cos\left(2\pi \cdot 4f_c t + \pi\right) \tag{4.145}$$

Equation (4.145) indicates that the process of raising the input signal to the multiplier to the fourth power results in a DC component, a signal at $2f_c$ that, depending on the value of i, has a phase of $\pi/2$ or $3\pi/2$ (i.e., a modulated signal) and a signal of frequency $4f_c$ with a fixed phase shift of π (i.e., devoid of modulation). The PLL locks onto the $4f_c$ component and minimizes the variations resulting from noise. The output of the PLL is divided by 4 to give a received carrier of $\cos(2\pi f_c t + \pi/4)$. Unfortunately, the division of this signal results in a signal that has the desired frequency but is advanced in phase by 45° relative to the received carrier. As a result, it has to be passed through a −45° phase shifter to replicate the true received carrier phase. Further, in this case the division results in a fourfold phase ambiguity in the recovered carrier. To understand this we note that it is equally correct to write the PLL output signal as $\cos(2\pi \cdot 4f_c t + \pi)$, $\cos(2\pi \cdot 4f_c t + 3\pi)$, $\cos(2\pi \cdot 4f_c t + 5\pi)$, and $\cos(2\pi \cdot 4f_c t + 7\pi)$. However, dividing these signals by 4 yields, respectively, $\cos(2\pi f_c t + \pi/4)$, $\cos(2\pi f_c t + 3\pi/4)$, $\cos(2\pi f_c t + 5\pi/4)$, and $\cos(2\pi f_c t + 7\pi/4)$. As with BPSK, this ambiguity can be resolved with differential encoding.

The Costas Loop Method. The *Costas loop method* is basically equivalent to the multiply-filter-divide method, but with the VCO operating at the carrier frequency rather than n times the carrier. The block diagram of a Costas loop for a BPSK system is shown in Fig. 4.29. The input signal $s_1(t)$ is as given in Eq. (4.142). The VCO output frequency, which is

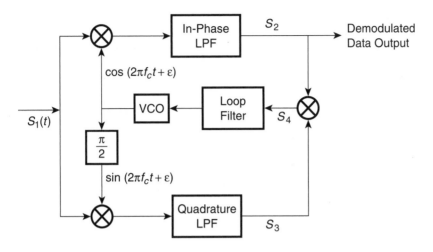

FIGURE 4.29   A BPSK carrier recovery Costas loop.

proportional to the output voltage of the loop filter, is assumed to be at the carrier frequency, but with a small phase error of ε, and thus given by $\cos(2\pi f_c t + \varepsilon)$. Using trigonometric identities, it can be shown that s_2, the output of the in-phase LPF, is given by

$$s_2 = \frac{1}{2}\cos\bigl(\pi(i-1)-\varepsilon\bigr) \tag{4.146}$$

and that s_3, the output of the quadrature LPF, is given by

$$s_3 = -\frac{1}{2}\sin\bigl(\pi(i-1)-\varepsilon\bigr) \tag{4.147}$$

The input s_4 to the loop filter is then given by

$$s_4 = s_2 \cdot s_3$$

$$= -\frac{1}{8}\sin\bigl(2\pi(i-1)-2\varepsilon\bigr)$$

$$= \frac{1}{8}\sin 2\varepsilon \tag{4.148}$$

The loop filter minimizes noise and other extraneous terms and the VCO is thus driven by a signal that, for small ε, is directly proportional to ε. This error term drives the VCO to change its average frequency until ε approaches zero. When this happens, the loop is said to be "in lock." We note from Eq. (4.148) that the loop error voltage s_4 driving the VCO is zero for a phase difference of 0 and 180°. As a result, phase lock can occur for two different phase angles of the incoming carrier. Thus, we see that the Costas loop creates the same 180° phase ambiguity with BPSK signals as is created with the multiply-filter-divide method. An interesting by product of the Costas method is that, if the in-phase LPF is designed to complement the modulator LPF for minimum ISI, then its output, s_2, is the demodulated data output.

The block diagram of a Costas loop for QPSK is shown in Fig. 4.30. It is somewhat more complex than that for BPSK, with a major difference being the addition of crossover arms that each incorporates a limiter. The operation of the loop can be understood from the following mathematical analysis:

Let the input signal $s_1(t)$ be as given in Eq. (4.144) and the output of the VCO assumed to be $\cos(2\pi f_c t + \varepsilon)$. Then, using trigonometric identities, it can be shown that s_2, the output of the in-phase LPF, is given by

$$s_2 = \frac{1}{2}\cos\left(\frac{\pi}{2}\left(i - \frac{1}{2}\right) - \varepsilon\right) \tag{4.149}$$

and that s_3, the output of the quadrature LPF, is given by

$$s_3 = -\frac{1}{2}\sin\left(\frac{\pi}{2}\left(i - \frac{1}{2}\right) - \varepsilon\right) \tag{4.150}$$

When s_2 is passed through the Limiter I, the output is constrained to be a fixed positive or negative value representing the demodulated output. Mathematically this removes the ε component from the phase. Another way to look at this is that the only way s_2 can give equal magnitude outputs for all four values of i is for the argument of the sin function to be $\pi / 4$, $3\pi / 4$, $5\pi / 4$, or $7\pi / 4$. This can only be so if ε equals 0. Thus s_4, the output of Limiter I, is given by

$$s_4 = \frac{1}{2}\cos\left(\frac{\pi}{2}\left(i - \frac{1}{2}\right)\right) \tag{4.151}$$

and, by similar logic, s_5, the output of Limiter Q, is given by

$$s_5 = -\frac{1}{2}\sin\left(\frac{\pi}{2}\left(i - \frac{1}{2}\right)\right) \tag{4.152}$$

The upper input to the subtractor, s_6, is given by

$$s_6 = s_2 \cdot s_5 \tag{4.153}$$

Substituting Eqs. (4.149) and (4.152) into Eq. (4.153), we get

$$s_6 = -\frac{1}{8}\sin\left(\pi\left(i - \frac{1}{2}\right) - \varepsilon\right) - \frac{1}{8}\sin\varepsilon \tag{4.154}$$

Similarly, it can be shown that s_7 is given by

$$s_7 = -\frac{1}{8}\sin\left(\pi\left(i - \frac{1}{2}\right) - \varepsilon\right) + \frac{1}{8}\sin\varepsilon \tag{4.155}$$

Finally, we have s_8, the output of the subtractor, given by

$$s_8 = s_6 - s_7$$

$$= -\frac{1}{4}\sin\varepsilon \qquad (4.156)$$

The subtractor output is fed to the loop filter, which minimizes the impact of noise and controls the loop dynamics. The error voltage fed to the VCO is proportional to ε for small values of ε. This voltage adjusts the VCO output signal phase in the proper direction and by the proper amount so as to drive the loop into a stable locked mode. The QPSK carrier recovery Costas loop has the same fourfold ambiguity in the recovered carrier as its multiply-filter-divide counterpart. This is easily demonstrated by simply replacing the assumed carrier at the VCO output by either $\cos(2\pi f_c t + \pi/2 + \varepsilon)$, $\cos(2\pi f_c t + \pi + \varepsilon)$, or $\cos(2\pi f_c t + 3\pi/2 + \varepsilon)$ in the preceding analysis. For all cases it will be found that the subtractor output is given by Eq. (4.156). It should be noted that, when the loop locks onto the frequency $2\pi f_c t + \pi/2$, or $2\pi f_c t + 3\pi/2$, the data inputted into the I channel in the modulator is outputted from the Q channel in the demodulator and vice versa, with the data inverted in the case of $2\pi f_c t + \pi/2$.

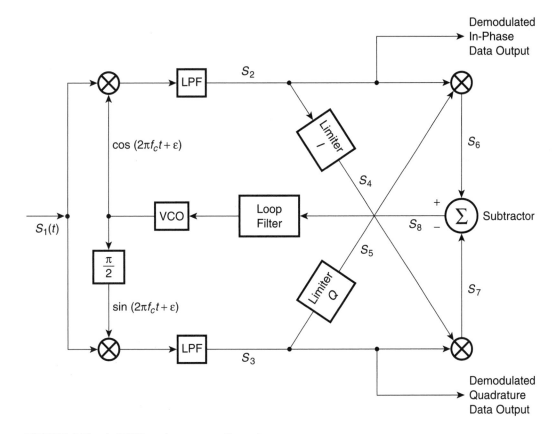

FIGURE 4.30   A QPSK carrier recovery Costas loop.

The Decision-Directed method. The multiply-filter-divide method and the Costas loop method are commonly used with BPSK and QPSK, where small phase errors in the recovered carrier do not seriously degrade performance. However, for high-order QAM where, due to the compactness of the signal states, even small phase errors can impair performance, the third method mentioned, the *decision-directed method,*[11] is much more suitable and thus the preferred method. As the analysis of the operation and performance of this method (and variations thereof) is highly complex, it will not be presented here. A broad overview of the concepts behind the method, however, follows.

The general approach is to assume that a recovered carrier, close in phase to the desired, is available, and to make decisions on the data symbols as if the phase error is zero. Then, based on the averaging of these symbols over many time samples, slowly direct corrective changes to the recovered carrier phase. In these circuits, the outputs of the *I* and *Q* rails are feed to high-resolution analog to digital converters having much more levels than those of the transmitted data signals. The outputs of the converters are used to re-create a constellation diagram. The points on this diagram are then digitally compared with those of an ideal diagram, the coordinates of which are stored in memory. Fig. 4.31 shows decoded states of a 16-QAM system in circles and the ideal states as black points. Noise causes the decoded constellation states to be not precise points, but to be spread out in a disk-shaped region around an average position. The decoded states are rotated from ideal

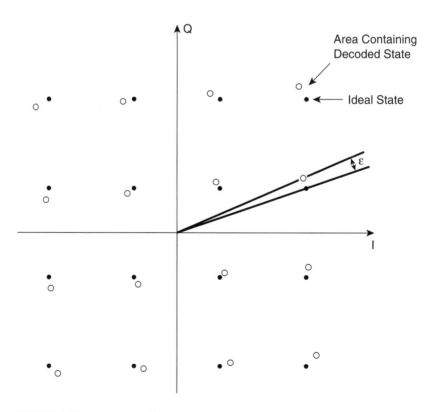

FIGURE 4.31   The impact of a $\varepsilon°$ phase error on a 16-QAM constellation.

by ε radians. The situation shown indicates the need for a clockwise rotation of the decoded signal constellation. The constellation comparator therefore generates an error signal that is low-pass filtered and applied to the carrier recovery VCO that adjusts the carrier frequency and phase slowly so as to rotate the decoded constellation in a clockwise direction until ε approaches zero. It is rotated slowly because this improves performance in the presence of noise. To assist the acquisition process and to increase the capture range of the loop, frequency sweeping can be used. As with the methods discussed for QPSK carrier recovery, the decision directed method results in the same fourfold ambiguity in recovered carrier phase for $(L \times L)$-QAM systems. This is because, due to the fourfold symmetry of their constellations, the constellation comparator has no way of knowing which of the four possible phases that result in a locked mode is correct.

4.3.9.4 Timing Recovery

The decision threshold units in the various demodulators discussed previously require a clock synchronized to the start and stop instances of the incoming symbols in order to trigger the sample instants. This clock is normally extracted from the data signal at the output of the low-pass filter following demodulation. Since the spectra of such signals are continuous, typically filtered sin x / x in shape, and containing no discrete $1/\tau_B$ symbol rate component, a nonlinear operation on it is required to extract the symbol frequency. Two methods commonly used for *timing recovery* or, as it's often referred to, *clock recovery*, are the *square and filter method* and the *delay and multiply method.*

Square and Filter Method. A block diagram of a timing recovery circuit using this method is shown in Fig. 4.32(a). The input signal is fed to a squaring device that performs the nonlinear operation. It can be shown mathematically that the result of squaring the input signal is to create a spectrum as shown in Fig. 4.32(b), which, though it continues to have a continuous component, also contains discrete spectral lines at multiples of $1/\tau_B$. A bandpass filter following the squarer is centered at $1/\tau_B$ and its output is fed to a PLL that locks onto it and outputs the desired timing signal. Despite the filtering effects of the bandpass filter and the PLL, it is not possible to eliminate the noise and continuous spectrum immediately surrounding the desired spectral component. As a result, the recovered timing signal contains an error component. This is referred to as timing jitter and, if significant, can degrade system performance.

Delay and Multiply Method. A block diagram of a timing recovery circuit using this second method is shown in Fig. 4.33. First, the input signal is fed to a simple unclocked threshold detector to create a signal with a wideband spectrum. A nonlinear operation is then performed on the output of the detector by multiplying it with a version of itself delayed by $\tau_B /2$. The output of the multiplier consists of a signal with a spectrum similar to that shown in Fig. 4.32(b). This signal is fed via a bandpass filter centered at $1/\tau_B$ to a PLL, which provides the desired timing signal.

4.3.9.5 Differential Encoding/Decoding

As indicated in Section 4.3.9.3, carrier recovery is not fully sufficient to re-create the transmitted signal because of ambiguities that usually exist as to the exact phase

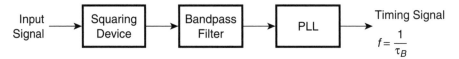

(a) Block Diagram of Square and Filter Timing Recovery Circuit

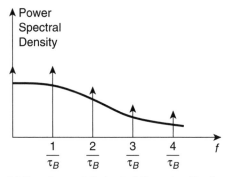

(b) Spectrum at Output of Squaring Device

FIGURE 4.32   Square and filter timing recovery method.

orientation of the received signal set relative to the recovered carrier. This problem is aggravated by varying phase delay introduced in the transmission path because of atmospheric anomalies. To remove phase ambiguities, *differential encoding* is applied to the modulator and the corresponding *differential decoding* applied to the demodulator. Differential encoding operates by representing some or all of the information bits that define a signaling state as a change in the phase of the transmitted carrier, not as a component in defining an absolute location of the carrier in the constellation diagram. In the demodulator, the operation of the differential decoder is simply the inverse of the encoder. Thus absolute phase information is unnecessary and any phase ambiguity introduced via carrier recovery is of no consequence.

In QPSK differential encoding, where two successive bits define a signaling state, both of these bits are differentially encoded. One encoding scheme is to have the four possible pairs of information bits represented by the phase shifts shown in Fig. 4.34(a). The corresponding constellation diagram is shown in Fig. 4.34(b). Note that the phase state representing the present state, though shown in the first quadrant, could have been in any

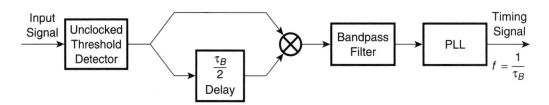

FIGURE 4.33   Block diagram of delay and multiply timing recovery circuit.

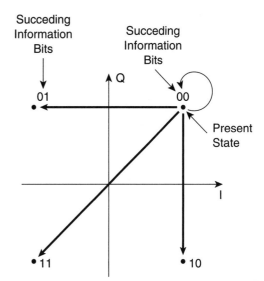

Information Bits	Phase Shift
00	0
01	$\pi/2$
11	π
10	$3\pi/2$

(a) Carrier Phase Shift Versus Information Bits

(b) Constellation Diagram Representation of Phase Shift Relative to Present Phase State

FIGURE 4.34   Carrier phase shift in differentially encoded QPSK.

of the four quadrants. Note also that the changes are Gray coded. Despite this, because of the nature of differential coding, the bit error rate in an *M*-ary PSK differentially encoded system is twice that of a similar Gray coded *M*-ary PSK system with no differential encoding.[12] This error, expressed as a multiplication factor, is referred to as a coding penalty, in this case a penalty of 2.

For multiple amplitude and phase shift keying systems such as the high-order QAM systems described previously, an effective differential coding technique is the one outlined by Weber.[12] For 2^{2n}-QAM systems differentially encoded with this technique, the first two information bits of the *2n* bits that represent a signaling state are differentially encoded and represent the change in quadrant. The remaining bits define the location within the quadrant and are Gray coded. Thus, for 64-QAM, for example, where six information bits define a transmitted state, the first two information bits are differentially encoded while the remaining four define the location of the state within a quadrant. With this approach, the coding penalty for 2^{2n}-QAM decreases as the number of modulation states increases, going from 1.67 for 16-QAM to 1.27 for 256-QAM.

Differential encoders and decoders can be implemented using logic circuits or lookup tables, with the decoding operation being the inverse of the encoding one.

4.3.9.6   Differential Phase Shift Keying (DPSK)

In the preceding section we saw that differential encoding can be used to resolve the impact of phase ambiguity in the recovered carrier in coherent detection systems. Differential encoding can serve another purpose, however. It can be used to permit the demodulation of phase modulated systems without the need of a recovered carrier. The simplest such system is *differential phase shift keying (DPSK)*, which is basically BPSK, but where the value of the current bit determines not the absolute value of the phase but whether or not the phase is changed relative to the phase in the previous signaling interval. Normally, in DPSK systems, a 1 leads to no change in phase and a 0 leads to a 180° change. A circuit for the generation of DPSK signals is shown in Fig. 4.35. A typical input message bit stream, $b(t)$, is shown in Fig. 4.35(a), as well $b'(t)$, the resulting differentially encoded sequence, and $b'(t - T)$, the encoded sequence delayed by T, a single bit period. The first digit of $b'(t)$ shown is arbitrary and is here assumed to be 1. Fig. 4.35(b) shows how $b'(t)$ is generated. In the logic circuit, when $b(t)$ and $b'(t - T)$ are the same, $b'(t) = 1$. When $b(t)$ and $b'(t - T)$ are different, $b'(t) = 0$. At the start of the encoding process, the first digit of $b'(t)$, of value 1, goes through the delay unit and is fed into the logic circuit along with the first digit of $b(t)$. The transition of $b'(t)$ through the logic circuit is represented by an arrow in Fig. 4.35(a). This results in a new $b'(t)$ of 1, and continuation of the process results in the $b'(t)$ sequence shown. The level shifter converts the 0 and 1 values of $b'(t)$ to $-V$ and $+V$ Volts, respectively, creating the bipolar signal $v(t)$, say, and this signal mixes with the carrier to create the output DPSK signal.

A circuit for recovering the message bit stream from the DPSK received signal is shown in Fig. 4.36(a). Note that the only difference between this demodulator and the standard BPSK demodulator is that reference signal into the correlator, instead of being the recovered carrier, is a delayed version of the received signal. Thus, during each symbol period we compare the received signal with its predecessor and look for a true correlation (i.e., in phase) or an anticorrelation (i.e., 180° out of phase). The output of the multiplier, $v_o(t)$, is given by

$$v_o(t) = v(t)\cos\omega_0 t \cdot v(t - T)\cos\omega_0(t - T) \tag{4.157a}$$

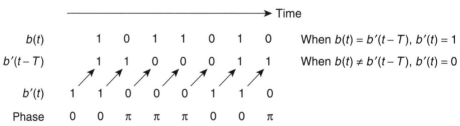

(a) Input and Output Signals of Logic Circuit and Resulting Phase of Carrier

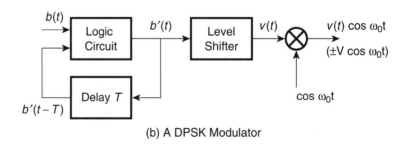

(b) A DPSK Modulator

FIGURE 4.35   DPSK signal generation.

$$= v(t)v(t-T)\frac{\cos\omega_0 T}{2} + v(t)v(t-T)\frac{\cos 2\omega_0(t-T)}{2} \tag{4.157b}$$

As shown in Fig. 4.36(a), $v_o(t)$ is passed through a low-pass filter. This action removes the second term on the right-hand side of Eq. (4.157b), leaving the first term, which is the waveform $v(t)v(t-T)$ multiplied by the constant $\cos \omega_0 T / 2$. In order that this first term be as large as possible, the relationship between T and ω_0 is chosen so that $T\omega_0 = 2n\pi$, where n is an integer, and hence $\cos \omega_0 T = +1$. Why generate $v(t)v(t-T)$? Because, in a noise-free environment, $v(t)v(t-T)$ is in fact the bipolar version of $b(t)$, the message bit stream. To see that this is so, one need only study Fig. 4.36(b), which shows, given that $b(t)$ is as indicated in Fig. 4.35(a), the resulting $v(t)$, $v(t-T)$, and the multiplication of these two waveforms, $v(t)v(t-T)$.

The simplicity with which DPSK is demodulated comes at a price. This price is a decrease in BER performance. To gain an intuitive feel as to why this is so, focus on the fact that bit determination is made on the basis of the signal received in two successive bit intervals. As a result, noise in one bit interval may result in errors in two bit determinations. The derivation of the probability of error of versus E_b / N_0 relationship for DPSK is complex and is not given here. The relationship is[10]

$$P_{eb} = \frac{1}{2}e^{-E_b/N_0} \tag{4.158}$$

For a probability of error of 10^{-3}, a plot of Eq. (4.158) shows that, to achieve the same probability of error, the E_b / N_0 required with DPSK is about 1.2 dB larger than that required with BPSK. For any probability of error of $\leq 10^{-4}$, the E_b / N_0 required with DPSK to achieve the same probability of error as BPSK is larger than that required with BPSK by a maximum of 1 dB. Thus the degradation in performance is relatively small in return for a simpler circuit and one that incurs no synchronization delay, the latter feature being very useful in bursty-type communications.

4.3.9.7   Differential Quadrature Phase Shift Keying (DQPSK)

In *differential quadrature phase shift keying (DQPSK)* modulation the incoming information bits are grouped into consecutive pairs called dibits and the dibits are used to modulate the carrier. Unlike QPSK, however, where they are used to determine the phase of the output, here they are used to determine the change in phase of the output. A possible mapping between dibits and output change in phase is shown in Fig. 4.37(a) along with other data that will be discussed later. A demodulator to address a signal modulated via this mapping is shown in Fig. 4.37(b). In this demodulator, the delay T introduced in the front end is equal to the dibit period (i.e., twice the information bit period). Also, as in the case of DPSK, the relationship between T and ω_0, the carrier frequency, is chosen so that $T\omega_0 = 2n\pi$, where n is an integer. The demodulator works as follows:

Let the incoming signal to the demodulator, resulting from the modulator output created by the kth dibit, be $v_k(t)$, where

$$v_k(t) = \sqrt{2}\cos(\omega_0 t + \varphi_k) \tag{4.159}$$

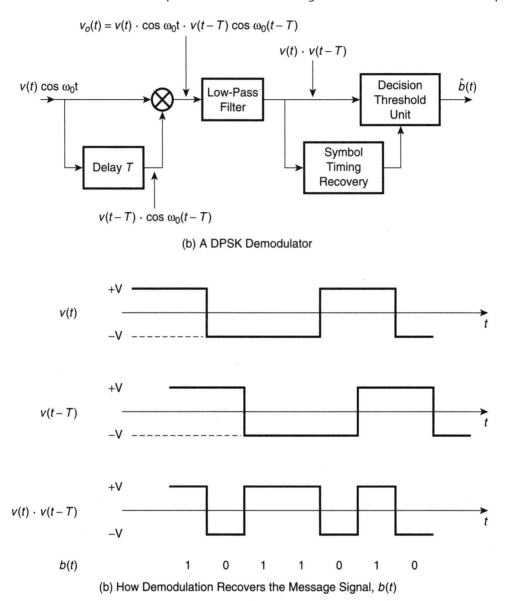

FIGURE 4.36   DPSK signal demodulation.

(There is no significance to the $\sqrt{2}$ multiplier. It has been added to simplify an analytical comparison later.)

Then the output of the upper multiplier, $v_{ku}(t)$, is given by

$$v_{ku}(t) = \sqrt{2}\cos(\omega_0 t + \phi_k) \cdot \sqrt{2}\cos\left(\omega_0(t-T) + \phi_{k-1}\right) \qquad (4.160)$$

Similarly, the output of the lower multiplier, $v_{kl}(t)$, where there is an additional $-\pi/2$ phase shift in the delayed path, is given by

$$v_{ku}(t) = \sqrt{2}\cos(\omega_0 t + \phi_k) \cdot \sqrt{2}\cos\left(\omega_0(t-T) - \frac{\pi}{2} + \phi_{k-1}\right) \qquad (4.161)$$

The mathematical expansion of the right-hand side of these equations and the elimination of terms at twice the carrier frequency gives the signals at the outputs of the low-pass filters following the multiplier. The signal at the output of the upper filter, I_k say, is given by

$$I_k = \cos(\phi_k - \phi_{k-1}) \qquad (4.162)$$

and the signal at the output of the lower filter, Q_k say, is given by

$$Q_k = \sin(\phi_k - \phi_{k-1}) \qquad (4.163)$$

We return now to Fig. 4.37(a), where possible values of I_k and Q_k are shown. We observe that the magnitudes of the possible values of I_k and Q_k are all the same (i.e., $1/\sqrt{2}$). Thus, the information content of I_k and Q_k is in their signs. For example, if I_k is positive and Q_k

Dibits	Phase Change $=$ $\phi_k - \phi_{k-1}$	$I_k=$ $\cos(\phi_k - \phi_{k-1})$	$Q_k=$ $\sin(\phi_k - \phi_{k-1})$
00	$+\dfrac{\pi}{4}$	$\dfrac{1}{\sqrt{2}}$	$\dfrac{1}{\sqrt{2}}$
01	$+\dfrac{3\pi}{4}$	$-\dfrac{1}{\sqrt{2}}$	$\dfrac{1}{\sqrt{2}}$
10	$-\dfrac{3\pi}{4}$	$-\dfrac{1}{\sqrt{2}}$	$-\dfrac{1}{\sqrt{2}}$
11	$-\dfrac{\pi}{4}$	$\dfrac{1}{\sqrt{2}}$	$-\dfrac{1}{\sqrt{2}}$

(a) Relationship between Modulating Dibits, Phase Change, and Quadrature Outputs of Demodulator

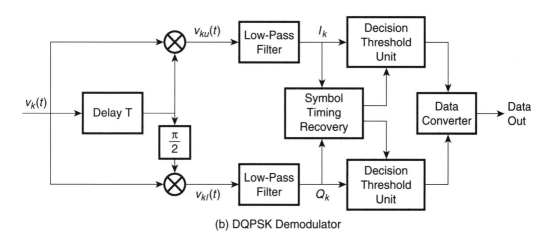

(b) DQPSK Demodulator

FIGURE 4.37   DQPSK modulation mapping and demodulation.

is positive, then, from Fig. 4.37(a), it is clear that the phase change was $+\pi / 4$ and thus the information dibit was 00. To re-create the original information stream, the outputs of the low-pass filters go through decision threshold units, and the outputs of these units are fed to a data converter. This converter knows the relationship between the I_k and Q_k signs and the dibit bits and thus outputs the original information bit stream.

Like DPSK, the derivation of the DQPSK probability of error versus the E_b / N_0 relationship is complex and also not provided here. The relationship is[13]

$$P_{eb} = erfc\left(\frac{4E_b}{N_0}\sin^2\left(\frac{\pi}{8}\right)\right)^{\frac{1}{2}} \tag{4.164a}$$

$$= 2Q\left(\frac{8E_b}{N_0}\sin^2\left(\frac{\pi}{8}\right)\right)^{\frac{1}{2}} \tag{4.164b}$$

A plot of Eq. (4.164b) shows that, for probabilities of error less than about 10^{-3}, to achieve the same probability of error, the E_b / N_0 required with DQPSK is about 2.3 dB larger than that required with QPSK.

4.4　NONLINEAR MODULATION, 4-LEVEL FSK

In nonlinear modulation, a linear relationship does not exist between the baseband signal and the modulated RF carrier. Nonlinear modulation is essentially *frequency shift keying* (*FSK*) modulation, where only the frequency of the carrier is varied as a function of the modulating signal. As a result, the modulated carrier has a constant envelope. However, since frequency is the rate of change of phase, FSK results in a different rate of change of phase from symbol to symbol, and thus a relative phase that is not constant over a symbol period, as is the case with ideal PSK. A general analytical expression for a *multilevel FSK* (*M-FSK*) modulated signal, where the symbol duration is τ_B and where the frequency is modulated to M discrete values, is given by

$$v_{i(FSK)} = A\cos(2\pi f_i t) \tag{4.165}$$

where $$0 \le t \le \tau_B$$

$$i = 1, 2, \ldots, M$$

A key determining factor in the spectral properties as well as the probability of error performance of an M-FSK system is its modulation index m, defined as the ratio of twice the peak frequency shift Δf of the carrier to the modulating symbol rate f_B. Thus mathematically we have

$$m = \frac{2\Delta f}{f_B} \tag{4.166}$$

Another important factor in the spectral properties of an M-FSK modulated signal is the nature of the phase transition at symbol transitions. Abrupt phase transitions, as would be created by switching between *M* independent oscillators, tend to result in large side lobes in the frequency domain, reducing the spectral efficiency of the system. Wireless M-FSK systems are thus normally designed to ensure smooth phase transitions. The modulation in such systems is referred to as *continuous-phase FSK (CPFSK)*. In Fig. 4.38(a) a 4-level PAM modulating signal is shown, and Fig. 4.38(b) is a sketch of the resulting CPFSK signal that shows phase continuity at the symbol boundaries. A common method for its generation is the feeding of the data signal to a single voltage-controlled oscillator (VCO), whose output frequency shift from its center frequency is proportional to the input voltage. When generated in this fashion, continuous phase occurs naturally. The constant envelope feature of M-FSK has the advantage of allowing amplification by highly nonlinear amplifiers, which, in general, provide higher output power than their linear counterparts. Unfortunately, given the same spectral efficiency, this advantage is normally more than offset by their typically poorer probability of error performance and higher sensitivity to interference, particularly in high spectral efficiency situations. As a result, only one specific M-FSK technique has found wide application in fixed wireless systems. This technique is 4-level CPFSK with noncoherent demodulation. Its specific advantage is ease of implementation leading to competitive costs where spectral efficiency of not greater than approximately 1.5 bits/s/Hz. is required.

As an alternative to generating an M-FSK signal with a VCO, it can be created via quadrature modulation, and some modern systems do just that. Since 4-FSK is our M-FSK modulation of interest, we will examine the generation of 4-FSK signals via quadrature

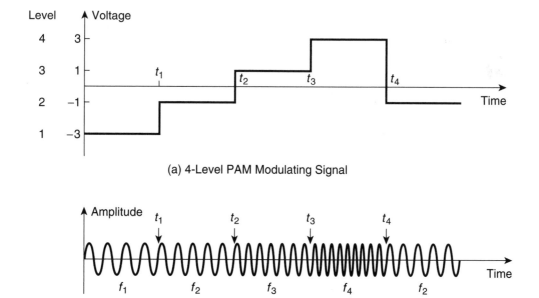

(a) 4-Level PAM Modulating Signal

(b) 4-Level CPFSK Modulated Signal

FIGURE 4.38   4-level CPFSK modulating and modulated signals.

modulation. Consider the quadrature circuit shown in Fig. 4.39(a), with in-phase sinusoidal modulating signal $m_i(t)$ and quadrature modulating sinusoidal signal $m_q(t)$, where

$$m_i(t) = \cos 2\pi f_1 t \tag{4.167}$$

and

$$m_q(t) = \cos(2\pi f_1 t + \pi / 2) = -\sin 2\pi f_1 t \tag{4.168}$$

Then the combined RF output signal, $S(t)$, is given by

$$S(t) = \cos 2\pi f_1 t \cos 2\pi f_c t - \sin 2\pi f_1 t \sin 2\pi f_c t \tag{4.169a}$$

$$= \cos 2\pi (f_c + f_1) t \tag{4.169b}$$

It can similarly be shown that if $m_q(t)$ lags $m_i(t)$ by $\pi / 2$; that is, if

$$m_q(t) = \cos(2\pi f_1 t - \pi / 2) = \sin 2\pi f_1 t \tag{4.170}$$

then

$$S(t) = \cos 2\pi (f_c - f_1) t \tag{4.171}$$

Thus, by feeding the quadrature multipliers with identical baseband sinusoidal signals of frequency f_1, we can create a discrete RF signal of frequency $f_c + f_1$ if $m_q(t)$ leads $m_i(t)$ by 90°, and of frequency $f_c - f_1$ if $m_q(t)$ lags $m_i(t)$ by 90°. If the multipliers are fed with signals of frequency $f_2 = 3f_1$, then we can create RF signals of frequency $f_c + 3f_1$ and $f_c - 3f_1$. Thus by the appropriate choice of modulating signals we can create RF signals at four discrete and equally spaced frequencies as shown in Fig. 4.39(b) (i.e., a standard 4-FSK signal of peak frequency shift $\Delta f = 3f_1$). A 4-FSK quadrature modulator is constructed by grouping the incoming bit stream into pairs to create dibits and using each of the four possible dibits to generate one of two modulating frequency and one of the two phase relationships. An example of a possible relationship between incoming bit pairs, modulating signals, and the resulting RF outputs is given in Fig. 4.39(c). Note that for a modulation index of 1 or less, which is likely to be the case with fixed broadband systems, the modulating "signals," which have a duration $1/f_B$, are really sinusoidal-shaped symbols with a length of a half a cycle or less. To create continuous phase 4-FSK, the $m_i(t)$ and $m_q(t)$ symbol streams are low-pass filtered, thus eliminating instantaneous transitions.

Like modulation, there is more than one method of demodulating M-FSK signals. Demodulators using all methods achieve the same result, however. They create a voltage at their output that is proportional to the instantaneous frequency deviation from the unmodulated carrier of the M-FSK signal at their input. A popular method, originally developed for analog modulated FM systems, is called *discrimination detection*. A second method is referred to as *differential detection, differential discrimination,* or *product detection.*

A commonly used discrimination detector is shown in Fig. 4.40(a). It is called a *balanced discriminator* and consists of two tuned circuits, one tuned to frequency $f_c + f_d$, say, and the other to frequency $f_c - f_d$, say, where f_c is the carrier frequency. These circuits feed envelope detectors, with the detector diode directions being such that the output voltage of

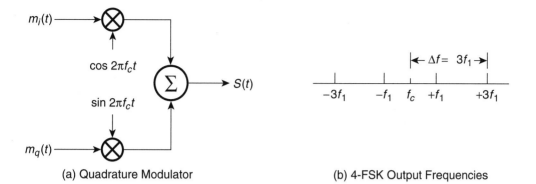

(a) Quadrature Modulator

(b) 4-FSK Output Frequencies

Input Bit Pair (Dibit)	$m_i(t)$	$m_q(t)$	Frequency of $S(t)$
00	$\cos 2\pi \cdot 3f_1 t$	$\sin 2\pi \cdot 3f_1 t$	$f_c - 3f_1$
01	$\cos 2\pi f_1 t$	$\sin 2\pi f_1 t$	$f_c - f_1$
11	$\cos 2\pi f_1 t$	$-\sin 2\pi f_1 t$	$f_c + f_1$
10	$\cos 2\pi \cdot 3f_1 t$	$-\sin 2\pi \cdot 3f_1 t$	$f_c + 3f_1$

(c) Relationship between Input Bit Pairs, Quadrature Modulating Symbols, and Resulting RF Frequencies in 4-FSK Quadrature Modulation

FIGURE 4.39   FSK quadrature modulation.

the higher tuned circuit, V_1, say, is positive and of the lower tuned circuit, V_2, say, is negative, as shown in Fig. 4.40(b). These outputs are summed to create the combined response $V_1 + V_2$ that is also shown in Fig. 4.40(b). With the careful choice of component values, the slope of the combined response over the bandwidth of the input signal can be made quite linear. Because the discriminator output is sensitive not only to the frequency of the input signal but also to its amplitude, variations in amplitude are normally removed by an amplitude limiter prior to discrimination. Since FSK is a constant amplitude modulation scheme, this limiting has no adverse impact on the FSK content of the signal.

A block diagram of the differential detector is shown in Fig. 4.41. It will be obvious that it's very similar to the DPSK demodulator discussed in Section 4.3.9.6. The only difference is in the delay time, T. Here T is chosen such that $2\pi f_c T = \pi / 2$. The reason for this choice will become clear shortly. For an input signal, $S_i(t)$, equal to $\cos 2\pi f_c t$, the output of the multiplier, $S_{om}(t)$, is given by

$$S_{om}(t) = \cos 2\pi f_i t \cdot \cos 2\pi f_i (t - T) \qquad (4.172a)$$

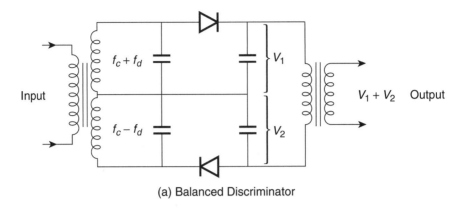

(a) Balanced Discriminator

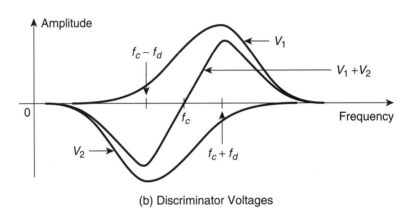

(b) Discriminator Voltages

FIGURE 4.40   Discrimination detection.

$$= \cos 2\pi f_i T + \cos(4\pi f_i t - 2\pi f_i T) \qquad (4.172b)$$

The low-pass filter following the multiplier removes the second term on the right-hand side of Eq. (4.172b). Thus the output of the filter, $S_{of}(t)$, is given by

$$S_{of}(t) = \cos 2\pi f_i T \qquad (4.173)$$

Letting the input frequency, f_i, be stated as $f_c + f_m$, then Eq. (4.173) can be restated as

$$S_{of}(t) = \cos 2\pi (f_c + f_m)T \qquad (4.174a)$$

$$= \cos(2\pi f_c T + 2\pi f_m T) \qquad (4.174b)$$

Recalling that T is chosen such that $2\pi f_c T = \pi / 2$, and substituting this equality into Eq. (4.174b), we get

$$S_{of}(t) = \cos(2\pi f_m T + \pi / 2) \tag{4.175a}$$

$$= -\sin 2\pi f_m T \tag{4.175b}$$

Thus, the choice of T results in the output of the filter, $S_{of}(t)$, being proportional to $-\sin 2\pi f_m T$. Since for values of $x \ll \pi / 2$, $\sin x \simeq x$, then for values of $2\pi f_m T \ll \pi / 2$, $S_{of}(t)$ is proportional to $-2\pi f_m T$ and hence $-f_m$. As f_m is the deviation of the input signal from the carrier frequency f_c, then provided $2\pi f_m T \ll \pi / 2$, the differential detector creates an output of amplitude directly proportional to the input frequency deviation, but inverted in sign. Because, like discrimination detection, the output amplitude is sensitive not only to the frequency, but also to the amplitude of the input signal, an amplitude limiter is normally used prior to detection.

A simplified block diagram of a typical four-level CPFSK system is shown in Fig. 4.42. As with the linear systems described previously, it employs scrambling, descrambling, and demodulator timing recovery. It uses a VCO as its FSK modulator and a limiter- discriminator for detection. The received signal is down converted to an IF frequency prior to being fed to the limiter discriminator. Down conversion is covered in Section 4.5.1. As discrimination detection does not require a replica of the carrier, then obviously neither carrier recovery or differential encoding/decoding is required. The absence of these circuits and the overall simplicity of the modulation/demodulation process clearly simplify the design and hardware requirements. It will be observed that, in the modulator, the 4-level PAM signal is filtered prior to driving the frequency modulator. This is done to further limit the width of the generated spectrum. The combined transfer function of this premodulator filter and the postdiscriminator filter in the demodulator is designed to result in a spectrum at the input to the decision threshold unit that is raised cosine or very close to raised cosine. As a result, intersymbol interference is minimized. This design is usually accomplished with root raised cosine (RRC) filters in the modulator and demodulator. Alternatively, *Gaussian*-type filters[14] are used. A Gaussian premodulator filter, if narrow enough, can result in better out-of-band spectral performance as compared to RRC filtering but achieves this at the expense of ISI performance, which degrades slightly. When a Gaussian premodulator filter is used, the modulation is sometimes referred to as *Gaussian frequency shift keying (GFSK)*.

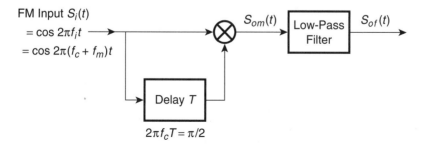

FIGURE 4.41   Differential detector/differential discriminator.

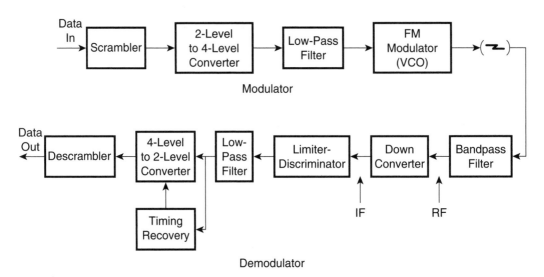

FIGURE 4.42   Block diagram of typical one-way 4-level CPFSK system.

In frequency modulation, the nonlinear relationship between the modulating signal and the modulated signal is such that individual frequency components of the modulating signal result in an infinite number of sidebands (i.e., signals on either side of the carrier at frequencies that are integer multiples of the modulating frequency). The levels of these sidebands depend on the shape of the pulses of the modulating signal, the phase transition between modulated symbols, and the modulation index. The derivation of the CPFSK spectrum is thus quite complicated and will not be addressed here. For the situation where the modulating pulses are rectangular, Lucky et al.[2] have derived a formula and presented normalized spectra for the 2-, 4- and 8-level cases. Their computed spectral densities for 4-level CPFSK and various values of modulation index are shown in Fig. 4.43. Note that they have presented spectra as a function of what they label Δ, where Δ equals the ratio of the frequency shift between adjacent levels divided by the modulating symbol rate. For 4-FSK, modulation index m as defined in Eq. (4.166) is equal to 3Δ. Since low-pass filtering of the modulating pulses results in a narrowing of the spectrum, the results of Lucky et al. can be regarded as an upper bound on spectrum width for a given modulation index.

As an alternative to complex computation, a simple rule exists that approximates the minimum transmission bandwidth, $W_{\min}$, required for FM systems. It's called *Carson's rule* and states that $W_{\min}$ is given by

$$W_{\min} \simeq 2(\Delta f + f_m) \qquad (4.176)$$

where f_m is the maximum modulating frequency and $W_{\min}$ contains at least 98% of the power associated with the FM signal. For an unfiltered n-level PAM modulating signal of symbol rate and f_B, the amplitude has a $\sin x / x$ form and is already down to 0.64 of its maximum at the Nyquist bandwidth of $f_B / 2$. Thus a practical estimate of f_m would appear

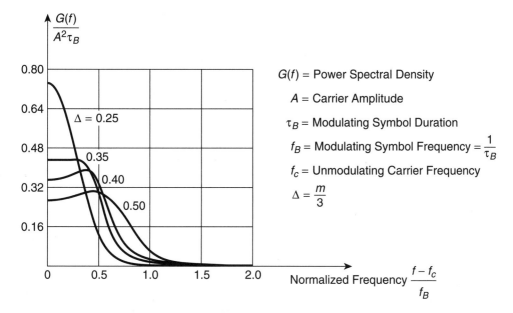

FIGURE 4.43   Spectral density of 4-level CPFSK. (By permission from Ref.2.)

to be $f_B / 2$. Substituting into Eq. (4.176) this estimate of f_m as well as the equivalence of Δf as given by Eq. (4.166), we get

$$W_{\min} \simeq (m+1)f_B \tag{4.177}$$

and for 4-FSK, where $f_b = f_B / 2$, we have

$$W_{\min(4L)} \simeq \frac{f_b}{2}(m+1) \tag{4.178}$$

where f_b is the system bit rate.

The interpretation of Carson's rule as indicated in Eq. (4.177) is in accordance with the interpretation of Mazo et al.[15] Comparing the minimum (double-sided) bandwidth as indicated by Eq. (4.177) with the (single-sided) spectra presented in Fig. 4.43 suggests that this approximation is a good one for 4-level CPFSK with unfiltered (rectangular) modulating pulses. For a spectral efficiency of 1 bit/s/Hz (the minimum normal regulatory requirement) or better, $W_{\min}$ needs to be less than f_b and thus, by Eq. (4.178), the modulation index m needs to be less than approximately 1.

The probability of bit error versus E_b / N_0 performance of 4-level CPFSK with discriminator detection is a function of the signal to noise ratio at the receiver input, the modulation index, the premodulation and postdetection low-pass filtering, and the prediscriminator bandpass filtering. Its theoretical computation is far more complex than that for the linear modulation methods addressed previously and will not be presented here. The derivation available in the literature that's closest to our interests and that results in a straightforward formula appears to be that of Mazo et al.[15] In the model they analyze, they assume a multilevel FSK transmitter with unfiltered modulating pulses. In the receiver they assume an RF filter of bandwidth W_{min}, ideal discriminator detection, and an *integrate-and-dump*,[16] circuit as the postdetection filter. Their derived equation for the probability of symbol error, P_{se}, is presented as their Eq. (4). Rewriting this equation for the specific case of 4-FSK and converting the parameters used to those given in this text, we get

$$P_{se(4L)} = \frac{1}{(2\pi\rho)^{\frac{1}{2}}} \frac{\cot\left(\frac{\pi m}{6}\right)}{\left(\cos\left(\frac{\pi m}{6}\right)\right)^{\frac{1}{2}}} \exp\left[-2\rho\sin^2\left(\frac{\pi m}{6}\right)\right],$$

(4.179)

$$\rho \gg 1, \quad m < \frac{3}{2}$$

where ρ = the RF signal-to-noise ratio in the minimum bandwidth $W_{min(4L)}$. In terms of E_b / N_0, ρ is given by

$$\rho = \frac{2}{m+1} \frac{E_b}{N_0}$$

(4.180)

Equation (4.179) is applicable where *clicks* or *spikes*[16] associated with the well-documented FM threshold effect are not significant. This is what drives the criteria that m be less than 3/2 and ρ be very much greater than 1.

For 4-FSK with Gray coding of the PAM symbols, one symbol error leads to one of two bits being in error. Thus the probability of Gray coded bit error $P_{be(4L)}$ is given by

$$P_{be(4L)} = \frac{1}{2} P_{se(4L)}$$

(4.181)

For 4-FSK systems transmitting standard bit rates in standard channel bandwidths, the typical minimum spectral efficiency required varies from about 1.2 to 1.3 bits/s/Hz. For example, a spectral efficiency of 1.2 would be required for a system transmitting 8.45 Mb/s (4 E1s) in a 7-MHz channel. This spectral efficiency requires that $W_{min(4L)} = f_b / 1.2$. Substituting this value of $W_{min(4L)}$ into Eq. (4.178) leads to a modulation index m of 0.67. For systems where the modulating pulses are filtered, as is almost always the case, the minimum transmitted bandwidth will be less than implied by Eq. (4.178) and thus should allow a somewhat higher modulation index while still having the transmitted spectrum stay within the authorized bandwidth. In Fig. 4.44 a graph of probability of bit error $P_{be(4L)}$ versus E_b / N_0 for Gray coded 4-FSK with $m = 0.75$ is presented, based on Eqs. (4.179) and (4.181). It will be observed that for a probability of bit error of 10^{-6} the E_b / N_0 required is 15.3 dB, which is 4.8 dB greater than that required for QPSK. Normally, this deficiency relative to QPSK is partly but not fully eliminated by the higher transmitted power afforded by FSK.

In practice, the system designer optimizes the probability of error performance of 4-FSK by finding the combination of filtering and modulation index that provides best probability of error performance while containing the radiated spectrum just within the

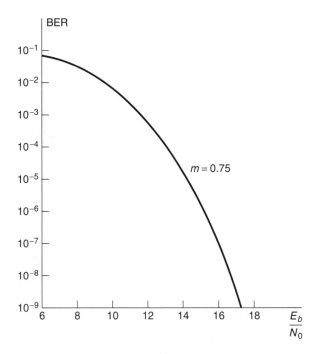

FIGURE 4.44   BER versus E_b / N_0 for 4-FSK with noncoherent
discriminator detection and $m = 0.75$.

spectral mask prescribed by the licensing authority. This optimization is carried out on either a hardware or a software model.

4.5 TRANSMISSION FREQUENCY COMPONENTS

The modems described previously typically do not operate at the frequency of actual RF
transmission but, instead, at a lower standard frequency. The most common modem operating frequency for fixed wireless systems is 70 MHz, but other frequencies such as 140
MHz are also used. The modem frequency is referred to as the *intermediate frequency*
(*IF*). The primary exception to the use of an IF signal in fixed wireless occurs in FSK systems where, if a VCO is used, the modulator typically operates at the desired RF. Even
here, however, an IF frequency is normally employed on the receive side. There are many
advantages to the modem being at a standard frequency. Among these are a standard design that greatly aids production efficiency, design at a frequency where components are
cost effective, and design that allows complex signal processing such as high-order bandpass filtering. To effect transmission at the desired RF frequency, components are added
to the modulator and demodulator to create the full transmitter and receiver. Fig. 4.45
shows a digital transmitter with these components, and Fig. 4.46 shows the associated receiver. In this section we discuss these components.

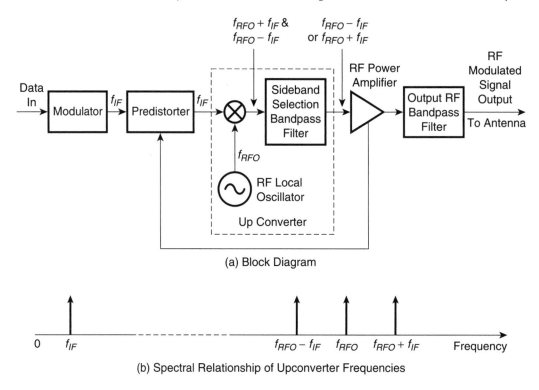

FIGURE 4.45   Digital transmitter.

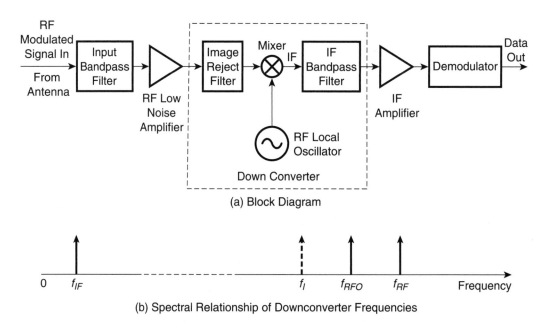

FIGURE 4.46   Digital receiver.

4.5.1   Transmitter Upconverter and Receiver Downconverter

The *transmitter upconverter*, shown in Fig. 4.45(a), translates the modulator output IF signal of frequency f_{IF} up to the desired RF frequency. It accomplishes this by mixing (multiplying) the IF signal with a RF local oscillator signal. If the IF signal is $m(t)\cos(2\pi f_{IF}t)$, where $m(t)$ represents the modulation on the carrier, and the RF local oscillator signal is $\cos(2\pi f_{RFO}t)$, then the output of the mixer $s_{RF}(t)$ is given by

$$s_{RF}(t) = m(t)\cos(2\pi f_{IF}t)\cdot\cos(2\pi f_{RFO}t)$$

$$= \frac{1}{2}m(t)\cos\left(2\pi\left(f_{RFO}-f_{IF}\right)t\right)+\frac{1}{2}m(t)\cos\left(2\pi\left(f_{RFO}-f_{IF}\right)t\right) \quad (4.182)$$

Thus the output of the mixer consists of two sideband signals, one below and one above the RF oscillator frequency by amounts equal to the IF frequency. The spectral relationship of the frequencies associated with upconversion is shown in Fig. 4.45(b). The designer chooses f_{RFO} so that one of these sideband frequencies is the one desired. The undesired sideband is eliminated by passing the output of the mixer through a bandpass filter centered on the desired frequency.

The *receiver downconverter*, shown if Fig. 4.46(a), does the opposite of the upconverter. It translates the received modulated RF signal, of frequency f_R, down to the IF frequency f_{IF}. Note, however, that if the desired received frequency is above the RF local oscillator frequency, then the process also translates to IF any signal at a frequency below that of the RF local oscillator that's offset by the IF frequency and vice versa. This undesired frequency is called the image frequency, f_I say, and any received interfering signal at or close to this frequency must be eliminated by a filter placed ahead of the mixer. Note also that even if an interfering signal is not present at the image frequency, there is always thermal noise there that, if not removed, is downcoverted and results in a doubling of the noise at the demodulator input. In Fig. 4.46(a) an image reject filter is shown just ahead of the mixer. This filter, however, is only necessary if the bandwidth of the input filter is such as to not reject the image frequency. The spectral relationship of the frequencies associated with downconversion is shown in Fig. 4.46(b).

Some receivers, for frequency agility purposes, or to allow the sharing of one RF local oscillator between the receiver and the accompanying transmitter, employ a double downconversion process and hence have two IF frequencies.

4.5.2   Transmitter RF Power Amplifier and Output Bandpass Filter

The transmitter RF *power amplifier* follows the upconverter or, in the case of VCO modulated 4-level FSK, the VCO, and its purpose is to provide a high level of output power so that adequate signal level is available to the receiver even with significant fading. These amplifiers today are all solid state, with bipolar transistors typically used below about 2 GHz and GaAs FETs typically used at about 2 GHz and above. The output power of fixed wireless transmitters vary from slightly less than a tenth of a Watt to several Watts, with maximum attainable power decreasing with frequency.

For FSK systems, the power amplifier is generally operated in a fully saturated mode. However, for linear modulation systems with signal states of varying amplitudes, linear amplification is essential to maintain acceptable performance, with higher and higher linearity required as the number of modulation states increases. The effects of non-linearity on such systems are a nonlinear displacement of signaling states in the phase plane and the regeneration of spectral sidelobes removed by prior filtering. The displacement of the signaling states degrades the error probability performance while the regenerated spectrum can cause interference to signals in adjacent channels. To avoid these effects the power amplifier must be capable of linearly amplifying all signaling states and thus amplifying the peak signal power. However, high order linear modulation results in signals where the ratio between peak power and average power can be several decibels. For example, for 16-QAM, 64-QAM, and 256-QAM, this ratio is 6 to 7 dB, 7 to 8 dB, and 8 to 9 dB, respectively. Thus, power amplifiers processing these signals must operate at an average power that is backed off from the peak linear power available by a minimum of the peak to average ratio of the amplified signal.

The output signal of the RF power amplifier is normally fed via a bandpass filter to the antenna. The bandwidth of this filter is typically wider than the signal spectrum and its purpose is to allow duplexing of this signal with an associated incoming signal (Section 7.5).

4.5.3   Power Amplifier Predistorter

In a number of wireless transmitter designs, the nonlinearity of the RF power amplifier is counteracted by employing a RF *predistorter,*[17] typically placed in the IF path as shown in Fig. 4.45. The predistorter's function is to purposely insert a nonlinearity into the IF signal that is the complement of the nonlinearity of the RF power amplifier. Though possible at RF, predistortion is normally done at IF as it is much more easily realized at this frequency, is able to account for the nonlinearity of the upconverter in addition to that of the RF amplifier, and it allows a standard one-frequency design. Many modern predistorters are adaptive. In such designs, circuits are added to the RF amplifier that continuously measure the nonlinearity at its output and feed this measurement back to the predistorter, which adjusts its nonlinearity in such a way as to minimize the RF output nonlinearity. By the use of predistorters, it is typically possible to increase transmitter output power by 1 to 2 dB relative to non-predistorted transmitters.

Baseband predistortion is also possible. In this case predistortion is accomplished by altering the modulating baseband signal voltages in such a fashion that the signaling states end up in their correct positions after passage through the RF nonlinearity.

4.5.4   The Receiver Front End

The term *receiver front end* is normally used to mean the combination of the receiver input bandpass filter, the low-noise amplifier, and the downconverter. The purposes of the input filter are to eliminate unwanted signals and, if no image reject filter is equipped just ahead of the mixer, to filter out frequencies at or close to the image frequency. Following the input filter there is normally a low-noise amplifier, which plays a large part in determining the overall noise performance of the receiver. The characteristic of the receiver that determines the signal-to-noise ratio presented to the demodulator is the receiver *noise figure*. For the receiver shown in Fig. 4.46, the noise figure describes the deterioration of the signal-to-noise ratio from the receiver RF input to the IF amplifier output due to presence of all the circuitry between these inputs and is given by

$$F = \frac{P_{Si} / P_{ni}}{P_{So} / P_{no}} \qquad (4.183)$$

where P_{Si} = RF input signal power
$\quad P_{So}$ = IF output signal power
$\quad P_{ni}$ = RF input noise power in a frequency band *df*
$\quad P_{no}$ = IF output noise power in a frequency band *df*
Thus

$$P_{So} / P_{no}(\text{dB}) = P_{Si}(\text{dBm}) - P_{ni}(\text{dBm}) - F(\text{dB}) \qquad (4.184)$$

where 1 dBm = 1 milliWatt.

The signal-to-noise ratio at the IF amplifier output, P_{So} / P_{no}, is also the signal-to-noise ratio at the demodulator input and thus that which determines the error rate performance. Equation (4.184) is important because it indicates that P_{So} / P_{no} can be determined from a knowledge of the signal-to-noise ratio at the receiver input and the receiver noise figure.

The thermal noise power in Watts available in a small frequency band *df* Hertz from a source having a noise temperature *T* degrees Kelvin (K) is given by

$$P_n = k \cdot T \cdot df \qquad (4.185)$$

where $k = 1.38 \cdot 10^{-23}$ Joules/degree Kelvin (Boltzmann's constant).

For a terrestrial wireless system, the source of thermal noise at the receiver input is the receiving antenna. Antenna noise temperature is normally assumed to be 290K. At this temperature, the antenna noise transferred to the receiver is given by Eq. (4.185) to be –174 dBm per Hertz of bandwidth. Thus the input thermal noise in the bit rate bandwidth f_b Hertz of a digital wireless system is given by

$$P_{nib}(\text{dBm}) = -174 + 10 \, \log_{10} f_b \qquad (4.186)$$

Substituting Eq. (4.186) into Eq. (4.184) gives the ratio of IF output signal power P_{So} to noise power in the bit rate bandwidth P_{nob} to be

$$P_{So} / P_{nob}(\text{dB}) = P_{Si}(\text{dBm}) + 174 - 10 \, \log_{10} f_b - F(\text{dB}) \qquad (4.187)$$

recognizing that

$$\frac{P_{So}}{P_{nob}} = \frac{E_b f_b}{N_0 f_b} = \frac{E_b}{N_0} \qquad (4.188)$$

where E_b is the energy per bit at the IF amplifier output/demodulator input and N_0 is the noise power spectral density at the IF amplifier output/demodulator input, then Eq. (4.187) can be restated as

$$E_b / N_0 (\text{dB}) = P_{Si}(\text{dBm}) + 174 - 10 \; \log_{10} f_b - F(\text{dB}) \qquad (4.189)$$

Thus, knowing the received input signal level, the bit rate, and the receiver noise figure, one can calculate the E_b / N_0 at the demodulator input and, from the appropriate probability of error versus E_b / N_0 relationship, the theoretical probability of error.

REFERENCES

1. Feher, K., *Digital Communications: Microwave Applications,* Prentice Hall, 1981.
2. Lucky, R. W., Salz, J., and Weldon, E. J., *Principles of Data Communication,* McGraw-Hill, 1968.
3. Bennett, W. R., and Davey, J. R., *Data Transmission,* McGraw-Hill, 1965.
4. Volertas, V., and Nossen, E., *Phase Modulation Techniques for Digital Communication Systems,* Proceedings of the International Telecommunications Exposition, Atlanta, Georgia, October 1977, pp. 474–480.
5. Morais D. H., and Feher, K., "Bandwidth Efficiency and Probability of Error Performance of MSK and Offset QPSK Systems," *IEEE Trans. Commun.,* Vol. COM-27, Dec. 1979, pp. 1794–1801.
6. Morais, D. H., and Feher, K., "The Effects of Filtering and Limiting on the Performance of QPSK, Offset QPSK and MSK Systems," *IEEE Trans. Commun.,* Vol. COM-28, Dec. 1980, pp. 1999–2009.
7. Wesel, R. D., *New Coding Techniques for Multicarrier Modulation,* University of California, Los Angeles, Department of Electrical Engineering, Final Report 1997–1998 for MICRO Project 97-204.
8. Burr, A., *Modulation and Coding for Wireless Communications,* Prentice Hall, 2001.
9. Winch, R. G., *Telecommunication Transmission Systems,* McGraw-Hill, 1993.
10. Sklar, B., *Digital Communications Fundamentals and Applications,* 2nd ed., Prentice Hall, 2001.
11. Shafi, M., and More, D. J., *Digital Implementation of a Simplified Carrier Recovery Loop,* Conference Record, ICC, Chicago, 1985, pp. 31.2.1–31.2.5.
12. Weber, W. J. III., "Differential Encoding for Multiple Amplitude and Phase Shift Keying Systems," *IEEE Trans. Commun.,* Vol. COM-26, March 1978.
13. Lindsey, W. C., and Simon, M. K., *Telecommunications Systems Engineering,* Prentice Hall, 1973.
14. Feher, K., *Wireless Digital Communications: Modulation and Spread Spectrum Applications,* Prentice Hall, 1995.
15. Mazo, J. E., Rowe, H. E., and Salz, J., "Rate Optimization for Digital Frequency Modulation," *Bell Syst. Tech. J.,* Nov. 1969, pp. 3021–3030,
16. Taub, H., and Schilling, D. L., *Principles of Communication Systems,* McGraw-Hill, 1971, pp. 366–369.
17. Ivanek, F., editor, *Terrestrial Digital Microwave Communications,* Artech House, 1989.

CHAPTER 5

THE FIXED WIRELESS PATH

5.1 INTRODUCTION

A fixed wireless system communicates between sites via the propagation of radio waves over a *path*. In an ideal world the path would linearly attenuate the transmitted signal by a fixed amount resulting in a predictable signal level at the receiver input. A digital system so deployed in "free space" could then be designed so that the received level resulted in an acceptable bit error rate and no further analysis would be required. In fact, ground-station to satellite paths behave close to such an ideal. For point-to-point terrestrial links, however, atmospheric anomalies often result in significant deviation from the ideal just postulated, and such deviation is referred to as *fading*. A typical fixed wireless link is shown in Fig. 5.1. As will be seen in succeeding sections, the maximum length of a wireless path for reliable communications varies depends on (a) the propagation frequency, (b) the antenna heights, (c) terrain conditions between the sites (in particular, the minimum clearance between the direct signal path and ground obstructions), (d) atmospheric conditions over the path, and (e) the radio equipment and antenna system electrical parameters. For systems operating in the 2- to 10-GHz bands, paths are typically 20 to 40 miles in length. For those operating in the 18- to 38-GHz bands, however, paths lengths are much more restricted, being typically in single digit miles, much of this restriction being due to fading resulting from rain.

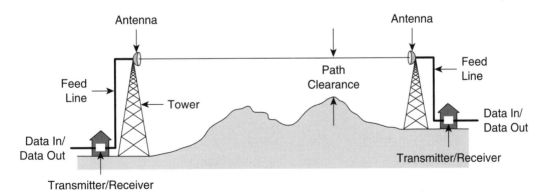

FIGURE 5.1   Typical fixed wireless link.

In this chapter we will examine propagation in an ideal environment and then study the various types of fading and how such fading impacts path reliability. As antennas provide efficient means to launch and receive radio waves, a brief review of their characteristics is in order as the first step in addressing ideal propagation.

5.2   ANTENNAS

5.2.1   Introduction

There are myriad antenna shapes and designs for wireless communication, and the variations are often a function of operating frequency. For the fixed broadband systems of interest, operating frequency is typically between about 2 and 40 GHz. Over this frequency range, only three types of antennas are typically used for point to point links; namely, the *parabolic antenna,* the *horn-reflector antenna,* and, more recently, *the flat plane* or *planar array antenna.* These antennas (more correctly "antennae") can be made to be highly *directive.* For point-to-point links this feature is desirable, with the transmitting antenna focusing its energy in a narrow beam that is directed to the receiving antenna, where it is collected in an efficient manner. For point-to-multipoint links, the outstations, like in point-to-point links, require directive antennas. For base stations, however, the antenna is required to transmit and receive signals over a wide sector and, sometimes, omnidirectionally. Omnidirectional antennas are usually of the *stacked dipole* or, as they are sometimes called, *collinear array* type. Sector antennas come in many forms. Some use a reflector behind a feed that creates a wide beam, some use a *lens-corrected horn,* and others use an array of mini-antennas combined electrically to produce a wide lobe.

Many antenna characteristics are important in designing wireless broadband systems. The most important of these will be reviewed followed by a brief description of those antennas commonly used in such systems.

5.2.2   Antenna Characteristics

Antenna gain is the most important antenna characteristic. It is a measure of the antenna's ability to concentrate its energy in a specific direction relative to radiating it isotropically (i.e., equally in all directions). The more concentrated the beam, the higher the gain of the antenna. It can be shown[1] that a transmitting antenna that concentrates its radiated energy within a small beam has an a gain G in the direction of maximum intensity with respect to an isotropic radiator of

$$G(dB) = 10 \cdot \log_{10}\left(\frac{4\pi A_e}{\lambda^2}\right) \qquad (5.1)$$

where A_e = effective area of the antenna aperture
        λ = wavelength of the radiated signal

The antenna's physical area A_p is related to its effective area by the following relationship:

$$A_e = \eta A_p \qquad (5.2)$$

where η is the efficiency factor of the antenna.

Equation (5.1) indicates that antenna gain is function of the square of the frequency ($\lambda f = c$), thus doubling the frequency increases the gain by 6 dB. It is also a function of the area of the aperture. Thus, for a parabolic antenna, for example, the gain is a function of the square of the diameter and so doubling the diameter increases the gain by 6 dB. The efficiency factor η in Eq. (5.2) accounts for the fact that the antenna is not 100% efficient. This is because the total incident power on the radiator is not radiated forward as per theory. Some of it is loss to spillover at edges, some is misdirected because the radiator surface is not manufactured perfectly to the desired shape and some is blocked by the presence of the feed radiator. Nominal values of η are 0.55 (55% efficient) for parabolic antennas and 0.75 for the horn-reflector type. The operation of an antenna in the receive mode is the inverse of its operation in the transmit mode. It therefore comes as no surprise that its receive gain, defined as the energy received by the antenna compared to that received by an isotropic absorber, is identical to its transmit gain. Figure 5.2 shows a plot of antenna gain versus angular deviation from its direct axis, which is referred to as its *pole* or *boresight*. It shows the mainlobe, where most of the power is concentrated. It also shows sidelobes and backlobes, which can cause interference into or from other wireless systems in the vicinity.

The *beamwidth* of an antenna is closely associated with its gain. The higher the gain of the antenna, the narrower the width of the beam. Beamwidth is measured in radians or degrees and is usually defined as the angle that subtends the points at which the peak field power is reduced by 3 dB. Fig. 5.1 illustrates this definition. Antennas used in broadband wireless systems have beamwidths that vary from a fraction of a degree to several degrees. The narrower the beamwidth, the more interference from external sources, including nearby antennas, is minimized. This improvement comes at a price, however, as narrow beamwidth antennas require precise alignment, and being larger, have a higher wind load, necessitating increased support structure mechanical stability. This is because a very small shift in alignment results in a measurable decrease in the transmitted power directed at a receiving antenna and the received power from a distant transmitting antenna.

The *front-to-back ratio* is another important antenna characteristic. It is defined as the ratio, usually expressed in decibels, of its maximum gain in the forward direction to the maximum gain in the region of its backward direction, the latter being the maximum backlobe gain. This ratio is particularly critical in repeater systems, where the same

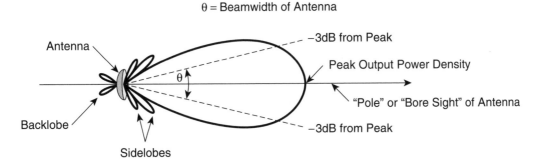

FIGURE 5.2   Antenna gain versus angular deviation from its pole.

frequencies are used in both directions for a station. Unfortunately, the front-to-back ratio of an antenna in a real installation can vary widely from that in a purely free space environment. This variation is due to foreground reflections in a backward direction of energy from the main transmission lobe by objects in or near the lobe. These backward reflections can reduce the free space only front-to-back ratio by 20 to 30 dB.

The polarization of an antenna refers to the alignment of the electric field in the radiated wave. Thus, in a *horizontally polarized* antenna, the electric field is horizontal and in a *vertically polarized* antenna, the electric field is vertical. When a signal is transmitted in one polarization, only a small fraction may be converted to the other polarization due to imperfections in the antenna and the path. Thus, to improve adjacent channel discrimination most standard frequency assignments are such that adjacent channels operate on different polarizations. The ratio of the power received in the desired polarization to the power received in the undesired polarization is referred to as the *crosspolarization discrimination*. Cross polarization typical varies from about 25 to 40 dB depending on the antennas and the path.

Finally, we note that an antenna related parameter often regulated for fixed wireless systems is the *equivalent isotropically radiated power* (*EIRP*). This power is the product of the power supplied to the transmitting antenna, P_t, say, and the gain of the transmitting antenna, G_{ta}, say. Thus, we have

$$EIRP = P_t \cdot G_{ta} \tag{5.3}$$

5.2.3   Typical Broadband Wireless Antennas

The *parabolic antenna* is the most commonly used type with fixed broadband wireless systems. It consists of a parabolic reflector radiated with RF energy from a feed source located at the focus of the reflector. In its simplest form the parabolic reflector is the only reflector used, and the feed source radiates directly onto it. The feed source consists of a waveguide that opens in the form of an enlarging taper in such a way as to match the impedance of the waveguide to that of free space. Such a feed source is, in fact, a member of the horn antenna class and is commonly referred to as the antenna feedhorn. A diagram of a cross section of such a parabolic antenna is shown in Fig. 5.3(a). As can be seen, the antenna's operation is very similar to that of a flashlight, RF waves following most of the rules as optics. One problem associated with directly radiating feedhorns is that their support structure blocks part of the beam they radiate and thus degrade performance. Parabolic antennas vary in diameter, from as low as about 1 foot for those operating at the high end of the common frequency bands (30 to 40 GHz), to 10 feet or more for those operating at the low end of the bands (2 to 11 GHz). The performance of a parabolic antenna radiated directly from a feedhorn can be enhanced by adding a *shield*, which takes the form of a forward projecting short cylinder attached to its circumference. The shield is normally covered by non–radiation-absorbing material called a *radome*. A shielded antenna will typically have slightly higher gain, slightly narrower beamwidth, and measurably higher (in the range of 10 to 20 dB) front-to-back ratio than its unshielded counterpart. Another version of the parabolic antenna that achieves these enhanced performance characteristics without the use of a shield is a dual-reflector type as shown in Fig. 5.3(b). Here the feedhorn is located at the apex of the main reflector and radiates a hyperbolic

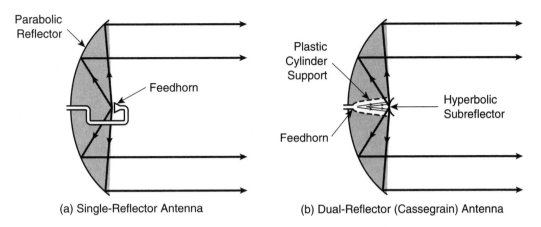

(a) Single-Reflector Antenna   (b) Dual-Reflector (Cassegrain) Antenna

FIGURE 5.3   Cross sections of single- and dual-reflector parabolic antennas.

subreflector that in turn radiates the main reflector. Such a dual reflector type is called a *Cassegrain* antenna. The subreflector is normally rigidly supported by a special plastic cylinder and so there is minimum radiated beam blockage.

Figure 5.4 shows the measured horizontal radiation pattern of an Andrew Corporation 6-foot-diameter parabolic antenna equipped with a shield when operating at 6.175

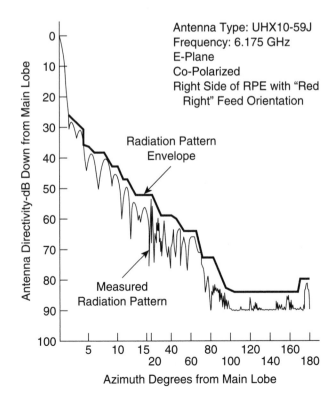

FIGURE 5.4   The horizontal radiation pattern of a 6', 6-GHz shielded parabolic antenna. *(Courtesy of Andrew Corporation.)*

GHz. It also shows the horizontal radiation pattern envelope, this envelope representing the "worst peaks" envelope of the radiation patterns of a number of such antennas. This antenna has a gain of 43.2 dB and, from the figure, we see that the beamwidth is very narrow indeed. In fact, it is specified as being 1.1°. The measured front-to-back ratio is seen to be just over 80 dB at close to 180 azimuth degrees from boresight.

The *horn-reflector antenna* is a high-performance device typically used in the 4- to 11-GHz bands on high-capacity systems. It consists of a vertically focused feedhorn tapering outward from its focal point that radiates RF energy onto a small section of a parabolic surface that then reradiates the energy outward. A simplified diagram of a cross section of such an antenna is shown in Fig. 5.5. This type of antenna has excellent side and backlobe suppression, resulting in a front-to-back ratio that's typically greater than 70 dB. Its impedance match to its waveguide feed is excellent, resulting in minimal reflected waves in the waveguide run feeding it. It exhibits a much wider bandwidth than parabolic antennas and can be used in both the vertical and horizontal polarization modes in the 4- to the 11-GHz bands. Its electrical characteristics are superior to its parabolic counterpart. However, for the same gain it is typically more expensive, larger, heavier, and more difficult to install.

The *lens-corrected horn antenna* is often used as a sector antenna in point-to-multipoint access networks. Figure 5.6 is a simplified diagram of such an antenna. As the diagram indicates, it consists of a horn usually of a simple wedge shape and somewhat similar to a parabolic reflector feedhorn, but with a lens, made from an appropriate dielectric material, placed at the exit of the horn to focus the radiated beam.

The *stacked dipole* or *collinear array*, as indicated previously, provides omnidirectional coverage. It should be noted that it does not actually provide true omnidirectional coverage, since such coverage would be equal in all directions and hence isotropic. What it provides is omnidirectional coverage in the horizontal (or close to horizontal) plane only. In the vertical plane its radiated energy diminishes as it propagates in a more and more

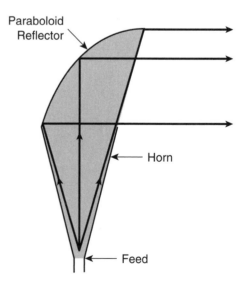

FIGURE 5.5   Cross section of a horn-reflector antenna.

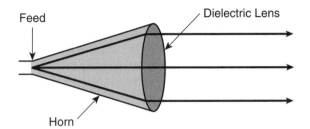

FIGURE 5.6   Cross section of a lens-corrected horn antenna.

vertical direction until it's essentially zero in the truly vertical direction. The simplest omnidirectional antenna is a single half-wave dipole. Its basic form is shown in Fig. 5.7(a) and its horizontal and vertical radiation pattern is shown in Fig. 5.7(b). The theoretical gain of this antenna is only 2.15 dB. Fortunately, by simply stacking identical dipoles into a single structure in the vertical plane and feeding them simultaneously with a phase-sensitive network, higher gain is achieved. When stacked vertically the resulting antenna is called a stacked dipole or collinear array. A stacked dipole with four dipole elements has a gain of approximately 6 dB, and one with eight elements has a gain of about 9 dB.

The *planar array* or *flat panel antenna* is a directive antenna whose key attribute is its flat profile. Its use has been spurred by the need to provide an environmentally unobtrusive presence in many metropolitan areas. Its primary application is in local access connections, either in point-to-point links or at the outstations of point-to-multipoint links. The physical structure of a planar antenna is one of dipole antennas arranged in a planar array, for example, as shown in Fig. 5.8. Such an arrangement provides two dimensions of control, allowing a beam, highly directive in both the horizontal and vertical coordinates, to be produced. Though each dipole in the array has an omnidirectional radiation pattern, they are cleverly connected together via a network that results in directivity in a forward direction along a line perpendicular to the plane of the array. Unfortunately, the pleasing aesthetics of the planar array typically comes at the expense of lower gain relative to a parabolic antenna of equal aperture.

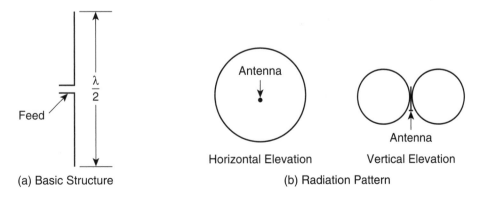

FIGURE 5.7   Half-wave dipole antenna.

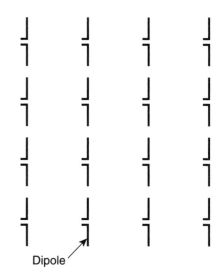

FIGURE 5.8 Typical dipole planar
array.

Dipole

5.3   FREE SPACE PROPAGATION

As indicated previously, the propagation of a signal over a wireless path is affected by
both atmospheric anomalies and the intervening terrain. Absent any such interfering effects we have signal loss only as a result of free space. *Free space loss* is defined as the
loss between two isotropic antennas in free space.

Consider the point source shown in Fig. 5.9 radiating isotropically a signal of power
P_t into free space. As the surface area of a sphere of radius d is $4\pi d^2$, then the radiated
power density $p(d)$ on a sphere of radius d, centered on the point source, is given by

$$p(d) = \frac{P_t}{4\pi d^2} \tag{5.4}$$

If a receiving antenna with an effective area A_{ef} is located on the surface of the sphere,
then the power received by this antenna, P_r, will be equal to the power density on the
sphere times the effective area of the antenna; that is,

$$P_r = \frac{P_t A_{ef}}{4\pi d^2} \tag{5.5}$$

To determine free space loss we need to know the power received by an isotropic antenna.
But, in order to know this, Eq. (5.5) indicates that we need to know the effective area of an
isotropic antenna. By definition, the gain G of an isotropic antenna is 1. Substituting this
value of gain into Eq. (5.1) gives

$$A_{ef} = \frac{\lambda^2}{4\pi} \tag{5.6}$$

Substituting Eq. (5.6) into Eq. (5.5), we determine that P_r, the power receiver by an isotropic antenna, is related to P_t, the power transmitted by an isotropic radiator, by the following relationship:

$$P_r = \frac{P_t}{\left(\dfrac{4\pi d}{\lambda}\right)^2} \tag{5.7}$$

The denominator of the right-hand side of Eq. (5.7) represents the free space loss, L_{fs}, experienced between the isotropic antennas. It is usually expressed in its logarithmic form; that is,

$$L_{fs}(dB) = 20\ \log_{10}\left(\frac{4\pi d}{\lambda}\right) \tag{5.8}$$

Substituting in Eq. (5.8) the well-known relationship $\lambda f = c$, where c is the speed of electromagnetic propagation and equals 3×10^8 meters/second for free space transmission, we get

$$L_{fs}(dB) = 32.4 + 20\ \log_{10} f + 20\ \log_{10} d \tag{5.9a}$$

where f is the transmission frequency in MHz
         d is the transmission distance in km

and

$$L_{fs}(dB) = 36.6 + 20\ \log_{10} f + 20\ \log_{10} d \tag{5.9b}$$

where f is the transmission frequency in MHz
         d is the transmission distance in miles

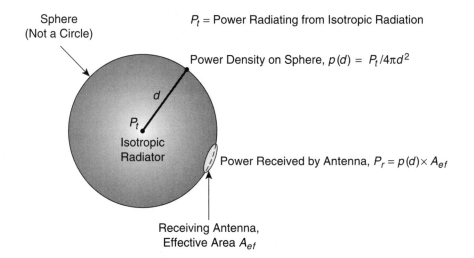

FIGURE 5.9   Power density radiated from an isotropic radiator and power received by an antenna.

To get a sense of the magnitude of free space loss typical of wireless broadband links, consider a 6-GHz path 30 miles long and a 38-GHz path 3 miles long. For the 6-GHz path the free space loss would be 141.7 dB, and for the 38-GHz path the loss would be 137.7 dB.

5.4   RECEIVED SIGNAL LEVEL

With a relationship to determine free space loss one is now able to also set out a relationship for the determination of the receiver input power P_r in a typical wireless link as shown in Fig. 5.1, assuming no loss to fading or obstruction. Starting with the transmitter output power P_t at the transmitter antenna port, one simply accounts for all the gains and losses between the transmitter output and the receiver input. These gains and losses, in decibels, in the order incurred are as follows:

> L_{tf} = loss in transmitter antenna feeder line (coaxial cable or waveguide depending
> on frequency)
> G_{ta} = gain of the transmitter antenna
> L_{fs} = free space loss
> L_a = free space absorption
> G_{ra} = gain of the receiver antenna
> L_{rf} = loss in the receiver antenna feeder line

Thus

$$P_r = P_t - L_{tf} + G_{ta} - L_{fs} - L_a + G_{ra} - L_{rf} \tag{5.10a}$$

$$= P_t - L_{sl} \tag{5.10b}$$

where $L_{sl} = L_{tf} - G_{ta} + L_{fs} + L_a - G_{ra} + L_{rf}$ is referred to as the *section loss* or *net path loss*.

EXAMPLE 5.1   COMPUTATION OF RECEIVED INPUT POWER

A microwave link has the following typical parameters:

Path length, $d = 25$ miles

Operating Frequency, $f = 6$ GHz

Transmitter output power, $P_t = 30.0$ dBm (1 Watt)

Loss in transmitter antenna feeder line, $L_{tf} = 3$ dB

Transmitter antenna gain, $G_{ta} = 40.5$ dB

Free space absorption, $L_a = 0.5$ dB

Receiver antenna gain, $G_{ra} = 37$ dB

Loss in receiver antenna feeder line, $L_{rf} = 2.5$ dB

What is the receiver input power?

Solution

By Eq. (5.9b) the free space loss, L_{fs}, is given by

$$L_{fs} = 36.6 + 20 \log_{10} 6000 + 20 \log_{10} 25 = 140.1 \text{ dB}$$

Thus, by Eq. (5.10a), the receiver input power, P_r, is given by

$$P_r = 30 - 3 + 40.5 - 140.1 - 0.5 + 37 - 2.5$$
$$= -38.6 \text{ dBm}$$

Received input power P_r resulting from free space loss only is normally designed to be significantly higher than the minimum receivable or threshold level R_{th} for acceptable probability of error. This power difference is engineered in so that if the signal fades, due to various atmospheric and terrain effects, it will fall below its threshold level for only a small fraction of time. This built-in level difference is called the link *fade margin* and is given by

$$\text{Fade margin} = P_r - R_{th} \qquad (5.11)$$

Fade margin will be addressed in some detail in following sections. However, to give some sense of its magnitude, it is typically designed to be between about 30 to 40 dB, depending on the path reliability desired. When only thermal noise is considered as the degrading factor to bit error rate performance during fading, then the fade margin is referred to as the *thermal fade margin* (*TFM*). For this situation R_{th} is defined as the signal level, distorted only by thermal noise, that results in a probability of bit error, or *bit error rate* (*BER*) of 10^{-3}. It should be noted that R_{th} corresponding to a BER of 10^{-3} is a dynamic threshold, used for outage calculations (Section 5.8) and "hands-off" field measurements in a normal fading environment. Another threshold used by industry is a static threshold and corresponds to a BER of 10^{-6}. This threshold, which is sometimes referred to as the *receiver sensitivity*, is measured manually in the factory or in the field in a nonfading environment by inserting an attenuator in the signal path and reducing the received level until the associated BER is achieved.

A useful first-order measure of the performance of a wireless link is its *system gain*. System gain is defined as the difference between transmitter output power and the receiver dynamic threshold level and is typically on the order of 80 to 110 dB. Since, by Eqs. (5.10b) and (5.11), thermal fade margin can also be expressed as system gain minus net path loss, then the larger the system gain, the larger the thermal fade margin for a given path.

5.5 FADING PHENOMENA

5.5.1 Atmospheric Effects

As wireless signals travel along a path from a transmitting to a receiving antenna, they are impacted by atmospheric factors that often distort their transmission relative to true free space propagation. These factors can be grouped into three main categories—namely, refraction, reflection, and absorption. As will be seen, the first impacts transmission primarily below approximately 10 GHz, the second impacts all transmission, and the third impacts transmission primarily above 10 GHz.

5.5.1.1   Refraction

In ideal free space propagation it is assumed that a radio signal travels in a straight line. However, a terrestrial radio signal seldom travels in a truly straight line. This is because atmospheric refraction causes it to bend slightly from its ideal straight-line path. This bending is due to changes in the atmospheric refractive index with height. The *refractive index* η of a given medium, which is function of density, determines the speed, v, of an electromagnetic wave through that medium. Specifically,

$$v = \frac{c}{\eta} \tag{5.12}$$

where c = speed of light in a vacuum.

All electromagnetic waves are bent via refraction when they pass from a medium of one refractive index to another. Figure 5.10 is a simplified depiction of the refraction of a radio wave as it travels from a high-density (high refractive index) atmosphere to a low-density (low refractive index) atmosphere. As rays at the edge A of the wave front AB enter the less dense area, they start to travel at a faster speed than rays still traveling in the more dense area. As a result, when the wave front AB is fully in the less dense medium, the new wave front $A'B'$ is tilted downward as AA' is longer than BB'. Most of the time, the refractive index of the atmosphere decreases gradually and uniformly with height as a result of the density of air and its vapor content decreasing gradually and uniformly with height. This thus has the effect of curving a wave in a gradual downward direction as it travels from a transmitter to a receiver. For given antenna heights, this allows the wave to be transmitted further than if it traveled in a straight line by simply tilting the antennas slightly upward. Figure 5.11 shows this path-lengthening effect.

The downward curvature resulting from refraction in a normal atmosphere can be visualized by assuming that the radio wave does in fact travel in a straight line but over a section of the earth's surface that has an effective radius that is greater than that of the true earth. The magnitude of the curving of a transmitted signal over a path is characterized by the path's *earth radius factor, k*, where

$$k = \text{effective earth radius/true earth radius} \tag{5.13}$$

For a value of k of 4/3, for example, a propagation length approximately 15% longer than straight-line propagation is achievable. During normal atmospheric conditions, k ranges

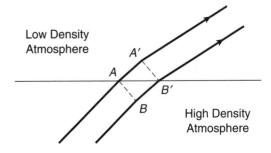

FIGURE 5.10   Refraction of a radio wave.

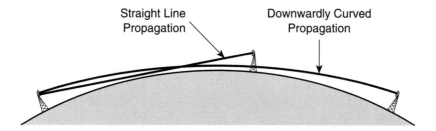

FIGURE 5.11   Lengthened path due to refraction-induced downward curving of radio wave.

from about 1 in dry, elevated areas to 4/3 in typical inland areas, to 2 to 3 in hot and humid coastal areas. However, in abnormal atmospheric conditions, it can vary significantly from its normal value, from a low of about 1/2 to a positive high of infinity and even to where it becomes a negative value. For values of *k* less than 1, the atmosphere is said to be *substandard* or *subrefractive*, whereas for values of *k* greater than 3 or negative, it is said to be *superstandard* or *superrefractive*.

The effects of differing values of *k* on a radio path are depicted in Fig. 5.12. Any value of *k* greater than 1 results in a downward curving of the transmitted signal as shown in Fig. 5.12(a). From a fictitious straight line radio signal point of view, as *k* increases above 1 the earth's curvature appears to be flattening, as shown in Fig. 5.12(b), until, when *k* equals infinity, it appears to be totally flat, permitting the radio wave to travel in

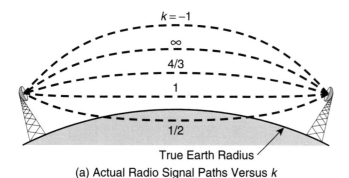

(a) Actual Radio Signal Paths Versus *k*

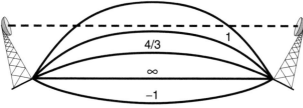

(b) Effective Earth Profiles Versus *k*

FIGURE 5.12   Effect of differing values of *k* on radio paths.

parallel with it until obstructed. Further downward bending of the wave leads to a depressed earth's surface or negative earth curvature from a straight line radio signal point of view and hence a negative value of k. When k equals 1, no refraction occurs and the radio signal travels in a straight line over the curved earth. For values of k less than 1, the transmitted signal curves upward, causing the earth's curvature from a straight-line signal point of view to appear to be bulging, possibly to the extent that it obstructs the path of the radio signal. Values of k less than 1 arise when atmospheric density increases with height instead decreasing, as is normally the case, and can occur when there is heavy fog or extremely cold air over less dense air near the ground, as occurs with a cold front passage.

In addition to the gradual bending of the direct beam described previously, refraction can result in additional signal impairment—namely, ducting, multiple refractive paths, and scintillation.

Ducting is an atmospheric phenomenon that can occur when values of k go negative. In this state, the atmosphere can act like a horizontal duct or waveguide, which, if located at the general level of the direct radio signal, can trap it within its boundaries. Ducts can be ground based or elevated.

In ground ducts the region close to the ground has a high refractive index, which changes rapidly to a low refractive index with height. If a transmitted signal finds itself in such an environment it is bent downward with a radius smaller than that of the earth. Thus, if the receiving antenna is located beyond the point where the signal hits the earth a fade large enough to cause complete outage would likely occur, unless the signal is fortuitously reflected and picked up by the antenna. Figure 5.13(a) shows a signal trapped in a ground duct with a possible reflected ray. Ground ducts tend to occur most frequently over or near

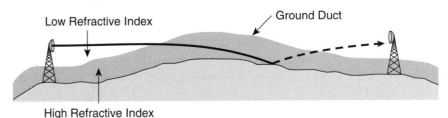

(a) Radio Signal Trapped in a Ground-Based Duct

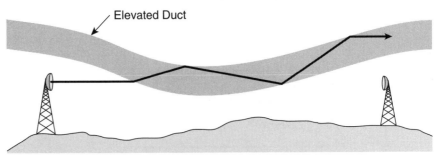

(b) Radio Signal Trapped in a Strong Elevated Duct

FIGURE 5.13   Ducting phenomena.

large expanses of water as a result of either evaporation or advection. Evaporation can lead to an unusually rapid decrease in water vapor density with height resulting in a high-humidity, high-density, high-refractive-index zone below a region of drier, low-refractive atmosphere. Such a phenomenon is most likely to occur in the afternoon, as a result of the cumulative effect of solar heating during the preceding period of the day. Ducts formed via convection are typically about 15 meters in height. Advection is the movement of one type of air over another. If dry air above land is blown over moist air above an expanse of water, a region of low density, low refractive index is created above a region of high humidity, high refractive index. Such advection usually occurs in the evening when breezes tend to blow from land to sea. Advection ducts are normally about 25 meters high.

Abnormal atmospheric conditions, such as a rise in temperature with increasing height, can also result in elevated ducts. Strong elevated ducts, more prevalent in tropical climates, can trap radio signals within their boundaries over relatively long distances, sometimes far beyond the horizon. Figure 5.13(b) shows how a signal can be trapped in such a duct. Once within the duct, it's reflected off the duct's boundaries and propagates within the duct much like how light travels in a fiber optic glass cable. A prediction method for calculating path attenuation due to superrefraction and ducting is given in CCIR Report 569-3.[2]

Of all the atmospheric anomalies that impact digital wireless transmission performance below about 10 GHz, none, in general, has greater impact on reliability than those that result in *multiple refractive atmospheric paths* between the transmitter and receiver. These anomalies occur when there is minimal normal atmospheric turbulence, thus permitting the formation of nonuniform vertical distributions of temperature and humidity. This in turn leads to significant variability, or, as it's sometimes referred to, stratification, in the refractive index vertical profile. This situation permits energy, leaving the radiating antenna at different angles, to travel to the receiving antenna via slightly different paths, each with a different length and hence different time delay. This stratification can be considered as a mild form of ducting. The creation of multiple refractive atmospheric paths typically begins to occur on clear, calm, hot summer evenings and builds up during the night, peaking in the early morning hours. Thereafter, as the morning unfolds, winds and rising convection currents tend to mix the atmosphere thoroughly, leading to little or no vertical or horizontal variations in the refractive index gradient in the vicinity of a typical path. As a result, only a single dominant path then tends to exist between transmitter and receiver. Figure 5.14 shows a radio link with multiple refractive atmospheric paths.

Scintillation[3] of a radio signal over a line-of-sight path is caused by fluctuations in the refractive index over a section of the atmosphere that the signal traverses. The

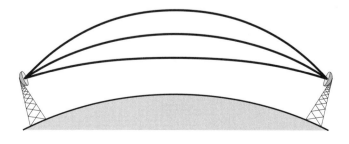

FIGURE 5.14 Link with multiple refractive atmosphere paths.

fluctuations result in a slight defocusing and refocusing of the signal. As a result, the received signal consists of several components arriving from slightly different directions, each of a different and continuously varying amplitude and phase. Summed together, these signals result in a composite signal that itself varies rapidly and randomly in amplitude and phase. Fortunately, fluctuations in received power due to the scintillation phenomenon are normally quite small, with a typical standard deviation of 0.6 dB.[4]

5.5.1.2   Reflection

In addition to traversing down their direct path, radio signals can reach the receiving antenna via atmospheric reflections. Signals can be reflected off ground ducts and elevated ducts described previously. The boundaries of these ducts are usually relatively stable, normally moving slowly in an up and down direction. Signals can also be reflected off *sheets*. Sheets are high-altitude undulating layers that are several miles long and constantly changing, often in a rapid fashion, in height and location. Figure 5.15(a) shows a path with signals reflected off an elevated duct, and Fig. 5.15(b) shows a path with a signal reflected off a sheet.

5.5.1.3   Rain Attenuation and Atmospheric Absorption

As a radio signal propagates down its path it may find itself subjected, in addition to the possible effects of refraction and reflection, to the attenuating effects of rain and absorption by atmospheric gases, primarily water vapor and oxygen.

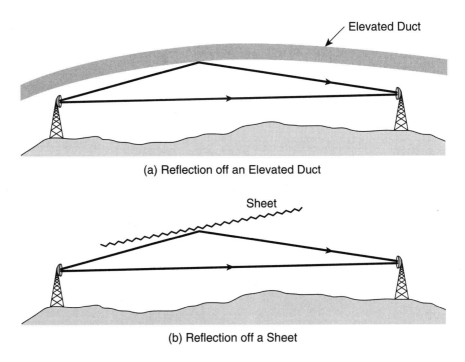

(a) Reflection off an Elevated Duct

(b) Reflection off a Sheet

FIGURE 5.15   Atmospheric reflections.

Raindrops attenuate a radio signal by absorbing and scattering the radio energy, with this effect becoming more and more significant as the wavelength of the signal decreases toward the size of the raindrops. Figure 5.16, which is from Rogers et al.[5] shows attenuation in dB/km versus frequency for several rain conditions. As can be seen from the figure, attenuation increases with propagation frequency and rain intensity. At 6 GHz, the attenuation due to a heavy cloudburst is about 1 dB/km. For a 40-km path, this would imply an attenuation of about 40 dB if the cloudburst extended across the entire path. Fortunately, the width of cloudburst cells tends to be on the order of kilometers, so maximum attenuation due to rain at these frequencies tend to be less than 10 dB, an amount easily accommodated by built in fade margin. In general, rain attenuation is not a major problem at or below about 10 GHz. However, this situation changes measurably as the frequency increases and normally must be factored in at frequencies of 10 GHz and above. For example, at 23 GHz, maximum attenuation due to rain is in the order of 20 dB/km. Thus, for a 23-GHz link, with a 40-dB fade margin and operating in a region where heavy cloudbursts occur, path length may have to be limited to no more than a couple of kilometers, if high reliability is to be maintained. The results in Fig. 5.16 are theoretically derived for spherical raindrops. Real raindrops, however, are somewhat flattened as they fall through the atmosphere. As a result, they have a smaller size in the vertical plane than in the horizontal one. This in turn leads to the attenuation of a vertically polarized wave being less

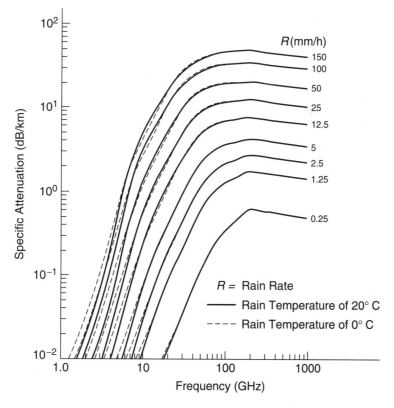

FIGURE 5.16 Attenuation due to rain. *(From Ref. 5, reproduced with the permission of the Minister of Public Works and Government Services Canada, 2003.)*

than that of a horizontally polarized one. For a more precise estimate of rain attenuation that factors in the polarization of the signal, the following formula can be applied:

$$\gamma_R = kR^\alpha \tag{5.14}$$

where γ_R = rain attenuation (dB/km)
        R = rain rate (mm/h)
        $k,\ \alpha$ = tabulated coefficients for both vertical and horizontal polarization as provided in ITU recommendation ITU-R P.838[6] and as reproduced in Table 5.1.

TABLE 5.1   Regression Coefficients for Estimating Specific Attenuation in Eq. (5.14)

Frequency (GHz)	k_H	k_V	α_H	α_V
1	0.0000387	0.0000352	0.912	0.880
2	0.000154	0.000138	0.963	0.923
4	0.000650	0.000591	1.121	1.075
6	0.00175	0.00155	1.308	1.265
7	0.00301	0.00265	1.332	1.312
8	0.00454	0.00395	1.327	1.310
10	0.0101	0.00887	1.276	1.264
12	0.0188	0.0168	1.217	1.200
15	0.0367	0.0335	1.154	1.128
20	0.0751	0.0691	1.099	1.065
25	0.124	0.113	1.061	1.030
30	0.187	0.167	1.021	1.000
35	0.263	0.233	0.979	0.963
40	0.350	0.310	0.939	0.929
45	0.442	0.393	0.903	0.897
50	0.536	0.479	0.873	0.868
60	0.707	0.642	0.826	0.824
70	0.851	0.784	0.793	0.793
80	0.975	0.906	0.769	0.769
90	1.06	0.999	0.753	0.754
100	1.12	1.06	0.743	0.744
120	1.18	1.13	0.731	0.732
150	1.31	1.27	0.710	0.711
200	1.45	1.42	0.689	0.690
300	1.36	1.35	0.688	0.689
400	1.32	1.31	0.683	0.684

From ITU Recommendation ITU-R P.838-1. Reproduced with the prior authorization of the ITU as copyright holder. The sole responsibility for selecting this extract for reproduction lies with the author and can in no way be attributed to the ITU. The complete volume of the ITU material, from which this table is extracted, can be obtained from:

International Telecommunications Union
Sales and Marketing Division
Place des Nations—Ch-1211 GENEVA 20 (Switzerland)
Telephone: +41 22 730 61 41 (English) / +41 22 730 61 42 (French) / +41 22 730 61 43 (Spanish)
Telex: 421 000 uit ch / Fax: +41 22 730 51 94
E-mail: sales@itu.int / http://www.itu.int/publications

Figure 5.17, which is from ITU Recommendation ITU-R P.676-5[7] shows attenuation in dB/km due to water vapor and dry air as a function of frequency in a low to moderate humidity region. For fixed wireless systems the highest frequency band currently authorized for use is in the 60-GHz range. From the figure it will be observed that attenuation due to water vapor has a first peak at approximately 22 GHz with a value of about 0.18 dB/km. It then declines with increasing frequency up to about 31 GHz, after which it increases back to about 0.18 dB/km in the region of 60 GHz. Attenuation due to dry air is

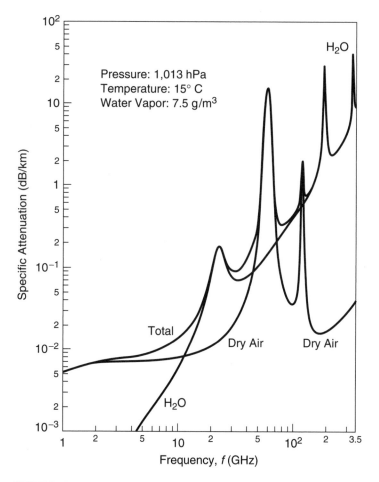

FIGURE 5.17 Specific attenuation due to atmospheric gases. *From ITU Recommendation ITU-R P.676-5. Reproduced with the prior authorization of the ITU as copyright holder. The sole responsibility for selecting this extract for reproduction lies with the author and can in no way be attributed to the ITU. The complete volume of the ITU material, from which this figure is extracted, can be obtained from International Telecommunication Union, Sales and Marketing Division, Place des Nations—Ch-1211 GENEVA 20 (Switzerland), Telephone: +41 22 730 61 41 (English) / +41 22 730 61 42 (French) / +41 22 730 61 43 (Spanish), / Telex: 421 000 uit ch / Fax: +41 22 730 51 94, E-mail: sales@itu.int / http://www.itu.int/publications*

below 0.01 dB/km for frequencies up to 20 GHz, then increases somewhat exponentially to a peak of approximately 15 dB/km in the region of 60 GHz. Paths at frequencies close to 22 GHz seldom exceed about 15 km in length due to rain attenuation and limited transmitter output power. Thus the maximum likely loss due to atmospheric absorption will be from water vapor and will be on the order of 3 dB, measurable, but not significant relative to likely rain loss. Paths close to 60 GHz, unfortunately, have the worst of two worlds. Rain attenuation can approach 40 dB/km and is additive to dry air attenuation of approximately 15 dB/km. As a result, fixed wireless systems operating at frequencies in this region and in locations subject to any appreciable rainfall are usually limited in path length to a couple of kilometers or less.

5.5.2   Terrain Effects

The propagation of radio signals through the atmosphere is affected by the terrain in or close to its path in addition to the atmospheric effects discussed previously. Potential propagation affecting terrain features are trees; hills; sharp points of projection; reflection surfaces such as ponds, lakes, and seas; and human-made structures such as buildings and towers. These features can result in either reflected or diffracted signals arriving at the receiving antenna.

5.5.2.1   Terrain Reflection

Reflected signals arriving at a receiving antenna are combined with the direct signal to form a composite signal that, depending on the strengths and phases of the reflected signals, can significantly degrade the desired signal. Fig. 5.18 shows a radio path with a direct signal and a reflected signal off a lake. The strength of a reflected signal at a receiving antenna depends on the directivity of both the transmitting and receiving antenna, the heights of these antennas above the ground, the length of the path, and the strength of the reflected signal relative to the incident signal at the point of reflection. The phase of a reflected signal at a receiving antenna is a function of the reflected signal path length and the phase shift, if any, incurred at the point of reflection.

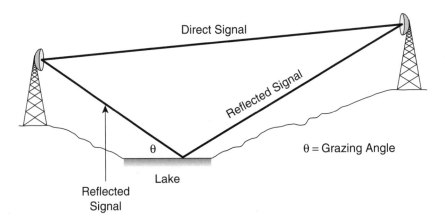

FIGURE 5.18   Ground reflected signal.

The strength and phase of a reflected signal relative to the incident signal at the point of reflection are strongly influenced by the composition of the reflecting surface, the curvature of the surface, the amount of surface roughness, the incident angle, and the radio signal frequency and polarization. If reflections occur on rough surfaces, they usually do not create a problem as the incident and reflected angles are quite random. Reflections from a relatively smooth surface, however, depending on the location of the surface, can result in the reflected signal being intercepted by the receiving antenna. The transfer function of a reflective surface is referred to as its *reflection coefficient*, *R*. For a highly reflective surface (a specular reflection plane) vertically polarized signals show more frequency dependence than horizontally polarized ones. However, at frequencies between about 2 and 10 GHz, variation with frequency for both polarizations is sufficiently small that the following general observations can be made:

- For horizontally polarized signals, reflection imparts a phase shift equal to, or very close to, 180° regardless of the grazing angle (see Fig. 5.18).
- For vertically polarized signals, reflection impart a phase shift of about 170° or greater when the grazing angle is very small (less than 1°), but as the grazing angle increases the imparted phase shift decreases rapidly to zero, being in the region of 30° for a grazing angle of 10°.
- For horizontally polarized signals, the magnitude of the reflection coefficient is unity for grazing angles between zero and a few degrees and then slowly decreases to about 0.8 at the maximum grazing angle of 90°. Thus for practical fixed wireless links, where grazing angle rarely, if ever, exceeds a couple of degrees, the magnitude of the reflection coefficient can be treated as unity.

- For vertically polarized signals, the magnitude of the reflection coefficient is unity for zero grazing angle, but falls off rapidly, to a value of about 0.9 at 0.4° grazing angle, and on to a minimum in the region of 0.1 at a grazing angle of about 6 to 7°. It then increases back up to about 0.8 at a grazing angle of 90°.

A method to calculate effective surface reflection coefficient is given in ITU Recommendation ITU-R P.530.[8]

5.5.2.2   Fresnel Zones

The proximity of terrain features to the direct path of a radio signal impacts their effect on the composite received signal. *Fresnel zones* are a way of defining such proximity in a very meaningful way. In the study of the diffraction of radio signals, the concept of Fresnel zones is very helpful. Thus, before reviewing diffraction, an overview of Fresnel zones is in order.

The first Fresnel zone is defined as that region containing all points from which a wave could be reflected such that the total length of the two segment reflected path exceeds that of the direct path by half a wavelength, $\lambda / 2$, or less. The *n*th Fresnel zone is defined as that region containing all points from which a wave could be reflected such that the length of the two segment reflected path exceeds that of the direct path by more than $(n - 1) \lambda / 2$ but less than $n\lambda / 2$. The boundary of a Fresnel zone surrounding the direct path in the plane of the path is an ellipsoid whereas it is a circle in the plane perpendicular to the path. Figure 5.19 shows the first and second Fresnel zone boundaries on a line-of-sight path. The perpendicular distance F_n from the direct path to the outer boundary of the nth Fresnel zone is approximated by the following equation:

$$F_1 = 72.1 \left[\frac{d_{11}d_{12}}{fd} \right]^{\frac{1}{2}} \tag{5.15}$$

where d_{n1} = distance from one end of path to point where F_n is being determined
d_{n2} = distance from other end of path to point where F_n is being determined
d = length of path = $d_{n1} + d_{n2}$
λ, d_{n1}, d_{n2} and d are measured in identical units

Specifically, F_1 is given by

$$F_1 = 17.3 \left[\frac{d_{11}d_{12}}{fd} \right]^{\frac{1}{2}} \quad \text{meters} \tag{5.16a}$$

where d_{11}, d_{12} and d are in km and f is in GHz, and given by

$$F_1 = 72.1 \left[\frac{d_{11}d_{12}}{fd} \right]^{\frac{1}{2}} \quad \text{feet} \tag{5.16b}$$

where d_{11}, d_{12} and d are in miles and f is in GHz.

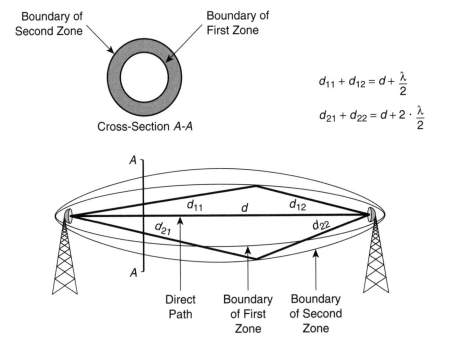

$$d_{11} + d_{12} = d + \frac{\lambda}{2}$$

$$d_{21} + d_{22} = d + 2 \cdot \frac{\lambda}{2}$$

FIGURE 5.19   First and second Fresnel zone boundaries.

For a typical 40 km path operating at 6 GHz, the maximum value of F_1, which occurs at the middle of the path, is, by Eq. (5.16a), 22.4 meters. However, for a typical 5-km, 23-GHz path, F_1 maximum is only 4 meters.

Most of the power that reaches the receiver is contained within the boundary of the first Fresnel zone (more on this in the next section). Thus, terrain features that lie outside this boundary, with the exception of highly reflective surfaces, do not, in general, significantly affect the level of the received signal. For a signal reflected at this boundary that acquires a $\lambda / 2$ phase shift as a result of the reflection, its total phase shift at the receiving antenna relative to the direct signal is λ and hence the reflected signal is additive to the direct signal. However, for a signal that experiences zero phase shift, as is possible with a vertically polarized reflected signal with a large incident angle, the total relative phase shift at the receiving antenna is $\lambda / 2$ and hence results in a partial cancellation of the direct signal. The maximum increase possible in signal strength resulting from a reflected signal is 6 dB. However, the maximum loss possible is in theory an infinite number of decibels, and in practice can exceed 40 dB.

5.5.2.3   Diffraction

So far in explaining propagation effects it has been tacitly assumed that the energy received by a radio antenna travels as a beam from transmitting to receiving antenna that's just wide enough to illuminate the receiving antenna. This is not exactly the case, however, as waves propagate following the *Huygens' principle*. Huygens showed that propagation occurs along a wavefront, with each point on the wavefront acting as a source of a secondary wavefront known as a wavelet, with a new wavefront being created from the combination of the contributions of all the wavelets on the preceding front. Importantly, secondary wavelets radiate in all directions. However, they radiate strongest in the direction of the wavefront propagation and less and less as the angle of radiation relative to the direction of the wavefront propagation increases until the level of radiation is zero in the reverse direction of wavefront propagation. The net result is that, as the wavefront moves forward it spreads out, albeit with less and less energy on a given point as that point is removed further and further from the direct line of propagation. This sounds like bad news. The signal intercepted at the receiving antenna, however, is the sum of all signals directed at it from all the wavelets created on all wavefronts as the wave moves from transmitter to receiver. At the receiver, signal energy from some wavelets tend to cancel signal energy from others depending on the phase differences of the received signals, these differences being generated as a result of the different path lengths. The net result is that, in a free space, unobstructed environment, half of the energy reaching the receiving antenna is canceled out.

Signal components that have a phase difference of $\lambda / 4$ or less relative to the direct line signal are additive. Signals with a phase difference between $\lambda / 4$ and $\lambda / 2$ are subtractive. In an unobstructed environment, all such signals fall within the first Fresnel zone. In fact, the first Fresnel contains most of the energy that reaches the receiver. Consider now what happens when an obstacle exists in a radio path within the first Fresnel zone. Clearly, under this condition, the amount of energy intercepted at the receiving antenna will differ from that intercepted if no obstacle were present. The cause of this difference, which is the disruption of the wavefront at the obstruction, is called *diffraction*. If an obstruction is raised in front of the wave so that a direct path is just maintained, the power reaching the receiver will be reduced, whereas the simplistic narrow beam model would suggest that full received signal would be maintained. A positive aspect of diffraction is that if the obstruction is further raised, so that it blocks the direct path, the signal will still be intercepted at the receiving antenna, albeit at a lesser and lesser level as the height of the obstruction increases further and further. Under the simplistic narrow-beam model, one would have expected complete signal loss.

Figure 5.20(a) shows "unobstructed" free space propagation, where path clearance is assumed to exceed several Fresnel zones. Figure 5.20(b) shows diffraction around an obstacle assumed to be within the region of the first Fresnel zone but not blocking the direct path. Figure 5.20(c) shows diffraction around an obstacle blocking the direct path. For simplicity, an expanding wave that has progressed partially down the path is shown in all three depictions, to the point of obstruction in the case of the obstructed paths. All wavelets on the wavefront shown in the unobstructed case are able to radiate a signal that falls on the receiving antenna. In the case of the obstructed paths, however, the size of the wavefronts and hence the number of wavelets radiating signals that fall on the receiving antenna are decreased.

For analytical purposes, diffracting terrain is normally classified as being one of three types—namely, smooth (spherical) earth, rounded obstacle, and knife edge. In reality, however, terrain property normally falls somewhere between these hypothetical types. Figure 5.21 shows, for each of these types of diffracting terrain, an example of diffraction around such terrain that's blocking a path. Like loss due to reflection, diffraction loss is a function of the nature of the diffracting terrain, varying from a minimum for a single knife-edge obstacle to a maximum for smooth earth. Much research and analysis of diffraction loss has been conducted over several decades leading to well accepted formulas[9] for its estimation. For the situation where the clearance between an obstacle and the direct path is equal to F_1, the first Fresnel zone clearance at the obstacle location, there is, in fact, a received signal gain. This gain varies from about 1.5 dB for the knife-edge case to 6 dB for the smooth-earth case. When the clearance decreases to $0.6F_1$, the received signal strength is unaffected by the obstruction, regardless of its type. For the grazing situation, where the clearance between the obstacle and the direct path is reduced to zero, diffraction loss varies from approximately 6 dB for the knife-edge case to approximately 15 dB for the smooth-earth case.

5.5.2.4 Path Clearance Criteria

Radio paths must be designed to minimize as much as possible the likelihood of significant loss in received signal level due to obstructions in the path. In so designing, the designer must take into account the fact that the effective clearance, F, between the obstruction and the direct path varies as the earth radius factor, k, varies. For a path where the normal value of k is 4/3, for example, then whenever k changes to a value less than 4/3, F diminishes, resulting in increased obstruction loss if the obstruction lies within the first Fresnel zone. Figure 5.22 shows the impact of varying k on clearance F. In this example, for $k = 4/3$, F is equal to F_1. When k decreases to 1/2, F is reduced to zero, whereas when k increases to 4, F doubles to $2F_1$. Note also that if the antennas are adjusted for maximum signal strength when k equals 4/3, then when k equals 1/2, the peak transmitted

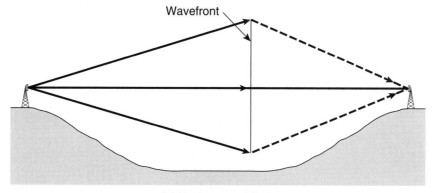

(a) Unobstructed Path

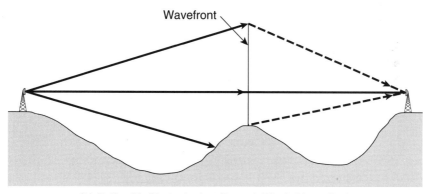

(b) Path with Obstacle that Doesn't Block Direct Path

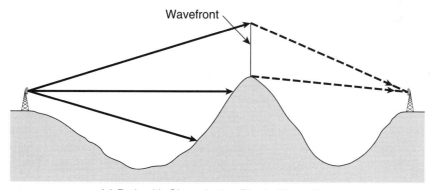

(c) Path with Obstacle that Blocks Direct Path

FIGURE 5.20   Graphical representation of diffraction.

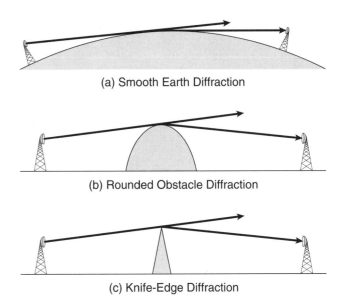

(a) Smooth Earth Diffraction

(b) Rounded Obstacle Diffraction

FIGURE 5.21   Various
types of diffraction.   (c) Knife-Edge Diffraction

signal passes above the receiving antenna and a slightly weaker direct path signal (a function of the transmitting antenna directivity) arrives at the receiving antenna. The receiving antenna further attenuates this signal as it's positioned to provide maximum gain to the k equals 4/3 signal. When k is 4, the peak direct signal now passes below the antenna and again the antennas will not be optimally aligned, and thus a slight reduction in direct path signal strength will result.

A number of planning procedures have been developed in different countries and by different regulating agencies in terms of the clearance of the controlling obstruction, relative to F_1, so as to guarantee, to a large extent, that diffracted signals do not result in significant signal loss under most atmospheric conditions. Among such procedures is one

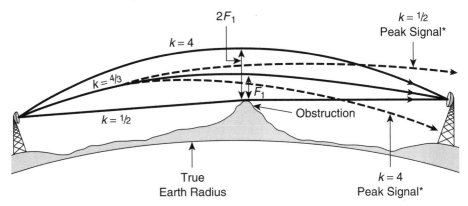

*If antennas adjusted for maximum signal strength when $k = 4/3$

FIGURE 5.22   Impact of varying k on clearance.

developed by the ITU.[8] This procedure, for paths with operating frequencies of about 2 GHz and above (the case for fixed wireless broadband links), is as follows:

Step 1: Determine antenna heights that allow clearance of 1.0 F_1 over the highest obsta-
cle at the median value of the single point (i.e., local) k factor (available via ITU Recommendation ITU-R P.435[10] or at $k = 4/3$ if k factor data is unavailable.

Step 2: Obtain the value of k_e (99.9%) from Fig. 5.23 for the path length in question (k_e is the effective value of k for the path, and k_e [99%] is the value of k_e exceeded 99.9 % of the worst month).

Step 3: For the value of k_e (99.9%) determined in Step 2, determine antenna heights that that allow clearance of
   (a) $0.0F_1$ (i.e., no clearance) if the climate is temperate and there is a single iso-
lated path obstruction
   (b) $0.3F_1$ if the climate is temperate and the path obstruction is extended along a portion of the path
   (c) $0.6\ F_1$ if the climate is tropical and the path length greater than 30 km

Step 4: Use the larger of the antenna heights obtained by Steps 1 and 3.

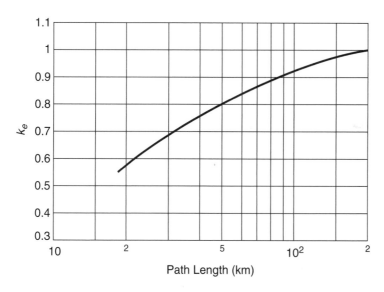

FIGURE 5.23   Value of k_e, exceeded for approximately 99.9% of the worst month (continental temperate climate). *From ITU Recommendation ITU-R P.530-10. Reproduced with the prior authorization of the ITU as copyright holder. The sole responsibility for selecting this extract for reproduction lies with the author and can in no way be attributed to the ITU. The complete volume of the ITU material, from which this figure is extracted, can be obtained from International Telecommunication Union, Sales and Marketing Division, Place des Nations—Ch-1211 GENEVA 20 (Switzerland), Telephone +41 22 730 61 41 (English) / +41 22 730 61 42 (French) / +41 22 730 61 43 (Spanish), Telex: 421 000 uit ch / Fax: +41 22 730 51 94, Email: sales@itu.int / http://www.itu.int/publications*

As the cost of a radio tower tends to increase significantly with height, and as high towers stand out environmentally, the radio path designer typically chooses antenna heights that result in these or similar criterion just being met or being met with only a small margin.

5.5.3   Signal Strength versus Frequency Effects

Fading is the variation in strength of a received radio signal due to the atmospheric and terrain effects in the radio's path as discussed in Sections 5.5.1 and 5.5.2. Fading is normally broken down into two main types—namely, *flat fading* and *frequency selective fading*. In flat fading the signal is attenuated uniformly across the frequency band occupied by the signal, whereas in frequency selective fading, the attenuation varies with frequency across the occupied band. These two types of fading can occur separately or together, though the latter is rare. Exactly when fading resulting from variations in atmospheric conditions is likely to occur cannot be predicted with any accuracy. Such fading, however, can be predicted on a statistical basis, given a statistical knowledge of the atmospheric conditions in the vicinity of the radio path.

5.5.3.1   Flat Fading

Beam Bending Fading.   The fading resulting from beam bending which in turn results from slow changes in the *k* factor is flat fading. As discussed previously, the impact of such possible fading is normally minimized by choosing antenna heights that result in the meeting or exceeding of path clearance criteria.

Duct Entrapment Fading.   Fading as a result of duct entrapment is also flat fading. Unfortunately, other than keeping the path length short to minimize the probability of entrapment, no effective countermeasures exist.

Rain Fading.   Finally, fading caused by rain is also flat fading. Effective countermeasures are a high fade margin, a path length short enough to guarantee a low probability of the maximum rain rate[11] resulting in path attenuation that exceeds the fade margin, and the use of vertical polarization on paths operating above 10 GHz, if possible. As indicated in Section 5.5.1.3, rain attenuation to vertically polarized signals is less than that to horizontally polarized ones. Thus, choosing vertical polarization reduces rain induced outages (by from 40 to 60%), compared to horizontal polarization.

5.5.3.2   Frequency Selective Fading

Ground Reflection Fading.   If a radio path passes over highly reflective ground or water, and the antenna heights allow a reflected path between the antennas, reflection fading is likely and, under certain circumstances, can be substantial. Ground reflection fading is frequency selective. The reflected signal will have, at any instant, a path length difference and hence a time delay, τ say, relative to the direct path that's independent of frequency. However, this delay, measured as a phase angle, varies with frequency. Thus, if the propagated signal occupies a band between frequencies f_1 and f_2, say, then the relative delay in radians at frequency f_1 will be $2\pi f_1 \tau$, the relative delay at f_2 will be $2\pi f_2 \tau$, and the difference in relative phase delay between the component of the reflected signal at f_1 and that at f_2 will be $2\pi(f_2 - f_1)\tau$. Because the reflected signal will have a phase shift relative to the direct signal that's a function of frequency, then when both these signals are combined, a different resultant signal relative to the direct signal will be created for each frequency. Depending on the value of τ and the signal bandwidth $f_2 - f_1$, significant differences can exist in the composite received signal amplitude and phase as a function of frequency as compared to the undistorted direct signal. Further, the closer the reflected signal is in amplitude to the direct signal, the larger the maximum signal cancellation possible; this occurs when the signals are 180° out of phase.

Normally, on a path with a highly reflective surface, the path designer will first try to shift the point where a reflected signal can be intercepted, by a receiving antenna, away from the highly reflective surface. This is done by adjusting the relative heights of the antennas. Alternatively, if an appropriate obstruction exists, the designer may locate the height of one antenna so that the obstruction denies the reflected signal a line-of-sight to the receiving antenna. Highly directive antennas may help somewhat as they reduce the strength of the received reflected signal relative to that resulting from less directive antennas. Improvement due to antenna discrimination may also be achieved by tilting one or both antennas slightly upward. However, in so doing the strength of the direct signal is slightly decreased. For overwater paths with an unblocked reflected signal, if path geometry is such that the grazing angle is greater than about 0.2°, then the use of vertical polarization can significantly reduce the maximum possible depth of reflection induced fading as compared to the horizontally polarized case. This is because, as pointed out in Section 5.5.2.1, the amplitude of the reflection coefficient for vertically polarized signals falls off rapidly beyond this grazing angle, reducing the cancellation effect at the receiver. For a horizontally polarized signal, however, the magnitude of the reflection coefficient remains very close to unit, resulting in the possibility of significant signal cancellation. ITU Recommendation ITU-R P.530-10[8] indicates that, for overwater paths at frequencies above about 3 GHz, it is advantageous to use vertical polarization, noting that "at grazing angles greater than about 0.7°, a reduction in the surface reflection of 2–17 dB can be expected over that at horizontal polarization."

If implementation of the aforementioned techniques, where possible, does not result in acceptable path performance, then the normal method used to improve reliability is space diversity reception.

Atmospheric Multipath (Rayleigh) Fading.   Atmospheric refractive multipaths are the dominant factor in fading for paths operating at about 8 GHz or below, assuming adequate clearance has been provided to minimize beam bending and obstruction related fading. This being so, going forward in this text, unless indicated otherwise, the term *multipath fading* will be used to mean fading resulting from atmospheric multipaths. For the same reason given for ground reflection fading, multipath fading is frequency selective. To gain a first-order insight into multipath fading, let us consider a simple two-ray model, with a direct ray and a refracted ray. If τ is the relative time delay between the direct and the refracted ray, then the relative phase delay between these rays differs by $2\pi(f_2 - f_1)\tau$ between f_1 and f_2, the signal band edge frequencies. The amplitude response versus frequency of this two-ray transfer function will be repetitive over a frequency difference $\Delta f = 1/\tau$, since this results in a phase change of 2π radians. If the bandwidth of the signal is much less than $1/\tau$, then the amplitude and phase variation across it will be small and the resulting multipath fading can be treated as flat fading. As Δf approaches about $1/10\tau$, however, the frequency selective effects become significant and cannot be ignored. Figure 5.24, which is from Rummler et al.,[12] shows a typical scan of a multipath fading event on a typical 30-MHz bandwidth, 6-GHz radio channel. It will be observed that the fade depth varies from about 25 dB to about 65 dB within the band. The group delay, which is the rate of change of phase and is constant with frequency for a linear nondistorting transfer function, similarly shows significant variation across the band. Inband fade depth variation similar to that shown here is quite common on paths subject to heavy multipath fading.

Multipath fades can be quite deep, often exceeding 40 dB. Figure 5.25, which is from Ivanek,[13] shows measurements of multipath fading versus time on a specific path. As can be seen, these fades take place fast, typically on a scale of seconds, and occur randomly. Fortunately, fade duration is inversely proportional to fade depth. For example, on a 4-GHz system with a path length of 30 to 35 miles, the median duration of 20 dB fades is about 30 seconds, whereas the median duration of 40 dB fades is about 3 seconds.[14] Figure 5.26, which is also from Ivanek,[13] shows measurements of fade depth versus the time that fades exceed that depth made on the same path from which the data in Fig. 5.25 was derived. The plot of Fig. 5.26 shows, for a one-month observation period, the time during which the fade depth specified is exceeded, except for some observations with fading less than about 10 dB, which were removed. It will be noted that for fade depth deeper than about 20 dB, the fade time decreases by a factor of about 10 for a 10 dB increase in fade depth. Such a distribution of received signal power versus time the power is exceeded is a Rayleigh distribution, as described in Chapter 3. As a result, multipath fading is often referred to as *Rayleigh fading*. Measurements of multipath fading indicate that the amount of discrete fades, as well as the percentage of time a fade exceeds a given level, tends to

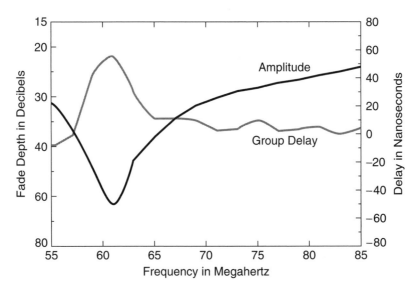

FIGURE 5.24　A typical scan of a multipath fading event in a 6-GHz radio channel. *(By permission from Ref. 12, © 1986 IEEE.)*

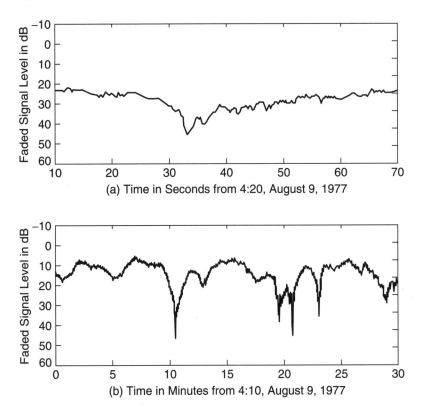

FIGURE 5.25   Time series of multipath fading at 6034.2 MHz on the 42.5 km path from Atlanta to Palmetto, Georgia. *(Reprinted with permission from* Terrestrial Digital Microwave Communications, *by Ferdo Ivanek, editor, Artech House, Norwood, MA, USA, www.artechhouse.com.)*

increase with path length and frequency. During periods of heavy rainfall, the right atmospheric conditions do not normally exist for multipath fading. Thus, at any given time, if fading is present, it is usually multipath or flat, but not both simultaneously.

5.5.3.3   Multipath Fading Channel Model

In this section a highly accepted model for simulating the frequency selective effects of multipath propagation is reviewed. Such models are very useful in evaluating the performance of digital radio links. Depending on their bandwidth and modulation complexity, such links can be highly sensitive to the frequency selective distortion resulting from multipath fading. This is because the severe amplitude and group delay distortion resulting from multipath fading causes severe intersymbol interference.[15] This in turn results in degraded bit error rate performance beyond that which would be generated by an equivalent flat fade.[16] It is important to recognize that the purpose of defining a channel model is only to describe the frequency response of the channel, not the physical propagation characteristics of the channel. This allows the channel model to be considerably less

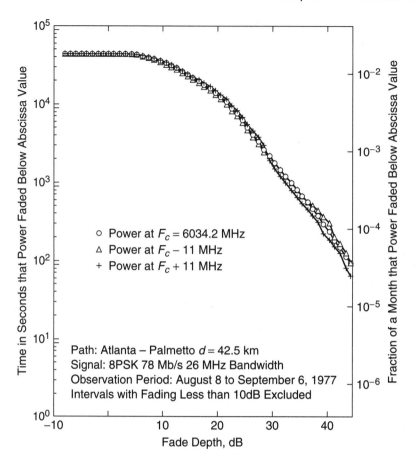

FIGURE 5.26   Distribution of received power levels for selected frequency components of a digital signal. *(Reprinted with permission from* Terrestrial Digital Microwave Communications, *by Ferdo Ivanek, editor, Artech House, Norwood, MA, USA, www.artechhouse.com.)*

complex than the actual propagation channel. With a multipath fading model, it becomes possible to estimate the fraction of time (i.e., the probability) that the propagation conditions on the path will be such that the radio link cannot meet a defined minimum acceptable performance, normally defined as a bit error rate (BER) of 10^{-3}.

Many multipath fading channel models have been proposed. However, over time, one model, the simplified three-ray (three-path) model, initially proposed by Rummler,[17] has found considerable favor, and this is the only model that will be reviewed here. For an excellent overview of multipath fading models, the reader is referred to Rummler et al.,[12] which has provided the basis of this review.

The simplified three-ray model defines the complex voltage transfer function $H(j\omega)$ of a multipath channel by the relationship

$$H(j\omega) = a\left[1 - be^{-j(\omega-\omega_0)\tau}\right] \tag{5.17}$$

where a is a flat loss term

the quantity in brackets suggest interference between two rays with relative delay τ that produces a minimum in the response (a notch) at ω_0, with both ω_0 and ω being measured from a common reference, usually the channel midfrequency

b is a term related to the notch depth relative to a

The flat fade power loss term in decibels, A, is given by

$$A = -20\log_{10} a \qquad (5.18)$$

and the relative notch depth in decibels, B, is given by

$$B = -20\log_{10}(1-b) \qquad (5.19)$$

Thus, $A + B$ equals the total fade depth at the response minimum which is the center of the notch. Figure 5.27, which is from Rummler et al.,[12] shows the amplitude response of $H(j\omega)$ for $\tau = 6.3$ ns.

When the amplitude b has a range from 0 to 1 and τ positive, the response is said to be *minimum phase*. When b continues to have a range from 0 to 1, but τ is negative, the response is said to be *nonminimum phase*. When b is greater than 1 and τ is positive, the response is also said to be nonminimum phase.

At first glance Eq. (5.17) suggests a two-ray model, the signal from one path being a and the signal from the other being $abe^{-j(\omega-\omega_0)\tau}$. In fact, this model is sometimes called the two-ray model with flat attenuation. However, this response can also be visualized as a three-ray model. In this visualization, there is a direct ray, ray 1 say, that is unfaded; a second ray, ray 2 say, of similar but varying negative and positive amplitude and close enough in delay to ray 1 that their composite amplitude response, of value a relative to that of ray 1, is flat over the channel width (no frequency selective fading); and a third ray, ray 3, at a relative

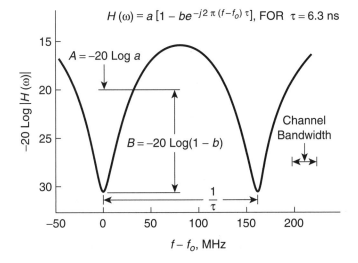

FIGURE 5.27   Attenuation of the modeling function used in the simplified three-ray model. *(By permission from Ref. 12, © 1986 IEEE.)*

delay τ to rays 1 and 2 that provides the frequency shaping of $H(j\omega)$. Figure 5.28 depicts the simplified three-ray model, recognizing that this illustration does not depict the physical path.

The strong acceptance of the simplified three-ray model is in part due to the fact that, in addition to its simplicity, it has been found to provide a very good fit to almost all measured responses in standard digital radio channels. When this model is being used, the delay parameter τ is usually fixed at a value that results in the period of $H(j\omega)$ (which equals $1/\tau$) being large compared to the channel bandwidth under consideration. In the original development of this model, τ was chosen so that the period of $H(j\omega)$ was six times the channel bandwidth under consideration. This led to a value of τ of 6.3 ns. This value has now been accepted as "standard" by many, but some still apply the factor of six rule, which gives a value of τ that's related to the specific channel bandwidth under consideration. Note that since the model is not representative of the physical path, but purely a mathematical way to predict the path transfer function $H(j\omega)$, no physical interpretation should be placed on τ.

To calculate the outage time of a specific radio link due to multipath fading, one must find a way to characterize the radio performance in the presence of such fading. The channel model provides a vehicle for this characterization, the radio performance being defined in terms of the model function parameters. Since the frequency selective effect of multipath fading on a signal is often referred to as dispersion, such a characterization is called the radio's *dispersion signature* and is defined for a specific BER performance threshold, typically 10^{-3}. Applying the simplified three-ray model, the radio is characterized in the laboratory via a path simulator that emulates Eq. (5.17) with $a = 1$ (achieved by a high fixed direct signal level that minimizes the effect of thermal noise), while varying b and ω_0 to obtain a BER of 10^{-3}. Thus the signature is a graph of fade notch depths B for a BER of 10^{-3} versus the frequency from the channel center.

A radio receiver BER performance in the presence of multipath fading differs depending on whether the propagation condition is minimum phase or non-minimum phase.

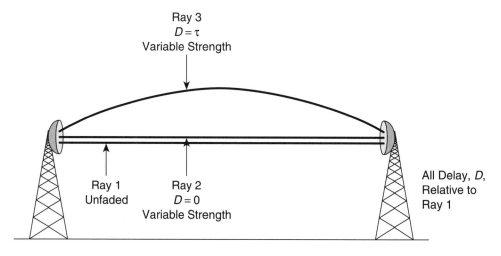

Note: This model is not representative of the physical path. It is a model to predict
      the path transfer function H(ω) only.

FIGURE 5.28   Depiction of the simplified three-ray model.

Fig. 5.29, which is from Rummler et al.,[12] shows the minimum phase and nonminimum phase dispersion signature of a 16-QAM radio for a BER of 10^{-3}. For all points below the signature, the radio BER is higher than 10^{-3}. It will be noted that the minimum phase signature is narrower than that for nonminimum phase. It thus requires greater notch depths, particularly at the band edges, to be degraded to a BER of 10^{-3} compared to the nonminimum phase case. Thus, for the radio characterized here, minimum phase performance is clearly superior to nonminimum phase. This is normally the case.

Careful consideration of the statistics of the parameters of the three-path model[17] has shown that the probability of location of the notch frequency is uniformly distributed over the width of the signature. Thus, for both the minimum phase and nonminimum phase condition, the probability of BER being greater than 10^{-3} is obtained by the integration of the area under the curve weighted by the probability at each frequency across the signature that the notch depth at that frequency will be exceeded. The probability of the BER being greater than 10^{-3} for the receiver as a whole is then taken as the weighed sum of the probability for each phase condition, equal weighting being common.

5.5.4 Cross-Polarization Discrimination Degradation Due to Fading

In order to double the capacity on a given transmission frequency, signals can be transmitted simultaneously on both the horizontal and vertical polarization. This is normally accomplished by using parabolic antennas with dual feeds, one for vertical polarization, the other for horizontal. Ideally, no energy transmitted on one polarization would be intercepted by the

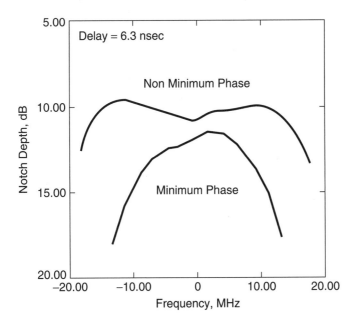

FIGURE 5.29   Signature curves for a 16-QAM digital radio at 10^{-3} bit error ratio. *(By permission from Ref. 12, © 1986 IEEE.)*

receiving antenna feed adjusted for reception of the signal on the other polarization. In practice, this is never the case, but under unfaded conditions, antenna systems can be deployed that result in the signal of the unwanted polarization being 30 to 40 dB below that of the intended polarization. This difference between wanted and unwanted signal level is referred to as *cross-polarization discrimination* (*XPD*). At these levels of XPD, the effect of interference on BER performance, even on high-level modulation systems, is negligible. Unfortunately, deep fades resulting from adverse path conditions can reduce XPD to unacceptable levels. These conditions are many and include heavy rain and strong frequency selective effects from refraction or ground reflection. The mechanisms causing XPD reduction are somewhat complex and will not be discussed here. It is of interest to note, however, that under conditions of multipath fading, XPD remains approximately constant for fade levels down to about 20 dB, but beyond that point, decreases decibel for decibel with the fade level.[13]

To minimize the impact of degraded XPD, modern digital wireless systems with cross-polarized signals employ adaptive *cross-polarization interference cancellers* (XPICs), the fundamentals of which we touch on in Section 6.4.

5.6   EXTERNAL INTERFERENCE

Often radio systems are subjected to interference from unwanted signals. In a given frequency band, channels are normally assigned one beside the other, with alternate channels being on alternate polarization, either horizontal or vertical. In some instances, where permitted by the licensing authority, operators double the capacity available from a given channel by operating on both horizontal as well as vertical polarization. Thus, a receiver operating at the end of a path may be subjected to interference from other channels over the same path. Also, depending on the geography of a path, it may be interfered with by channels on other paths of the same multihop system to which it belongs or channels on independent radio systems operating in the same general geographic area. Finally, it may be interfered with by satellite, radar, or any other type of systems radiating a signal at or very close to the receiver frequency. Because receivers usually have selectivity that increases rapidly outside the passband, only co-channel and adjacent channel signals normally pose a threat to system performance. Discrimination to an adjacent channel signal is achieved via receiver selectivity and the cross-polarization discrimination of the receiving antenna. Discrimination to a co-channel signal is only possible if it's cross-polarized, in which event the antenna provides the discrimination. Figure 5.30 shows examples of intrasystem interference in a three-link system. A receiver at a repeater site, tuned to frequency f_1, is shown being interfered with by an adjacent channel signal transmitted on the same path as the desired signal and two co-channel signals transmitted from adjacent paths. A receiver at an end terminal, tuned to frequency f_2, is shown being interfered with by an "overshoot" signal transmitted from two links away.

Radio manufacturers characterize their product's resistance to interference by the product's *threshold to interference* (*T/I*) ratio. It is usually provided for co-channel and adjacent channel and is normally the ratio in decibels of the 10^{-6} BER static threshold level to the interference level that degrades the BER static threshold 1 dB. It is measured with an interfering signal that is of the same type as the desired signal. Typically, the higher the modulation complexity, the more sensitive a receiver is to interference. In countries that keep a current database of all operating radio frequencies and signal radiated power, frequency search companies, utilizing T/I ratios, try to find for their client's frequencies that can be used in a specific path that will not interfere with or be interfered

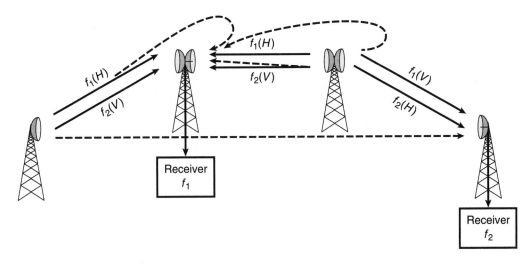

FIGURE 5.30   Example of adjacent channel and co-channel interference.

by other radio systems in the region. Thus interference is a problem that's normally reviewed and hopefully eliminated or minimized at the planning stage.

5.7   OUTAGE AND UNAVAILABILITY

An *outage event* is normally said to occur in a wireless receiver in any period of time when its BER exceeds 10^{-3}. Outage events where the BER exceeds 10^{-3} for periods of less than 10 seconds are referred to as periods of *unacceptable performance*. A loss of frame or synchronization also triggers an outage event. When outage events last for more than 10 consecutive seconds, traffic is usually disconnected and a radio link is said to be *unavailable*. The period of unavailability ends at the start of the period where the BER improves to better than 10^{-3} and stays better than 10^{-3} for ten consecutive seconds. Any one-second period where the BER exceeds 10^{-3} is termed a *severely errored second* (*SES*). Any one-second period where the bit error rate exceeds 10^{-6} but is less than 10^{-3} is termed a *burst errored second* (burst ES). Any one second period where the *residual bit error rate* (*RBER*) exceeds 10^{-10} but is less than 10^{-6} is termed a *dribbling errored second* (dribbling ES). Finally, when the RBER is less than 10^{-10}, the link is said to be in an *error free state*.

Unavailability is meant to be a measure of the time that a system is not available for service. Long-duration interruptions, such as those caused by equipment failure, as well as relatively short ones, such as those caused by rain fading or duct entrapment, contribute to unavailability. Unavailability objectives are normally expressed as a percentage of a year and refer to two-way system communications. Often, instead of specifying unavailability, *availability* is specified, availability being the time that a system is available for service. Thus the availability percentage of a link is 100 minus the unavailability percentage.

Multipath-induced outages generally consist of periods lasting less than 10 seconds and are therefore contributors primarily to unacceptable performance. Objectives for unacceptable performance are defined unidirectionally and are usually measured over a one-month period in ITU regions and annually in North America. As multipath fading is seasonal it results in a *worst month performance*. Rain-induced outages, on the other hand, generally consist of periods lasting minutes and are therefore contributors to unavailability.

Fixed wireless users in North America sometimes specify their link outage requirement in terms of *path reliability*. Path reliability is usually defined as the percentage of time annually that there is no outage due to fading. Thus, if P is the probability of outage due to fading, then the reliability, R, assuming outages are due only to fading, is given by

$$R = (1 - P) \cdot 100\%$$ (5.20)

Many wireless links in long-haul systems are designed to meet a reliability of 99.9999%. This translates to an outage time of only 32 seconds per year!

5.8   OUTAGE ANALYSIS

5.8.1   Rain Outage Analysis

To determine the fade margin required for a given rain related path availability requires knowledge of the rain-induced path attenuation exceeded for a given percentage of the time. The latter, in turn, requires knowledge of the probability distribution of rain attenuation as a function of frequency, polarization, and distance in the general geographic vicinity of the path. ITU Recommendation ITU-R PN.837[11] gives data on the rain rate $R_{0.01}$ exceeded 0.01% of the time. Figures 5.31(a), (b), and (c) show worldwide rain regions provided in this recommendation and Table 5.2 shows rain rates exceeded $x\%$ of the time for the various regions in Fig. 5.31(a), (b), and (c). ITU Recommendation ITU-R P.530[8] gives a procedure to use this data to compute the path attenuation $A_{0.01}$ exceeded 0.01% of the time on a non cross-polarized path. The procedure is as follows:

Step 1: Obtain $R_{0.01}$ from Table 5.2 or local sources of long-term measurements.
Step 2: Compute the specific rain attenuation γ_R for the rain rate $R_{0.01}$ via Eq. (5.14).
Step 3: Compute the effective path distance d_{eff} of the link, this distance being given by

$$d_{eff} = d \frac{1}{1 + d / d_0}$$ (5.21)

where d is the actual path length in km
where, for $R_{0.01} \leq 100$ mm/h

$$d_0 = 35 e^{-0.015 R_{0.01}}$$ (5.22)

and where, for $R_{0.01} > 100$ mm/h, use the value 100 mm/h in place of $R_{0.01}$.

Step 4: Compute $A_{0.01}$ via the relationship

$$A_{0.01} = \gamma_R d_{eff} \text{ dB}$$ (5.23)

Note that if the fade margin is equal to $A_{0.01}$, then this results in a rain related path availability of 99.99% and perhaps four to six rain outage events totaling 53 minutes per year.

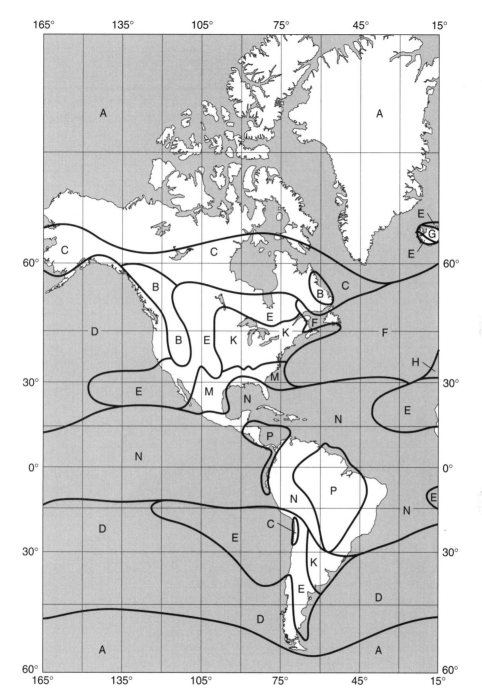

FIGURE 5.31(a)   ITU-R worldwide rain regions: the Americas and Greenland. *From ITU Recommendation ITU-R PN.837-1. Reproduced with the prior authorization of the ITU as copyright holder. The sole responsibility for selecting this extract for reproduction lies with the author and can in no way be attributed to the ITU. The complete volume of the ITU material, from which this figure is extracted, can be obtained from International Telecommunication Union, Sales and Marketing Division, Place des Nations—Ch-1211 GENEVA 20 (Switzerland), Telephone +41 22 730 61 41 (English) / +41 22 730 61 42 (French) / +41 22 730 61 43 (Spanish), Telex: 421 000 uit ch / Fax: +41 22 730 51 94, Email: sales@itu.int / http://www.itu.int/publications*

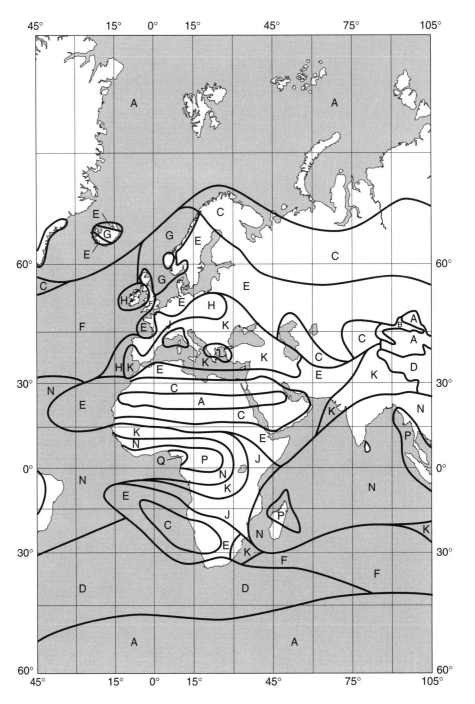

FIGURE 5.31(b)   ITU-R worldwide rain regions: Europe, Africa, and western Asia.

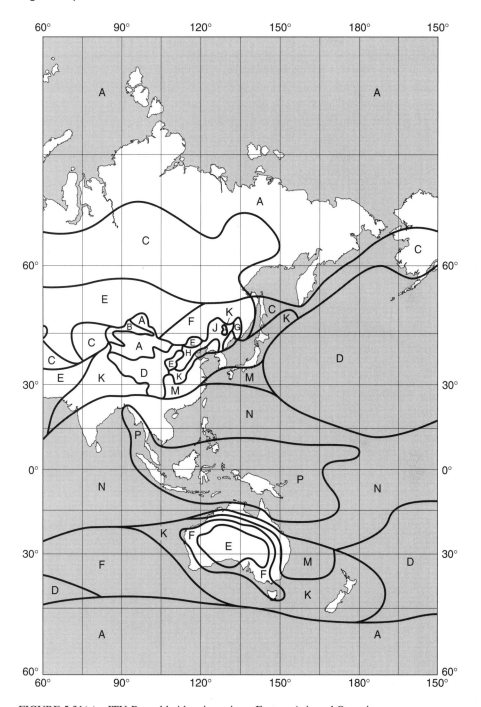

FIGURE 5.31(c)   ITU-R worldwide rain regions: Eastern Asia and Oceania.

TABLE 5.2 Rainfall Intensity Exceeded (mm/h) for Given Percentages of Time as a Function of Rain Region as Indicated in Fig. 5.31

Percentage of time (%)	A	B	C	D	E	F	G	H	J	K	L	M	N	P	Q
1.0	<0.1	0.5	0.7	2.1	0.6	1.7	3	2	8	1.5	2	4	5	12	24
0.3	0.8	2	2.8	4.5	2.4	4.5	7	4	13	4.2	7	11	15	34	49
0.1	2	3	5	8	6	8	12	10	20	12	15	22	35	65	72
0.03	5	6	9	13	12	15	20	18	28	23	33	40	65	105	96
0.01	8	12	15	19	22	28	30	32	35	42	60	63	95	145	115
0.003	14	21	26	29	41	54	45	55	45	70	105	95	140	200	142
0.001	22	32	42	42	70	78	65	83	55	100	150	120	180	250	170

From ITU Recommendation ITU-R PN.837-1. Reproduced with the prior authorization of the ITU as copyright holder. The sole responsibility for selecting this extract for reproduction lies with the author and can in no way be attributed to the ITU. The complete volume of the ITU material, from which this table is extracted, can be obtained from

International Telecommunications Union
Sales and Marketing Division
Place des Nations—Ch-1211 GENEVA 20 (Switzerland)
Telephone: +41 22 730 61 41 (English) / +41 22 730 61 42 (French) / +41 22 730 61 43 (Spanish)
Telex: 421 000 uit ch / Fax: +41 22 730 51 94
E-mail: sales@itu.int / http://www.itu.int/publications

TABLE 5.3   Factor *F* that A$_{0.01}$ Must Be Multiplied By to Determine A$_p$, the Path Attenuation Exceeded *p*% of the Time

a) Radio Links Located in Latitudes Equal to or Greater than 30° (North or South)					
p	1%	0.10%	0.05%	0.01%	0.005%	0.001%
F	0.12	0.39	0.52	1	1.28	2.14
b) **Radio Links Located in Latitudes Less than 30° (North or South)**					
p	1%	0.10%	0.05%	0.01%	0.005%	0.001%
F	0.07	0.36	0.53	1	1.19	1.44

The Recommendation also gives a procedure to compute A_p as a factor F of $A_{0.01}$, where A_p is the path attenuation exceeded p% of the time and where p lies in the range 0.001% to 1%. A_p is determined from the relationship $A_p = FA_{0.01}$, and two location-dependent formulas are given for F. Specifically, for radio links located in latitudes equal to or greater than 30° (north or south), the formula provided gives the values of F shown in Table 5.3(a) for various values of p. For radio links located in latitudes less than 30° (north or south), the formula provided gives the values of F shown in Table 5.3(b) for various values of p.

In addition to providing the preceding outage prediction method for non–cross-polarized paths, ITU Recommendation ITU-R P.530[8] also provides a method for predicting XPD outage due to intense rain for paths where both vertically and horizontally polarized signals are employed.

Typically, though the fade margin required for a given rain-related availability can be calculated manually, as demonstrated in Example 5.2, it is normally done by one of several easily accessible computer programs that store all relevant required location dependent data. It should be noted that some programs use R.F. Crane's rain outage model, which may give values for annual rain outage and availability that are slightly different from those given by the ITU model described previously.

EXAMPLE 5.2   FADE MARGIN REQUIRED TO ACHIEVE A GIVEN RAIN-RELATED AVAILABILITY

A 40-Ghz, 8-km (5 mile), vertically polarized path is located in the San Francisco Bay area (latitude 37° North). What is the fade margin required to achieve a 99.995 path availability?

Solution

The fade margin required to achieve a 99.995% path availability is the rain attenuation not exceeded 99.995% of the time and hence the rain attenuation exceeded 0.005% of the time, $A_{0.005}$. To find $A_{0.005}$ we must first, however, find $A_{0.01}$. We do this using the ITU-R P.530 procedure outlined previously.

Step1:   Obtain $R_{0.01}$, the rain rate exceeded 0.01% of the time. From Fig. 5.31(a) we determine that the San Francisco Bay Area lies in rain region D. From Table 5.2 we determine that $R_{0.01}$ in rain region D is 19 mm/h.

Step 2: Determine rain attenuation γ_R for $R_{0.01}$. From Table 5.1 (ITU-R table) we determine that for a vertically polarized, 40-GHz signal, $k_v = 0.310$ and $\alpha_v = 0.929$. Thus, by Eq. (5.14), we have

$$\gamma_R = 0.31 \times (19)^{0.929} = 4.78 \text{ dB/km}$$

(We note that, had we decided to conserve a few computation calories by determining γ_R from Fig. 5.16, the value found would have been approximately 5.5 dB/km, not too different from the ITU-R result.)

Step 3: Compute d_{eff}, the effective path distance. By Eq. (5.22), we have

$$d_0 = 35e^{-(0.015 \times 19)} = 26.3 \text{ km}$$

Thus, by Eq. (5.21), we have

$$d_{eff} = 8 \left[\frac{1}{1 + 8/26.3} \right] = 6.13 \text{ km}$$

Step 4: Compute $A_{0.01}$. By Eq. (5.23), we have

$$A_{0.01} = 4.87 \times 6.13 = 29.3 \text{ dB}$$

Now that we have $A_{0.01}$ we can compute $A_{0.005}$. As the latitude of the San Francisco Bay area is 37° and thus greater than 30°, we use Table 5.3(a) to determine the multiplication factor, F. From the table we determine that, for a path attenuation exceeded 0.005% of the time, $F = 1.28$. Thus

$$A_{0.005} = 1.28 \times 29.3 = 37.5 \text{ dB}$$

Thus, the fade margin required to achieve a 99.995% path availability is 37.5 dB. We note that an availability of 99.995% implies an unavailability of 0.005% which equates to an annual rain outage time of 26 minutes.

5.8.2   Multipath Fading and Interference-Induced Outage Analysis

A method commonly used to predict multipath fading and interference induced outage on a microwave radio link is the *composite fade margin* (*CFM*) method. In this method, a composite fade margin is determined that consists of individual components that factor in the effects of thermal noise, the dispersive (RF spectrum distortion) component of multipath fading, and interference. Then, applying well-researched statistics on multipath fading occurrence, the probability of this fade margin being exceeded is determined, and hence the outage of the link. Outage estimates using the *CFM* method show good agreement with observations made in the field.

Breaking down the *CFM* into its individual components requires that multipath fading be regarded as contributing two separate components: one associated with flat fading at the channel center frequency and represented by the thermal (Gaussian) noise power in the absence of interference at the receiver input, N_t say, and another associated with the noise equivalence of the dispersiveness of the fading and represented by a Gaussian-type noise source, of power N_d say, also at the receiver input. Note that this fictitious noise power is not a fixed power, per se, but rather one with a fixed relationship to the unfaded signal, regardless of that signal level. Interferers contribute a component by considering them to be represented by an equivalent Gaussian-type noise source of noise power I at the receiver input. The *CFM* is constructed based on the assumption that the total outage time due to fading is the linear addition of the outage times due individually to flat fading, dispersive fading, and interference. It has been shown by Shafi and Rummler[13] that if the *flat fade margin, FFM* (alternatively called the *thermal fade margin,* since flat fading only increases the thermal noise contribution to the BER) is given by

$$FFM = \left(\frac{C_u}{N_t}\right) - \left(\frac{C_f}{N_3}\right)$$ (5.24)

where C_u/N_t is the unfaded carrier to thermal noise ratio
and C_f/N_3 is the faded carrier to thermal noise ratio that results in unacceptable
        performance, such performance here assumed to be a BER of 10^{-3}

if the *dispersive fade margin, DFM*, is given by

$$DFM = \left(\frac{C_u}{N_d}\right) - \left(\frac{C_f}{N_3}\right)$$ (5.25)

where C_u/N_d is the unfaded (undispersed) carrier to the fictitious dispersive noise ratio
and C_f/N_3 is the faded (dispersed) carrier to dispersive noise ratio that results in unacceptable
        performance, such performance here assumed to be a BER of 10^{-3}

and if the *interference fade margin, IFM*, is given by

$$IFM = \left(\frac{C_u}{I}\right) - \left(\frac{C_f}{N_3}\right)$$ (5.26)

where C_u/I is the unfaded carrier to interference noise ratio
and C_f/N_3 is the faded carrier to interference noise ratio that results in unacceptable
        performance, such performance here assumed to be a BER of 10^{-3}

then the *CFM* is given by

$$CFM = -10\log_{10}\left[10^{-FFM/10} + 10^{-DFM/10} + 10^{-IFM/10}\right]$$ (5.27)

Note that the ratios (C_f/N_3) shown in Eqs. (5.24), (5.25) and (5.26) are identical since this analysis assumes Gaussian-type noise sources for all three methods of performance degradation and hence identical *C/N* for a given BER.

The *CFM*, as defined mathematically, is the channel center frequency fade depth exceeded for the same number of seconds as the threshold BER, 10^{-3}, and is exceeded as a result of flat fading, dispersion, and interference. Similarly, the *DFM* is the channel center frequency fade depth exceeded for the same number of seconds as the threshold BER is exceeded as a result of dispersion only. Note that, despite what one may be tempted to conclude from a cursory glance at Eq. (5.25), the *DFM* of a receiver is a function of the receiver characteristics only and hence independent of the unfaded signal level. It is a measure of the receiver's ability to withstand signal dispersion regardless of the unfaded signal level. The *DFM* limits the extent to which the *CFM* can be increased by increasing the *FFM*. For example, if the *FFM* is increased to a value that's 6 dB greater than the *DFM*, then, assuming negligible interference, increasing the *FFM* further will do little to reduce outage since doing so will increase the *CFM* by less than 1 dB relative to the *DFM*.

Typically, relatively low bit rate systems (less than about 10 Mb/s) that occupy a bandwidth of less than about 10 MHz can be treated as systems subject to flat fading only. This is not because there is no dispersive effect on the receiver but rather because the *DFM* tends to be so large that the receiver performance is dominated by the flat fade component.

The *DFM* of a receiver is computed from the receiver's minimum phase and non-minimum phase dispersive signatures described in Section 5.5.3.3. Assuming equal weighting to each of these signatures in BER performance degradation, then the *DFM* can be shown to be given by[18]

$$DFM = 17.6 - \log_{10}\left(\frac{S_w}{158.4}\right) \tag{5.28}$$

where

$$S_w = \int_{-f_1}^{f_1} \left(e^{\frac{-B_m(f)}{3.8}} + e^{\frac{-B_n(f)}{3.8}} \right) df \tag{5.29}$$

and where $B_m(f)$ is the signature notch depth in decibels; minimum phase fades
            $B_n(f)$ is the signature notch depth in decibels; non-minimum phase fades
            $-f_1$ and f_1 define the edges of the signal-occupancy bandwidth

Determining a receiver's *DFM* using Eqs. (5.28) and (5.29) would appear to be a consuming task. Fortunately for manufacturers, test equipment is available that not only simulates the simplified three-ray model for different values of τ, the multipath delay and computes the receiver dispersive signature automatically, but also calculates *DFM*.

The *IFM* is that depth of fade to the point where, with only interferers as degraders, BER is degraded to 10^{-3}. Using Eq. (5.26), *IFM* can be found, noting that (a) *I* is the total power of interferers at the receiver input, after reducing each by the appropriate receiver selectivity, and (b) (C_f / N_3) is assumed identical for *IFM* and *FFM*, thus the *FFM*'s (C_f / N_3) can used, N_3 being the thermal noise in the receiver's noise bandwidth.[13]

For fade multipath depths greater than about 20 dB, Barnett[19] has shown that, in the average worst month, $P(F_m > F)$, the probability that the center frequency fade depth, F_m dB, is greater than F dB, follows a Raleigh distribution and is given by

$$P_{ns}(F_m > F) = P_0 10^{-F/10} \tag{5.30}$$

where P_0 is the *Vigants multipath fade occurrence factor*[20] and given by

$$P_0 = c \cdot (f / 4) \cdot d^3 \cdot 10^{-5} \tag{5.31}$$

and where c is the *climate-terrain factor* for average terrain roughness and is given by

$c = 0.25$ Mountainous terrain in dry climate
$= 1$ Average terrain in temperate climates
$= 4$ Over water and near gulf coasts or in damp climates or climates exhibiting strong thermal excursions such as found in deserts

f is the operating frequency in GHz
d is the path length in miles

For path length d in km, P_0 is given by

$$P_0 = 6cfd^3 \cdot 10^{-7} \tag{5.32}$$

All other pertinent factors being similar, paths over rough terrain fade less than paths over smooth terrain. Vigants[20] has suggested that this may be because stable atmospheric layering, necessary for multipath fading, is less likely to occur over rough terrain. To account for terrain roughness, Vigants[20] suggests that the factor c in Eqs. (5.31) and (5.32) be replaced by a new factor, here labeled c', and given by

$$c' = Xa \tag{5.33}$$

where X is the *climate factor* and is given by

$X = 0.5$ Dry climate (high-dry mountainous)
$= 1$ Average climate (continental temperature or midlatitude inland)
$= 2$ Coastal areas (maritime temperature, coastal or high humidity/temperature)

and a is the *roughness factor*, given by

$$a = \left(\frac{w}{50}\right)^{-1.3}, \qquad 20 \leq w \leq 140 \tag{5.34}$$

where w is the *terrain roughness* measured in feet, and is defined as the square root of the average square of the terrain height deviation from the mean terrain height, terrain height being measured above some reference level (sea level, for example), obtained from the path profile at one-mile intervals, with the ends of the path excluded.

Figure 5.32, which is from Ref. 21, shows worldwide climate-terrain factors, c. Note that it gives more values for c than originally proposed by Barnett, namely, values of 2 and 6. The value 6 has been added to account for those areas that, based on climate only, would have been assigned a value of 4 by Barnett's relationship but are better represented by a value of 6 to account for flat ($w < 20ft$) terrain conditions. Note also that Vigants's climate factor, X, is equal to the $\sqrt{c}$, the square root of his climate-terrain factor. Thus X can be determined from Fig. 5.32 by taking $X = \sqrt{c}$, except where $c = 6$, in which case $c = 4$ should be used since roughness will be accounted for by a, the roughness factor.

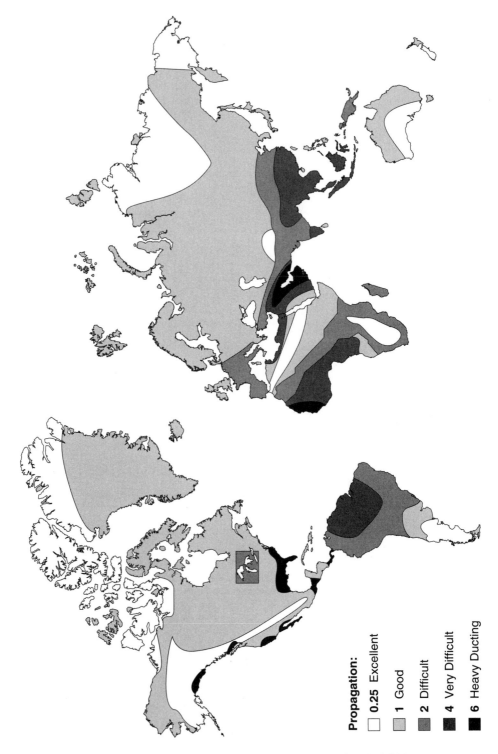

Propagation:

- ☐ **0.25** Excellent
- ☐ **1** Good
- ☐ **2** Difficult
- ■ **4** Very Difficult
- ■ **6** Heavy Ducting

FIGURE 5.32   Worldwide climate terrain factors, *c. (By permission from Ref. 21.)*

The probability of outage is, by definition, the probability that a BER of 10^{-3} is exceeded. This latter probability, however, is also equal to the probability that the center frequency fade depth exceeds the composite fade margin CFM. As a result, the average worst month outage probability for a nondiversity path, P_{nd}, is given by

$$P_{nd} = P(F_m > CFM) \qquad\qquad (5.35a)$$

$$= P_0 \times 10^{-CFM/10} \qquad\qquad (5.35b)$$

The predicted outage seconds (severed error seconds) in the average worst month is calculated by multiplying the average worst month probability of outage P_{nd} given by Eq. (5.35b) by the number of seconds in a month.

Since multipath fading takes place primarily in warm weather, Vigants[20] assumed that the length of the warm portion of the year, and hence the length of the multipath fading season, is proportional to the average annual temperature. Based on this assumption, he proposed that annual one-way outage for a nondiversity path, O_{nd}, be determined by multiplying P_{nd} by T_{fs}, the length of the multipath fading season; that is,

$$O_{nd} = P_{nd} \times T_{fs} \qquad\qquad (5.36a)$$

where T_{fs} is given by

$$T_{fs} = \left(\frac{t}{50}\right) \times 8 \times 10^6 \text{ s}, \qquad 35 \le t \le 75 \qquad (5.36b)$$

and where t denotes the average annual temperature in °F of the locality in question.

The average case is assumed to be where $t = 50°F$ and the fading season equals 3 months or 8×10^6 seconds. Fig. 5.33 shows worldwide average annual temperature, t.

The ITU, in Recommendation ITU-R P.530, suggests a similar but not identical process to calculate nondiversity outage on a radio link due to multipath fading. As in the CFM method it assumes that multipath fading consists of two separate components contributing to error rate degradation and hence outage: a flat fading component, and one due to dispersion, which it refers to as the selective fading component. It also factors in co-channel interference due specifically to a cross-polarized signal. However, unlike the CFM method, although mathematically equivalent to it, it computes the probability of outage due to each component separately, then adds these outages. Thus it gives the probability of total outage P_{ndt} as

$$P_{ndt} = P_{ns} + P_s + P_{XP} \qquad\qquad (5.37)$$

where P_{ns} is the probability of outage due to the nonselective component of fading
$\qquad P_s$ is the probability of outage due to the selective component of fading
and $\qquad P_{XP}$ is the probability of outage due to cross-polarized co-channel interference

Its P_{ns} is equivalent to the probability given by Eq. (5.35b) but with CFM replaced by FFM, and its P_s is equivalent to the probability given by Eq. (5.35b) but with CFM replaced by DFM. However, as different relationships to those given above for the CFM method are applied in determining these probabilities, the actual values of P_{ns} and P_s will be different to their CFM method equivalents. The ITU recommended process also outlines a procedure to convert

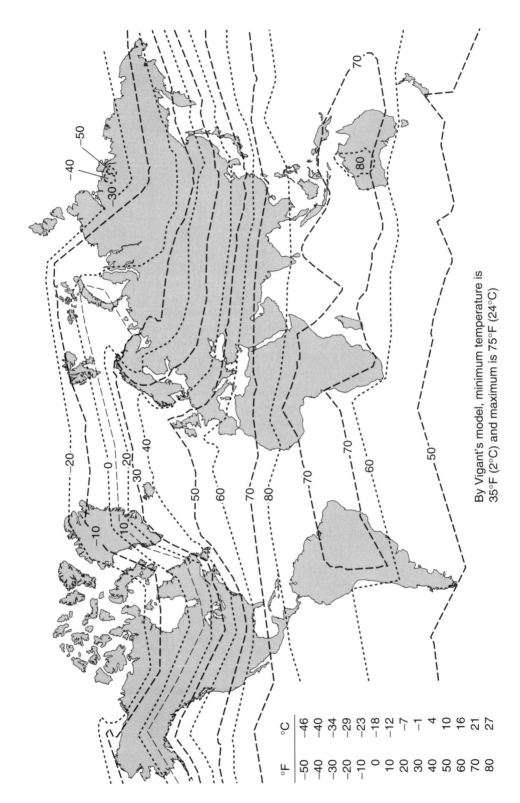

FIGURE 5.33 Worldwide average annual temperature (°F). *(By permission from Ref. 21.)*

By Vigant's model, minimum temperature is 35°F (2°C) and maximum is 75°F (24°C)

°F	°C
−50	−46
−40	−40
−30	−34
−20	−29
−10	−23
0	−18
10	−12
20	−7
30	−1
40	4
50	10
60	16
70	21
80	27

outage probability from average worst month to average annual. In general, the ITU recommended process[20] is a more complicated one than the process initially devised by Vigants.

As in the case of rain-induced outages, ITU Recommendation ITU-R P.530[8] provides a method for predicting XPD outage due to multipath activity for paths where both vertically and horizontally polarized signals are employed.

Finally, though nondiversity multipath fading and interference induced outage, as in the case for rain, can be calculated manually (as demonstrated in Example 5.3), it is normally computed via readily accessible computer programs. These programs use either the ITU-R P.530 or the Vigants model.

EXAMPLE 5.3   ANNUAL ONE-WAY OUTAGE OF A TYPICAL NONDIVERSITY MICROWAVE LINK

A nondiversity microwave link has the following properties:

Location: Madrid, Spain
Path length, $d = 30$ miles
Operating frequency, $f = 6$ GHz
$FFM = 40$ dB
$DFM = 50$ dB
$IFM = 51$ dB

What is the annual one-way outage in severed error seconds (SES)?

Solution

First we determine the climate terrain factor, c, and the average annual temperature, t, for a link in Madrid. From Fig. 5.32 we determine that $c = 2$ (difficult), and from Fig. 5.33 we determine that $t = 60°$F.

Next we compute the multipath fade occurrence factor, P_0, using Eq. (5.32). Thus,

$$P_0 = 2 \times (6/4) \times (30)^3 \times 10^{-5} = 0.81$$

We now compute the composite fade margin, CFM, using Eq. (5.27). Thus,

$$CFM = -10\log_{10}[10^{-40/10} + 10^{-50/10} + 10^{-51/10}] = 39.3 \text{ dB}$$

We now determine the worst month outage probability, P_{nd}, using Eq. (5.35b). Thus,

$$P_{nd} = 0.81 \times 10^{-39.3/10} = 9.52 \times 10^{-5}$$

Now the multipath fading season, T_{fs}, is determined using Eq. (5.36b):

$$T_{fs} = \left(\frac{60}{50}\right) \times 8 \times 10^6 = 9.6 \times 10^6 \text{ s}$$

Finally, using Eq. (5.36a), we have that the annual one-way outage, O_{nd}, is given by

$$O_{nd} = (9.52 \times 10^{-5}) \times (9.6 \times 10^6)$$

$$= 914 \text{ SES}$$

5.9   DIVERSITY TECHNIQUES FOR IMPROVED RELIABILITY

Diversity is a means whereby two or more versions of the same simultaneously transmitted information are intercepted at a receiver site. It is used to improve a link's resistance to the degrading effect of multipath fading when the performance from the nonprotected link is unacceptable. Three distinct types of diversity are employed in point-to-point, line-of-sight digital radio systems: *space diversity, angle diversity* and *frequency diversity*. Hybrid, quadruple, and other diversity schemes use combinations of these three distinct types. In space diversity one signal is transmitted, but at the receiver two or, in rare instances over difficult paths, three signals are received by using multiple antennas placed at differing vertical heights. Angle diversity is similar to space diversity but employs either two side-by-side antennas at the same vertical height with the feedhorns adjusted at slightly different angles or only one receiving antenna reflector fitted with two feedhorns adjusted at slightly different angles. Either approach results in signals at the receivers that travel over slightly different paths. Finally, in frequency diversity, the same information is transmitted from one antenna but on two separate frequencies and, at the receive end, one antenna intercepts these signals and feeds two separate receivers that demodulate the signals. Diversity works on the principle that signals received via different physical or electrical paths will be largely uncorrelated with respect to multipath-induced distortion. Thus, by clever choices at the receive end in how the original data stream is reconstructed, most errors resulting from multipath fading can be eliminated. Improvement in outage time on a radio link resulting from diversity is expressed as a *diversity improvement factor, I*, which is defined as the ratio of outage time without diversity to the outage time with diversity.

5.9.1   Space Diversity

Figure 5.34(a) shows a simplified one way vertical space diversity link. In such links, a number of different techniques are used at the receive end to reconstitute the original data stream errorlessly (i.e., without degrading its error performance). These techniques vary from combining the received signal at RF to combining the received signals at IF to selecting the appropriate data at baseband. However, RF combining is rarely used due to its high cost relative to IF combining. Fig. 5.34(b) shows an example of space diversity RF combining, Fig. 5.34(c) shows IF combining, and Figure 5.34(d) shows switching at baseband. Switching, either at RF or IF, is not employed on digital radio systems because the large number of data interruptions of such an approach would result in during fading. However, as will be discussed later, switching at baseband can be accomplished without data interruption. Combining either at RF or IF minimizes the amount of receiver hardware required as common equipment is used after the combiner. Baseband switching, however, requires two complete receivers to drive the switch. Following is a summary of some of the space diversity combining methods used on broadband digital links:

The *maximum power combiner (MPC)*,[22] also called the *equal gain combiner (EGC)*, was initially developed for analog systems, was among the first used in digital space diversity systems, and continues to be widely used. It is a device that can operate at either RF or IF. It shifts the phase of the signal entering through one antenna before

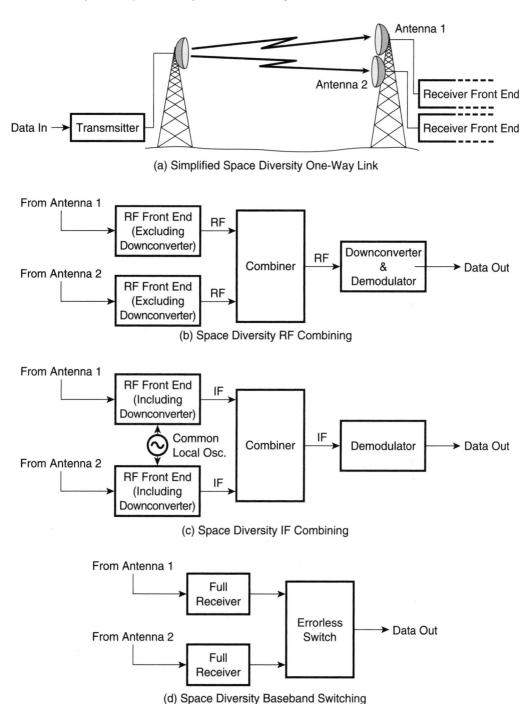

FIGURE 5.34   Space diversity configurations.

combining it, with equal weighting, with the signal entering through the other. The amount of phase shift applied is such as to bring together in phase the center frequency spectral components of the two received signals. The process is called co-phasing. As a result, the power of the combined received signal at the channel center frequency is maximized. Certain field results of outage times resulting from applying this method showed a factor of reduction that varied between one-half and one-and-one-half orders of magnitude relative to the nondiversity case.[23]

The *maximal ratio combiner (MRC)*, like the MPC, co-phases and sums, at RF or IF, the incoming signals. However, before summing, it weights the signals in an adaptive fashion so as to maximize its instantaneous output signal-to-noise ratio. As one would suspect, it provides somewhat better performance relative to the MPC.

The *minimum dispersion combiner,*[24] also an RF or IF type, combines the received signals in such a fashion as to minimize the in-band amplitude dispersion. It achieves this by monitoring the power present in two or more narrow bands within the combined channel bandwidth and adjusting the relative phase of the two received signals based on the monitored results. Unfortunately, if a fading situation should produce input signals at the two antennas that are similar over the frequency range of the channel, the signals will mutually cancel, leaving no useful combined output. This is avoided by switching to the maximum power combining state under such circumstances. Though more complex than the pure maximum power combiner, the minimum dispersion combiner is particularly attractive in high capacity and hence high bandwidth digital radio systems due to its superior minimization of in-band dispersion. Certain field results indicated diversity improvement factors of more than an order of magnitude when this method of combining was employed.[24]

The *diversity baseband switch* switches between the two receiver baseband outputs and on space diversity systems is usually designed to perform error-free switching. In such switching, the performance of each receiver is constantly monitored, and when distortion to amplitude, slope, and so on is detected on one stream the "anticipatory" switch selects data from the other stream before errors occur on either channel. It achieves errorless switching by using elastic storage to line up each data stream so that the same bit is available from either stream at exactly the same time. This way, no extra bits are added and neither are bits deleted.

In analog radio systems the space diversity improvement factor increases with antenna separation. However, because of the impact of selective fading, such a simple relationship does not exist for those digital systems where *DFM*, the dispersive fade margin, is sufficiently low as to dominate *CFM*, the composite fade margin. Rather, in such situations, theory suggests a maximum improvement for very small separation (on the order of a foot at 6 GHz), which then decreases rapidly to a constant factor as separation increases.[25] ITU Recommendation ITU-R P.530[8] outlines a procedure for determining appropriate antenna separation on space diversity systems. It also provides a method for predicting outage in space diversity digital systems. In this method, total average worst month outage probability, P_d, is given by

$$P_d = \left(P_{dns}^{0.75} + P_{ds}^{0.75} \right)^{4/3}$$

(5.38)

where P_{dns} is the nonselective (nondispersive) average worst month outage probability

and P_{ds} is the selective (dispersive) average worst month outage probability

Formulas for determining the factors in Eq. (5.38) are given in ITU-R P.530.[8] They are many and quite lengthy and will not be stated here. Note, however, that, unlike the non-protected case, the total outage probability, P_d, is not a linear addition of the outages due to the nonselective and selective components.

An alternative method for determining P_{dns} to that given in ITU-R P.530,[8] and, indeed, the first highly accepted method for determining P_{dns}, is that given by Vigants.[20] The nondispersive components of the *CFM* are the *FFM* and the *IFM*. We can thus define the *nondispersive (nonselective) fade margin, NDFM*, as

$$NDFM = -10\log_{10}\left[10^{-FFM/10} + 10^{-IFM/10}\right] \tag{5.39}$$

Vigants's formula, although developed for analog systems, is applicable to digital systems where the link's dispersive fade margin, *DFM*, is larger than the nondispersive fade margin, *NDFM*, by at least 6 dB say, so that the *CFM* is largely controlled by the nondispersive fade margin, *NDFM*. This is typically the case in most modern digital fixed wireless links where the *DFM* is large, on the order of 50 dB or greater, as a result of performance improvement techniques discussed in Chapter 6 or by using highly directive antennas that lower the multipath signal's amplitude. By Vigants, we have P_{sd-ns}, the space diversity, nonselective, average worst month outage probability given by

$$P_{sd-ns} = P_{nd} / I_{sd} \tag{5.40}$$

where P_{nd} is the nondiversity average worst month probability of outage as given by Eq. (5.35 b)

and I_{sd} is the *space diversity improvement factor*, given by

$$I_{sd} = 7 \times 10^{-5} s^2 f \frac{1}{d} 10^{(CFM-V/10)}, \qquad s \leq 50 \tag{5.41a}$$

where s = vertical separation of receiving antennas, in feet, center to center

f = frequency, in GHz

d = path length, in miles

V = absolute difference in gain between the two receiving antennas

For antenna separation in meters, and path length in km, I_{sd} is given by

$$I_{sd} = 1.2 \times 10^{-3} s^2 f \frac{1}{d} 10^{(CFM-V/10)}, \qquad s \leq 15 \tag{5.41b}$$

I_{sd} as defined above is equally applicable to links with receivers using either errorless baseband switches or IF dispersion sensing combiners.

EXAMPLE 5.4   ANNUAL ONE-WAY OUTAGE OF A TYPICAL SPACE DIVERSITY MICROWAVE LINK

Assume that the nondiversity path in Example 5.3 is converted to a space diversity one, with equal gain receiver antenna dishes, and that these dishes have a vertical separation of 25 feet. (a) Is the Vigants space diversity improvement formula applicable to this link? (b) If the answer to (a) is yes, what is the annual one-way outage?

Solution

(a) By Eq. (5.39), the nondispersive fade margin, *NDFM*, is given by

$$NDFM = -10 \log_{10}[10^{-40/10} + 10^{-51/10}] = 39.7 \text{ dB}$$

Since the dispersive fade margin, *DFM* = 50 dB, it is more than 6 dB greater than the *NDFM*. Thus, the composite fade margin, *CFM*, is largely controlled by the *NDFM*. In fact, as we determined in Example 5.3, the *CFM* is 39.3 dB, a value very close to the *NDFM* value of 39.7 dB. Thus, the Vigants space diversity improvement formula is indeed applicable to this link.

(b) First we determine the space diversity improvement factor, I_{sd}, from Eq. (5.41a). Thus,

$$I_{sd} = 7 \times 10^{-5} \times 25^2 \times 6 \times \left(\frac{1}{30}\right) \times 10^{(39.3-0)/10} = 74.5$$

By Eq. (5.40), the worst month space diversity outage probability is the worst month nondiversity outage probability divided by I_{sd}. It follows, therefore, that the annual one-way space diversity outage, O_{sd}, is the annual one-way nondiversity outage, O_{nd}, 914 SES as determined in Example 5.3, divided by I_{sd}. Thus,

$$O_{sd} = 914/74.5 = 12 \text{ SES}$$

5.9.2   Angle Diversity

As indicated previously, early researchers into space diversity found that the diversity improvement factor did not increase linearly with separation, but rather seemed to be independent of practical separation distances. At about the same time researchers began experiments with different types of diversity antennas at the same height and discovered large diversity improvement factors.[25] Further research led to the finding that by slightly offsetting the elevation adjustments of two identical antennas at the same height, or by slightly offsetting the elevation adjustments of two feedhorns of a single parabolic reflector, similar large diversity improvement actors were realized. This approach was therefore appropriately named angle diversity. The reason that such small differences in the composite signal received by the feedhorns can produce a large diversity improvement factor is as follows. The creation of a notch deep enough to cause a digital radio outage requires that the two predominant combined rays be of very nearly equal power and 180° out of phase at the notch frequency. However, in this condition, even a small reduction in the power level of one ray relative to the other can reduce the notch depth significantly. For example, a 1-dB reduction in the relative power level can reduce notch depth from 40 dB to less than 20 dB.[25] Thus, during multipath fading, if the received signal on one feed has a deep notch, it is unlikely that the signal on the other feed will simultaneously have a deep notch, since it is receiving signal components of slightly different magnitude. Angle diversity using one reflector significantly lowers the cost of diversity implementation relative to vertical separation diversity as it not only lowers the antenna cost but reduces antenna wind load on the supporting tower and potentially lowers the tower height. For links with *DFM* sufficiently high that their *CFM* is controlled primarily by their *FFM*, space diversity provides improved diversity performance relative to angle diversity. Thus, angle diversity is normally employed where installation limitations such as space, aesthetics, and tower loading, prohibit space diversity. Like space diversity systems, angle diversity systems employ maximum power combiners, minimum dispersion combiners, dispersion-sensing IF combiners, and diversity baseband switches.

ITU Recommendation ITU-R P.530[8] provides a technique for predicting outage using angle diversity. As with vertical space diversity, total outage probability, P_d, is defined in terms of nonselective and selective outage probability via the relationship given in Eq. (5.38).

5.9.3 Frequency Diversity

Figure 5.35 shows a simplified one-way frequency diversity link. It will be noted that, unlike space and angle diversity, it requires two transmitters. Further, it requires two complete receivers followed by an errorless switch. This arrangement is often less expensive than vertical space diversity. However, the opposite is usually the case compared to angle diversity, where only one transmitter, one receiving antenna, and, if RF or IF combining is employed, less than two full receivers are required. A major disadvantage of frequency diversity is that it utilizes twice the RF capacity for the protection of a single working channel (one-by-one protection) as compared to space or angle diversity. As a result, many licensing authorities restrict its use. Frequency diversity works well on digital systems by, ironically, taking advantage of the frequency selective nature of multipath fading. Since both signals travel over identical paths, then even with the diversity channels separated by two channel widths, which is the normal minimum, it's highly unlikely that both channels will simultaneously suffer an outage resulting deep notch. This is because the frequency separation between the notches is inversely proportional to the relative delay between the various propagation paths (see Section 5.5.3.2). Field measurements have shown that the probability of occurrence of a given value of relative delay decreases exponentially as that given value increases.[26] Thus, the long delay required for closely spaced deep notches has a very low probability. Figure 5.36, which is from Lin et al.,[25] shows a comparison of one-by-one frequency diversity improvement factors for a number of digital radio systems. It indicates improvement factors that vary from one to two orders of magnitude. Further, it shows that these factors are higher than those predicted by the standard analog radio model by about one order of magnitude. Note, however, that in these systems, developed in the early days of digital radio, the *DFM* was very likely lower than the *FFM* and thus the controlling parameter of the *CFM*. In systems where the *DFM* is larger than the *FFM*, and thus the *FFM* is the controlling parameter of the *CFM*, as is very often the case with well designed modern systems, the diversity improvement factor tends to closely match that of analog radio model.

A procedure for predicting outage on one-by-one frequency diversity digital systems is given in ITU Recommendation ITU-R P.530.[8] As with vertical space diversity and

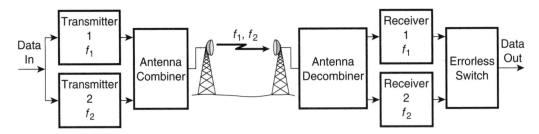

FIGURE 5.35   Simplified frequency diversity one-way link.

angle diversity, total outage probability, P_d, is defined in terms of nonselective and selective outage probability via the relationship given in Eq. (5.38). The formula given for determining $P_{fd\text{-}ns}$, the frequency diversity, nonselective, average worst month outage probability, is the analog radio formula given by Vigants.[20] As with Vigants's analog space diversity formula, it is applicable to digital systems where the *CFM* is largely controlled by the *NDFM* and is as follows:

$$P_{fd-ns} = P_{nd} / I_{fd} \qquad (5.42)$$

where P_{nd} is the nondiversity average worst month probability of outage as given by Eq. (5.35b)

and I_{fd} is the *frequency diversity improvement factor*, given by

$$I_{fd} = 50\Delta f \frac{1}{f^2 d} 10^{CFM/10} \qquad (5.43a)$$

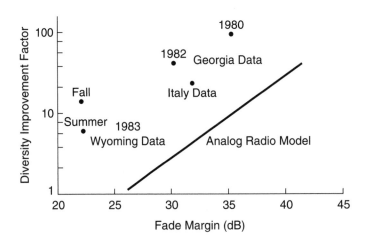

FIGURE 5.36   Comparison of one-by-one frequency diversity improvement factors for a number of digital radio systems. *(By permission from Ref. 25, © 1988 IEEE.)*

where Δf is the frequency separation in GHz
   f is the frequency in GHz
   d is the path length in miles

For path length d in km, then I_{fd} is given by

$$I_{fd} = 80\Delta f \frac{1}{f^2 d} 10^{CFM/10} \tag{5.43b}$$

ITU-R P.530[8] indicates that Eq. (5.43b), and hence Eq. (5.43a), apply only for the following ranges of parameters:

$2 \leq f \leq 11$ GHz
$30 \leq d \leq 70$, when d is in km
$19 \leq d \leq 44$, when d is in miles

$$\frac{\Delta f}{f} \leq 5\%$$

EXAMPLE 5.5   ANNUAL ONE-WAY OUTAGE OF A TYPICAL FREQUENCY DIVERSITY MICROWAVE LINK

Assume that the nondiversity path in Example 5.3 is converted to a frequency diversity one by adding a second frequency at 6.12 GHz. (a) Is the Vigants frequency diversity improvement formula applicable to this link? (b) If the answer to (a) is yes, what is the annual one-way outage?

Solution

(a) Yes, the Vigants frequency diversity improvement formula is applicable, as it has already been determined in Example 5.4 that the *CFM* is largely controlled by the *NDFM*.

(b) To determine the annual one-way outage, O_{fd}, we first we determine the frequency diversity improvement factor, I_{fd}, from Eq. (5.43a). Thus,

$$I_{fd} = 50 \times 0.12 \times \frac{1}{\left(6^2 \times 30\right)} \times 10^{39.3/10} = 47.3$$

By Eq. (5.42), the worst month frequency diversity outage probability is the worst month nondiversity outage probability divided by I_{fd}. It follows, therefore, that the annual one-way frequency diversity outage, O_{fd}, is the annual one-way nondiversity outage, O_{nd}, and 914 SES as determined in Example 5.3, divided by I_{fd}. Thus,

$$O_{fd} = 914 / 47.3 = 19 \text{ SES}$$

REFERENCES

1. Kraus, J. D., *Antennas*, McGraw-Hill, 1950.
2. Recommendations and Reports of the CCIR, XVI Plenary Assembly, 1986, Vol. V, *Propagation in Non-Ionized Media, Report 569-3*, ITU, Geneva, 1986.
3. Clifford, S. F., and Strohbehn, J. W., "The Theory of Microwave Line-of-Sight Propagation through a Turbulent Atmosphere," *IEEE Trans. on Antennas and Propagation,* 1970, 18, pp. 264–274.
4. Barnett, W. T., "Microwave Line-of-Sight Propagation with and without Frequency Diversity," *Bell System Technical Journal,* 1970, pp. 1827–1871.
5. Rogers, D. V., and Olsen, R. L., *Calculation of Radiowave Attenuation Due to Rain at Frequencies of up to 1000 GHz*, CRC Report No. 1299, Communications Research Centre, Dept. of Communications, Ottawa, Canada, Nov. 1976.
6. ITU Recommendation ITU-R P.838-1, *Specific Attenuation Model for Rain Use in Prediction Methods*, ITU, Geneva, 1999.
7. ITU Recommendation ITU-R P.676-5, *Attenuation by Atmospheric Gases*, ITU, Geneva, 2001.
8. ITU Recommendation ITU-R P.530-10, *Propagation Data and Prediction Methods Required for the Design of Terrestrial Line-of-Sight Systems*, ITU, Geneva, 2001.
9. ITU Recommendation ITU-R P.526-7, *Propagation by Diffraction*, ITU, Geneva, 2001.
10. ITU Recommendation ITU-R P.453-8, *The Radio Refractive Index: Its Formula and Refractivity Data*, ITU, Geneva, 2001.
11. ITU Recommendation ITU-R PN.837-1, *Characteristics of Precipitation for Propagation Modelling*, ITU, Geneva, 1994.
12. Rummler, W. D., Coutts, R. P., and Liniger, M., "Multipath Fading Channel Models for Microwave Digital Radio," *IEEE Communications Magazine,* Vol. 24, No. 11, November 1986, pp. 30–42.
13. Ivanek, F., editor, *Terrestrial Digital Microwave Communications*, Artech House, 1989.
14. Members of the technical staff, Bell Telephone Laboratories, *Transmission Systems for Communications*, Revised 4th ed., Bell Telephone Laboratories, Inc., 1971.
15. Siller, C. A. Jr., "Multipath Propagation: Its Associated Countermeasures in Digital Microwave Radio," *IEEE Communications Magazine,* Vol. 22, No. 2, February 1984.
16. Anderson, C. W., Barber, S. G., and Patel, R. N., "The Effect of Selective Fading on Digital Radio," *IEEE Trans. Commun.,* Vol. COM-27, Dec. 1979, pp. 1870–1876.
17. Rummler, W. D., "A New Selective Fading Model: Application to Propagation Data," *Bell System Tech. Jour.,* Vol. 58, No. 5, May/June 1979, pp. 1037–1071.
18. Bellcore Technical Reference TR-TSY-000752, *Microwave Digital Radio Systems Criteria*, October 1989.
19. Barnett, W. T., "Multipath Propagation at 4, 6, and 11 GHz," *Bell System Technical Journal,* Vol. 51, February 1972, pp. 321–361.
20. Vigants, A., "Space-Diversity Engineering," *Bell System Technical Journal,* January 1975, pp. 103–142.

21. Laine, R. U., *Digital Microwave Systems Application Seminar: Vol. 1–Digital Microwave Link Engineering*, Issue 9, Jan. 2003, Harris Corporation, Microwave Communications Division.

22. Rummler, W. D., *Modeling the Diversity Performance of Digital Radios with Maximum Power Combiners*, IEEE Conference on Communications, 1984, pp.657–660.

23. Documents CCIR Study Group Period 1982–1986, *Effects of Propagation on the Design and Operation of Line-of-Sight Radio-Relay Systems*, Draft Report, 784-1 (Annex 1. Docs, 9/2049/402).

24. Komaki, S., Tajima, K., and Okamoto, Y., "A Minimum Dispersion Combiner for High Capacity Digital Microwave Radio," *IEEE Trans. Commun.*, Vol. COM-32, No. 4, April 1984, pp. 419–428.

25. Lin, S. H., Lee, T. C., and Gardina, M. F., "Diversity Protections for Digital Radio—Summary of Ten-Year Experiments and Studies," *IEEE Communications Magazine,* Vol. 26, No. 2, February 1988, pp. 51–64.

26. Sasaki, O., and Akiyama, T., "Multipath Delay Characteristics on Line-of-Sight Microwave Radio System," *IEEE Trans. Commun.*, Vol. COM-27, No.12, December 1979, pp. 1876–1886.

CHAPTER 6

LINK PERFORMANCE OPTIMIZATION IN THE PRESENCE OF PATH ANOMALIES AND IMPLEMENTATION IMPERFECTIONS

6.1 INTRODUCTION

The probability of error analyses reviewed in Chapter 4 assumed an ideal linear path between the transmitter and receiver. However, as we have seen in Chapter 5, real-world terrestrial transmission often deviates from this ideal. Digital systems employing linear modulation methods are particularly susceptible to the in-band distortions created via multipath fading. Further, this susceptibility increases, in general, as the number of modulation states increases. Because spectrum is limited, there is constant pressure to improve spectral efficiency as a way to increase throughput. This in turn leads to systems with higher and higher numbers of modulation states. Such systems, in addition to being highly susceptible to in-band distortion, are also susceptible to their own implementation imperfections. This makes the attainment of error rate performance close to ideal difficult to achieve, even in a linear transmission environment, and results in a BER, at even high signal-to-noise ratios, that levels off at a residual value that may be higher than desirable. A number of highly effective techniques have been developed to address these susceptibilities, with some leading to BER performance better than the theoretical optimum in the absence of these techniques. As a result, by their application, the transmission of very high data rates at very high levels of spectral efficiency is possible. In this chapter, some of the more important of these techniques will be reviewed.

189

6.2 FORWARD ERROR CORRECTION CODING

6.2.1   Introduction

Error-control coding is a means of permitting the robust transmission of data by the deliberate introduction of redundancies into the data. One method of accomplishing this is to have a system that looks for errors at the receive end, and once an error is detected, makes a request to the transmitter for a repeat transmission of the bit sequence containing the error. In this method, called the *automatic request for repeat* or *ARQ* method, a return path is necessary. This method is widely used in computer networks and is starting to find application in fixed wireless systems optimized for bursty data transmission. However, when error-control coding is employed in fixed wireless systems optimized for continuous data transmission, it is normally *forward error correction (FEC)* coding. In this form of coding, not only is the receiver capable of detecting errors, it is capable of correcting errors without having to request retransmission. Use of FEC results in the reduction of the residual BER, usually by several orders of magnitude, and a reduction of the receiver 10^{-6} threshold level by about one to several dBs depending on the specific scheme employed. Fig. 6.1 shows typical error performance characteristics of an uncoded versus FEC coded system. The advantage provided by a coded system can be quantified by *coding gain*. The coding gain provided by a particular scheme is defined as the reduction in E_b / N_0 in the coded system compared to the same system but uncoded for a given BER

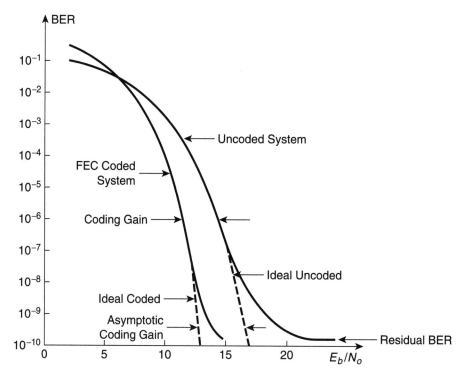

FIGURE 6.1   Typical error performance of an uncoded versus FEC coded system.

and the same data rate. Coding gain varies significantly with BER, as can be seen from Fig. 6.1, and above a very high level may even be negative. As BER decreases, the coding gain increases until it approaches a limit as BER approaches zero (zero errors). This upper limit is referred to as the *asymptotic coding gain* of the coding scheme.

FEC works by adding extra bits to the bit stream prior to modulation according to specific algorithms. These extra bits contribute no new message information. However, they allow the decoder, which follows the receiver demodulator, to detect and correct, to a finite extent, errors as a result of the transmission process. Thus, improvement in BER performance is at the expense of an increase in transmission bit rate. The simplest error detection-only method used with digital binary messages is the parity-check scheme. In the even-parity version of this scheme, the message to be transmitted is bundled into blocks of equal bits and an extra bit is added to each block so that the total number of 1s in each block is even. Thus, whenever the number of 1s in a received block is odd the receiver knows that a transmission error has occurred. Note, however, that this scheme can detect only an odd number of errors in each block. For error detection and correction, the addition of several redundant (check) bits is required. The number of redundant bits is a function of the number of number of bits in error that are required to be corrected.

FEC codes can be classified into two main categories—namely *block codes* and *convolution codes*. Further, these codes can be combined and decombined sequentially, creating *serial concatenated codes*, or combined so as to be decoded in an iterative fashion, creating *turbo codes*. The subject of FEC coding can be quite complex, is a significant field in its own right, and will not be addressed in detail in this text. Rather, what follows is a top-level overview of such coding that is typically applied in fixed wireless systems. For the reader interested in pursuing this subject in greater detail, a number of excellent texts are available.[1-3]

6.2.2 Block Codes

In systematic binary linear block encoding, the input bit stream to be encoded is segregated into sequential message blocks, each k bits in length. The encoder adds r check bits to each message block, creating a codeword of length n bits, where $n = k + r$. The code created is called an (n, k) block code, having a block length of n and a coding rate of k/n. Such a code can transmit 2^k distinct codewords, where, for each codeword, there is a specific mapping between the k message bits and the r check bits. The code is *systematic* because a part of the sequence in the codeword (in this case the first part) coincides with the k message bits. As a result, it is possible to make a clear distinction in the codeword between the message bits and the parity bits. The code is *binary* because its codewords are constructed from bits, and *linear* because each codeword can be created by a linear *modulo-2 addition* of two or more other code words. Modulo-2 addition is defined as follows:

$$0 + 0 = 0$$
$$0 + 1 = 1$$
$$1 + 0 = 1$$
$$1 + 1 = 0$$

The following simple example[2] will help explain the basic principles involved in linear binary block codes.

EXAMPLE 6.1   THE BASIC FEATURES AND FUNCTIONING OF A SIMPLE LINEAR BINARY CODE

Consider a (5, 2) block code where 3 check bits are added to a 2-bit message. There are thus four possible messages and hence four possible 5-bit encoded codewords. Table 6.1 shows the specific choice of check bits associated with the message bits. A quick check will confirm that this code is linear. For example, codeword 1 can be created by the modulo-2 addition of codewords 2, 3, and 4.

How does the decoder work? Suppose codeword 3 (10011) is transmitted, but an error occurs in the second bit so that the word 11011 is received. The decoder will recognize that the received word is not one of the four permitted codewords and thus contains an error. This being so, it compares this word with each of the permitted codewords in turn. It differs in four places from codeword 1, three places from codeword 2, one place from codeword 3, and two places from codeword 4. The decoder therefore concludes that it is codeword 3 that was transmitted, as the word received differs from it by the least number of bits. Thus, the decoder can detect and correct an error.

The number of places in which two words differ is referred to as the *Hamming distance*. Thus, the logic of the decoder in Example 6.1 is, for each received word, select the codeword closest to it in Hamming distance. The minimum Hamming distance between any pair of codewords, d_{min}, is referred to as the *minimum distance* of the code. It provides a measure of the code's minimum error-correcting capability and thus is an indication of the code's strength. In general, the error-correcting capability, t, of a code, defined as the maximum number of guaranteed correctable errors per codeword, is given by

$$t = \left\lfloor \frac{d_{min} - 1}{2} \right\rfloor \tag{6.1}$$

where $\lfloor i \rfloor$ means the largest integer not to exceed i.

An important subcategory of block codes is *cyclic block codes*. A code is defined as cyclic if any cyclic shift of any codeword is also a codeword. Thus, for example, if 101101 is a codeword, then 110110 is also a codeword, since it results from shifting the last bit to the first bit position and all other bits to the right by one position. This subcategory of codes lends itself to simple hardware encoding using linear shift registers with feedback. Further, because of their inherent algebraic format, decoding is also

TABLE 6.1   A (5, 2) Block Code

Codeword #	Codeword	
	Message Bits	*Check Bits*
1	00	000
2	01	110
3	10	011
4	11	101

(By permission from Ref. 2.)

accomplished with a simple hardware structure. The two most important types of block codes used in practical systems are the *Bose-Chaudhuri-Hocquenghem* (*BCH*) and the *Reed-Solomon* (*RS*) codes, both of which are cyclic codes.

6.2.2.1 BCH codes

BCH codes are a powerful class of cyclic codes that allow multiple error correction and provide a large choice of block lengths and code rates. The most commonly employed BCH codes are binary with the following parameters:

Codeword length: $n = 2^m - 1$ bits, where $m = 3, 4, \ldots$
Number of check bits: $n - k \leq mt$, where t is the number of correctable errors per codeword

6.2.2.2 Reed-Solomon (RS) Codes

Reed-Solomon codes are a subclass of BCH codes and are widely used in digital wireless systems. These codes are nonbinary cyclic block codes. In nonbinary block codes, the input bit stream is converted into symbols m bits long and these symbols are segregated into message blocks, each k symbols in length. The encoder then adds r check symbols, each also m bits long, creating a codeword of length n symbols. RS (n, k) codes exist for all n and k for which

$$0 < k < n < 2^m + 2 \tag{6.2}$$

However, for the most commonly used RS (n,k) code,

$$(n,k) = (2^m - 1, 2^m - 1 - 2t) \tag{6.3}$$

where t is the number of correctable symbol errors per codeword, and thus the number of parity symbols, $n - k$, equals $2t$.

Figure 6.2 shows an example of a RS codeword based on Eqs. (6.2) and (6.3) and where symbol length is 3 bits ($m = 3$) and $t = 1$ and hence $n = 7$ and $k = 5$.

For nonbinary codes, the distance between two codewords is defined as the number of places in which the symbols differ. For RS codes, the code minimum distance d_{min} is given by[4]

$$d_{min} = n - k + 1 \tag{6.4}$$

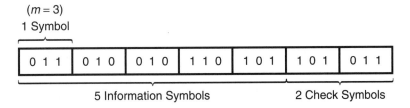

($m = 3$)
1 Symbol

| 0 1 1 | 0 1 0 | 0 1 0 | 1 1 0 | 1 0 1 | 1 0 1 | 0 1 1 |

5 Information Symbols   2 Check Symbols

FIGURE 6.2   Reed-Solomon codeword example.

Substituting d_{min} from Eq. (6.4) into Eq. (6.1), we get

$$t = \left\lfloor \frac{n-k}{2} \right\rfloor \qquad\qquad (6.5a)$$

$$= \left\lfloor \frac{r}{2} \right\rfloor \qquad\qquad (6.5b)$$

Reed-Solomon codes may be shortened by making a number of information desig-nated symbols at the encoder zero, the first i symbols say, not transmitting them, and then reinserting them at the decoder prior to decoding. A $(n - i, k - i)$ *shortened code* of the original code is created in such a fashion, with the same minimum distance, and hence the same correctable symbol error capability, of the original code.

RS codes achieve the largest possible d_{min} and, hence, the largest possible error-correcting capability of any linear code, given the same values of n and k. Further, they are especially effective in correcting long strings of errors normally referred to as burst er-rors. This is because for a given symbol being in error, the bit error correction perfor-mance of the code is independent of the number of bits in error in the symbol. Consider, for example, a RS (63, 57) code where there are 6 bits in a symbol. Such a code is capable of correcting any three symbols in a block of 63. Consider also what happens if an error burst of up to 13 contiguous bits occurs. These bits in error would be contained within three symbols regardless of when the sequence commenced and would thus all be cor-rected. Further, error bursts of between 14 and 18 contiguous bits may, depending on when the sequence commenced, be contained within three symbols. This capability pro-vides RS codes with a significant burst error-handling advantage over binary codes and helps explain their popularity. Note, however, that this advantage comes at a price. If the bits in error had been spread over more than three symbols in the 63-symbol block, then all the symbol errors and hence all the bit errors could not have corrected.

Additive Gaussian noise typically causes random errors. With random noise, the bit error probabilities are independent of each other. The longer the codeword length, the greater the probability that that number of random errors in a codeword will be the aver-age number of errors for that length and thus the more effective the code. Thus, to combat random errors, RS codes are usually have long codewords. Fig. 6.3(a)[5] shows perfor-mance curves for RS codes of rate 0.92 and codelengths n of values varying from 51 to 255. The probability of input symbol error, P_{SE}, is shown on the horizontal axis and the probability of an uncorrectable error, P_{UE} is shown on the vertical axis. It will be observed that the larger the value of n, the more effective the code. The symbol length $m = 8$, and hence the codeword length $n = 255$, is a popular choice. Fig. 6.3(b) shows this codeword as well as a sequence of five of the shortened (51, 47) codewords. Since the rates are the same, given a choice of using the (255, 235) code versus the (51, 47) option, one would clearly choose the former, assuming the added delay (latency) in transmission is tolerable.

Figure 6.4[5] shows random symbol error block performance for a RS (255, 235) code for $t = 1,3,5,8$ and 10. Note that for the $t = 10$ option, a probability of input symbol error $P_{SE} = 10^{-3}$ results in a probability of output uncorrectable error, P_{UE}, of approxi-mately 10^{-14}. This significant probability of error reduction comes at the expense of a data rate that's only a factor of 255/235, or 8.5%, larger than the uncoded rate. In fixed

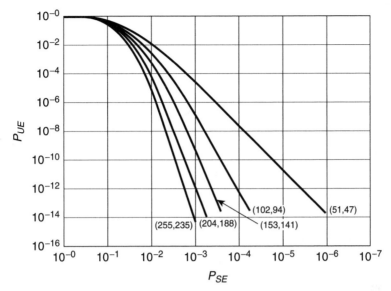

(a) Performance Curves for RS Codes of Rate 0.92

(By permission from Comtech AHA Corporation)

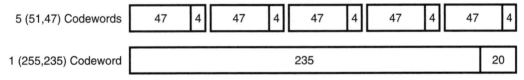

(b) Comparison of 1 (255,235) Codeword with a Sequence of 5 (51,47) Codewords

FIGURE 6.3   Comparison of RS codes of rate 0.92.

wireless systems, the FEC codes used typically add no more than about 10% to the information rate as adding more, though helpful in error correction, comes at a measurable reduction to spectral efficiency.

6.2.3 Convolution Codes

In convolution coding the message bit stream is encoded in a continuous fashion, rather than from message block to message block as in block coding. A generalized convolution encoder is shown in Fig. 6.5. It consists of a kK stage shift register and n modulo-2 adders. Register outputs (though not necessarily all) feed adders, with the choice of which register output feeds which adders determining the specifics of the encoder output. At each shift instant, k bits are shifted into the first k stages of the register and all bits already in the register are shifted k stages to the right. The outputs of the n adders are then sequentially sampled by the commutator to generate n *code bits*, or, as they are often called, *code symbols*. The sampling is repeated for each inputted k bit group. K is referred to as the

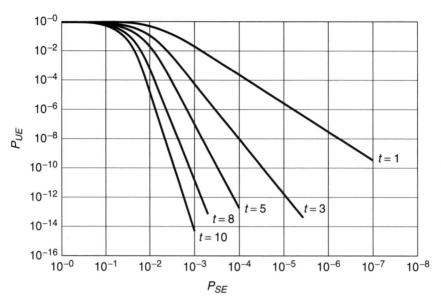

(By permission from Comtech AHA Corporation)

FIGURE 6.4   Random symbol block error performance for the RS(255, k) code for $k = 235$, $t = 10$, through $k = 253$, $t = 1$.

constraint length of the encoder and signifies the number of shifts, k bits at a time, over which a single information bit can influence the output of the encoder. Since for each input group of k message bits there are n code bits, the code rate is k/n. The choice of connections between the shift registers and the adders required to yield good distance properties is complicated and does not lend itself to straightforward solutions. However, computer searches have yielded good codes. Convolution encoders are said to possess

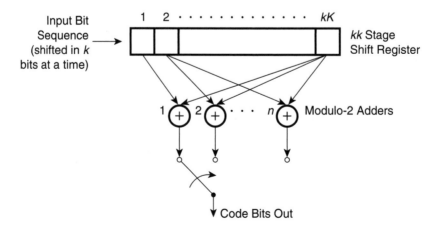

FIGURE 6.5   Generalized convolution encoder.

states. The states represent the system memory. The convention used here is that the current k bits are located in the leftmost stage, not on the wire feeding that stage. It should be noted, however, that is also common convention to show the current k bits as being on the input feed. Thus, in reviewing an encoder, it is important to establish the convention being used. With the convention being applied here, when k new bits are shifted into the register, what's then in "memory" are the additional $k(K-1)$ bits in the register. The resulting n code bits emitted from the encoder are not only a function of the k inputted bits, but also the $k(K-1)$ bits in memory. The number of states of the encoder equals the number of combinations of the bits in memory and thus equals $2^{k(K-1)}$.

To help in the understanding the functioning of convolution encoders, consider the very simple encoder shown if Fig. 6.6, where $k = 1$. This is referred to as a binary convolution encoder, the message bits being shifted into the encoder one bit at a time. As there is a two-stage shift register, the constraint length is K is equal to 2, the encoder possesses two states, and as there are two output bits for each input bit the rate is 1/2. In this encoder only one adder is used. Labeling the input bits as the i bits, the output bits out of the second register as the l bits, and those out of the adder as the m bits, we have

$$l_j = i_{j-1} \tag{6.6}$$

and $$m_j = i_j + i_{j-1} \text{ (modulo-2 addition)} \tag{6.7}$$

where $j = 1, 2, 3, \ldots$.

In Figure 6.6, for the purpose of clarity, the i, l, and m bits are shown in a way to represent the order of their movement into and out of the encoder. We assume that the register at the start contains all zeros. To demonstrate the functioning of the encoder with a real input, consider the input sequence $i_1 = 1$, $i_2 = 0$, $i_3 = 1$, $i_4 = 1$. Table 6.2 shows the register contents before and after each bit in the sequence is inputted. It also shows the resulting output sequences l_j and m_j as dictated by applying Eqs. (6.6) and (6.7) to the "new register contents." Thus, for the input bit sequence 1011, we get, at the output of the commutator, the output code bit sequence 01 11 01 10.

Convolution codes are linear. The sense in which it is linear is most easily explained by an example.

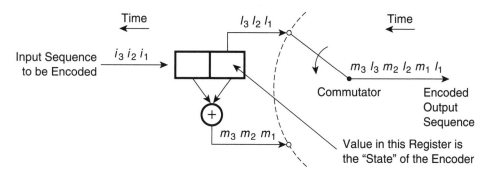

FIGURE 6.6   Simple convolution encoder.

TABLE 6.2   Relationship among Input Sequence, Register Contents, and Output Sequence for Encoder Shown in Fig. 6.6.

Input Bit Number	Input Bit Value	Old Register Contents	New Register Contents	l_j	m_j
1	1	00	10	0	1
2	0	10	01	1	1
3	1	01	10	0	1
4	1	10	11	1	0

EXAMPLE 6.2   DEMONSTRATION OF THE LINEAR FEATURE OF A CONVOLUTION CODE

Consider the output when the bit 1 moves through the encoder of Fig. 6.6 so that the input sequence is 10 (where no bit is defined we assume 0, hence the second "bit" is 0). For this sequence, the output is 01 11. Consider next the output sequence when, with the registers cleared, the bit 0 moves through the encoder, so that the input sequence is 00. For this sequence the output is 00 00. Now consider the bit 1 moving through again and then, finally, bit 1 moving through once more. Then, for input sequence 1011, the output sequence may be found by the *linear addition* of the time shifted individual output sequences as follows:

Input	Output
1	01 11
0	00 00
1	01 11
1	01 11

Linear addition
(Modulo-2 sum)   01 11 01 10 11
Output sequence   01 11 01 10

Thus the output due to 1011, by linear addition, is 01 11 01 10, the same as we obtained using Eqs. (6.6) and (6.7), demonstrating that this code is linear.

To determine the output sequence for a given input sequence it is not necessary to always do the computation as in Example 6.2. Instead, the output sequence can be determined with the aid of a *code tree*. The code tree for the encoder in Fig. 6.6 is shown in Fig. 6.7. The labels in the boxes, which are at the nodes of the tree, give the state of the encoder. The state represents the smallest amount of information that, together with the current input to the encoder, can predict the output of the encoder. For a rate $1/n$ convolution encoder, the state is represented by the contents in the rightmost $K - 1$ stages. Thus, in our example, where $K = 2$, the state is the content in the right register. The starting point of the code tree is to the left of node A and corresponds to the situation prior to the input of the first message bit. We start with both registers set to zero. The state of the encoder at node A is its state at time t_1, the time just after the first input bit is inputted to the first register and the second register is zero. If the first input bit is a zero, we follow the tree upward to node

B, reading the corresponding two output bits on the upper branch that is reached. However, if the input bit is a one, we go downward to node C, reading the output bits on the lower branch. The state at node B is the state at time t_2, the time just after the second input bit is inputted into the first register, and thus just after the first bit is inputted into the second. We proceed similarly for node C. This process continues from the new point reached (either node B or C), where if the second input bit is a zero, the tree at that node is followed upward, and if a one downward. Subsequent data bits move the position on the tree the right in the same manner. Thus, Fig. 6.7 can be used to determine the output sequence corresponding to any initial four-bit input sequence. The path for the input sequence 1011 is shown in dashed lines on the tree from which the output sequence is easily read.

A study of Fig. 6.7 shows that it repeats itself after two splits. Thus, from node E onward is the same as from node G onward, and from node D onward is the same as from node F onward. This repetition allows one to create a more compact diagram to determine the encoded sequences by merging node G with node E and node D with node F. This new diagram takes the form of a trellis, and as a result is called a *trellis diagram*. The trellis diagram equivalent of Fig. 6.7 is shown in Fig. 6.8. The nodes have been labeled to correspond to those in the tree and the output bits indicated on each branch. In this diagram, if the input bit is a zero we follow the solid line, if a one, the dashed line. Note that each horizontal line corresponds to one of the two states of the encoder and that the trellis continues ad infinitum. Once an encoding trellis is fully formed, it repeats itself in each succeeding time interval.

Decoding of convolution codes is quite complex, being accomplished by either *sequential decoding*[1,2] or *maximum likelihood sequence detection* (*MLSD*).[1,2] The latter, better known as *Viterbi decoding,* is the more elegant and the more easily accomplished and, hence, the more popular of the two. As there are no distinct codewords, the decoder decides between possible code sequences. In essence, the Viterbi algorithm, with a knowledge of the encoder trellis, starting from time t_2 and moving from one transition time to next, finds the path through this trellis that most closely resembles the received signal sequence. Since the paths through the trellis represent all possible transmit sequences, the algorithm searches every path through the trellis. However, by recognizing that at a given transition time, only one path to each given node on the trellis may be the correct one, it discards the unlikely paths at every node, keeping only the path that is closest to the received sequence. It thus restricts the number of paths under consideration to a manageable level. The path that is kept is called the *survivor path*. The algorithm used to determine the survivor path is most easily understood by way of an example.

EXAMPLE 6.3   DEMONSTRATION OF THE ALGORITHM TO DETERMINE THE SURVIVOR PATH

Let's assume that the input sequence 1011 that we considered earlier is coded and sent over a noise channel and, as a result, there is one error in the demodulated output, and hence decoder input, as shown in Table 6.3. The application of this decoding algorithm to the encoder trellis of Fig. 6.8 when the decoder input is as shown in Table 6.3 is shown in Fig. 6.9. The decoder works as follows:

At t_2 it computes the distance between the received code bits 01 and labels of the two branches leading to the two nodes shown at that time. The computed distance is shown in Fig. 6.9(a) above each label, and the total distance from t_1 is shown at the t_2 nodes.

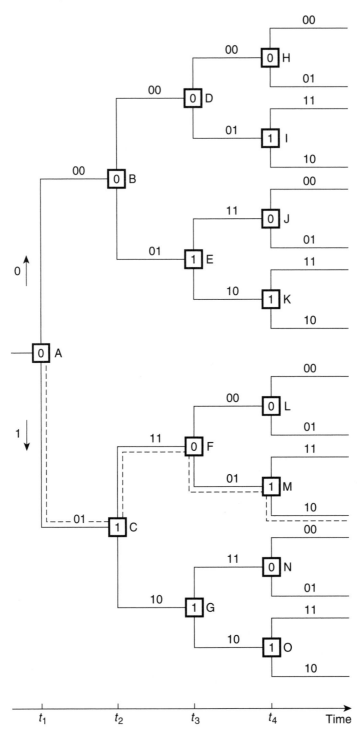

FIGURE 6.7　Code tree for encoder in Fig. 6.5.

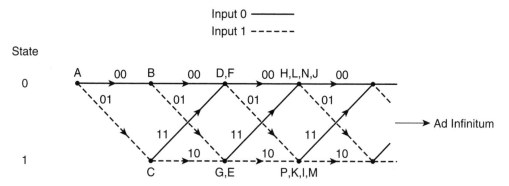

FIGURE 6.8   Trellis diagram for encoder in Fig. 6.5.

At t_3 it computes the distance between the received code 01 and the labels of the four branches leading to the two nodes at that time. It then adds these distances to the distances at the start of the branches. At each node it discards that branch with the higher total distance metric. The discarded branches are shown "crossed out" in Fig. 6.9(b).

An attempt is made to repeat the same process at time t_4, but now we have a problem. In Fig. 6.9(c), we observe that on the upper t_4 node both total metrics are identical, each being equal to two. When this happens, the rule is to arbitrarily throw out one of the two branches. Applying this rule, we throw out the upper branch. Note that in this figure we don't show previously discarded branches.

At t_5 we apply the process again. It results, as seen in Fig. 6.9(d), in the upper of the two branches to each node having the higher total metric distance and hence being discarded. Note what else happens, however. Since the branches discarded both originated at the upper t_4 node, clearly the chain of branches that feeds only t_4 can also be discarded. As a result, we end up with the surviving paths shown in Fig. 6.9(e), where from t_1 to t_4 there is only one common path. However, the final decoding rule says that if the deletion process leaves only one survivor path over some earlier data period, the input bit that would have led to that path is outputted as the decoded result. Thus for the period t_1 to t_2, the trellis is a dashed line and the output is therefore a 1. Similarly, for the period t_2 to t_3 we find that the output is 0 and for t_3 to t_4 it is 1. We cannot yet decode the fourth bit. More input bits are required to accomplish this.

In summary, we decoded correctly the first three bits encoded and sent, even though there was an error in one of the bits of the received sequence.

Example 6.3, though instructive, is based on a highly simplified encoder of constraint length K equal to 2. In practice encoders have constraint lengths of 3 or more, leading to more states and hence many more paths to process. In fact, the complexity of the Viterbi decoding increases exponentially with the code's constraint length. However, the larger the value of K, the larger the coding gains that are achievable. As a result, in most practical encoders, K is usually less than about ten in value. Though in our simple example we were able to decode the first three input bits after only four transition time periods, there is no way of knowing over how many transition time periods computations must be

TABLE 6.3   Encoder Input, Errored Demodulator Output Example

Time	t_1	t_2	t_3	t_4
Encoder input	1	0	1	1
Encoder output	01	11	01	10
Demodulator output/decoder input	01	01	01	10

↑
bit in error

made in order for a given information bit to be decoded. This being so, in order to ensure
a fixed, acceptable decoding delay, most decoders apply a *truncation window*, storing
only the portion of the survivor paths that fall within the window. Practice has shown that
with a truncation window of five or six times the constraint length, the degradation in
error correction capability is negligible.

As will be demonstrated later, the trellis diagram is a helpful tool in defining a measure of the convolution code's error-correcting ability. Unlike block codes, we cannot divide the stream of code bits into distinct codewords and must instead compare the distance
between defined code sequences. Thus, for convolution codes, it is the *free Hamming distance, d_f,* also called the *minimum Hamming distance*, that provides a measure of the code's
strength. This distance is the smallest distance between any pair of code sequences beginning and ending on the same state. Such sequences are the closest sequences that could be
confused by the decoder. It turns out that, because the code is linear, we need only consider
pairs of sequences that begin and end on the zero state (contents in the rightmost $K - 1$

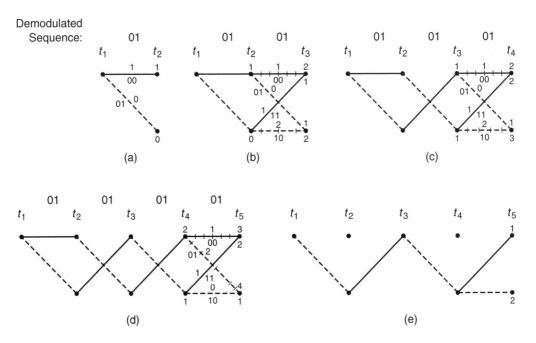

FIGURE 6.9   Viterbi decoding example based on trellis in Figure 6.8 and demodulated output as per Table 6.3.

registers at zero). One of these sequences is the all-zero sequence with a Hamming weight of zero. Hence the Hamming weight of the other (non-all-zero) sequence less the Hamming weight of the zero sequence is the Hamming weight of the non-zero sequence. Thus the free Hamming distance is the Hamming weight of the non-zero sequence. To demonstrate this concept, let's examine the trellis in Fig. 6.8 and start from the node A, which represents state 0, and return to state 0 at some later point. Clearly, in this case, the path with the smallest distance is path ACD, and the distance of this path is the distance of path AC, which is 1, plus the distance of path CD, which is 2, for a free Hamming distance, d_f, of 3.

To determine the error correcting capability of a convolution code we apply Eq. (6.1) but with d_{min} replaced by d_f, the free Hamming distance. Thus, the code characterized by the encoder shown in Fig. 6.6, where $d_f = 3$, as we have just shown above, is capable of correcting one channel error. Seems straightforward enough, except over what sequence is the code capable of correcting one error? In block codes the error-correcting capability is defined per codeword. However, with convolution codes, as there are no distinct codewords, error-correcting capability cannot be so sharply defined. All we can say is that the code can, with MLSD, correct t errors, as predicted by Eq. (6.1), within a few constraint lengths, where *few* here means 3 to 5, the exact length depending on how the errors are distributed.[1]

Convolution decoding can be accomplished with *hard decision decoding*, where the demodulator outputs either ones or zeros as the previous example demonstrates. Here, the path chosen is the one with the least Hamming distance from the received sequence. However, the decoding can be improved by employing *soft decision decoding*[1,2] algorithms. Here the demodulator output is greater than two levels, typically eight. Thus the output is still "hard" but more closely related to the analog version and thus contains more information about the original sequence. The metric used to compare paths is now the *Euclidean distance*,[1,2] the path chosen being the one with the least Euclidean distance from the received sequence. The Euclidean distance between sequences is, in effect, the root mean squared error between them. A simple example will aid in understanding the concept of Euclidean distance and its advantage over Hamming distance.

**EXAMPLE 6.4   DEMONSTRATION OF THE ADVANTAGE OF EUCLIDEAN
DISTANCE DECODING**

Consider an encoder that produces the four codewords in Table 6.1. Admittedly this is a block, not a convolution encoder, but it nonetheless serves the purpose of conveying in a straightforward fashion the basic concept and advantage of decoding using Euclidean distance versus Hamming distance as the decoding metric. Assume that codeword 2 (01110) is sent over a noisy channel in the form of the signal shown in Fig. 6.10(a), and, as a result, the demodulator analog output signal is as shown in Fig. 6.10(b). This analog output leads to a hard decision output of 10010 and, with eight-level quantization, a soft decision output as indicated on the figure.

Let's first assume that decoding is based on hard decisions. If $d_H(r,n)$ represents the Hamming distance between the hard decision outputs of the received signal and codeword n, then simple comparison yields $d_H(r,1) = 2$, $d_H(r,2) = 3$, $d_H(r,3) = 1$, and

$d_H(r,4) = 4$. Since $d_H(r,3)$ is the smallest Hamming distance, the decoder declares that the received codeword is codeword 3, i.e., 10011. It thus decodes in error.

Let's now assume that the decoder is using soft decisions, and that $d_E(r,n)$ represents the Euclidean distance between the soft decision outputs of the received signal and codeword n. We compute the squared Euclidean distance, $d_E^2(r,1)$, by determining the error between the soft decision output and codeword 1 for each of the five bits sent, squaring these errors, and then adding the squared values together. Since codeword 1 is 00000, its true output per bit would be $-1, -1, -1, -1, -1$, and thus the errors between its potential bit outputs and the received signal soft outputs are, sequentially, 1 3/7, 6/7, 6/7, 2, and 0. Thus, $d_E^2(r,1)$ is given by

$$d_E^2(r, 1) = (1\ 3/7)^2 + (6/7)^2 + (6/7)^2 + (2)^2 + (0)^2 = 7.51$$

Applying this same process to the other three codewords, we get

$$d_E^2(r, 2) = (1\ 3/7)^2 + (1\ 1/7)^2 + (1\ 1/7)^2 + (0)^2 + (0)^2 = 4.65$$

$$d_E^2(r, 3) = (4/7)^2 + (6/7)^2 + (6/7)^2 + (0)^2 + (2)^2 = 5.80$$

$$d_E^2(r, 4) = (4/7)^2 + (1\ 1/7)^2 + (1\ 1/7)^2 + (2)^2 + (2)^2 = 10.94$$

Since $d_E^2(r,2)$ is the smallest squared Euclidean distance, then $d_E(r,2)$ is the smallest Euclidean distance and hence the encoder chooses codeword 2, thus making the correct decision. In effect, the decoder's decision is based on the fact that it can't with much confidence decide what are the first three bits that have been sent, but it can with a high confidence decide that the last two bits sent are 10. Since only codeword 2 had these last two bits, it decides, correctly, that codeword 2 was sent.

For a channel with white Gaussian noise, soft decision decoding is optimum, with eight-level quantization typically increasing the coding gain by about 2 dB. It should be noted that, even though Example 6.4 was with a block code, soft decision decoding is not normally used with block codes. This is because determining the Euclidean distance of the received word from all 2^k possible code words becomes a prohibitively complex process for all but very short codes such as the one in our example.

Unlike soft decision decoding of long block codes, soft decision Viterbi decoding of convolution codes is relatively straightforward to implement, representing only a small increase in distance computation. All that is required is a change of distance metric from Hamming distance to the square of the Euclidean distance. To demonstrate this point, let's revisit the Viterbi decoding process shown in Fig. 6.9, but let us now assume soft decision decoding with eight-level quantization. For hard decision decoding, one can depict each of the four possible pairs of symbols at the demodulator output, and hence the decoder input, as one of the corners of a square on a plane, as shown in Fig. 6.11(a). For 8-level soft decision decoding, however, there are 64 possible pairs of received symbols. In this case, each of these pair can be depicted as one of the 64 cross points on an 8×8 planar grid as shown in Fig. 6.11(b). Note that for computation convenience the octal numbers 0 to 7 are used on each axis of the grid. With this numbering scheme we can use a pair of integers, each in the range 0 to 7, to designate any point in the set of 64 possible points. Let us now also assume that the first received symbols (at time t_1) of the decoding

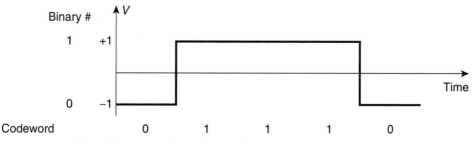

(a) Signal Sent Over Noisy Channel Representing Codeword 01110

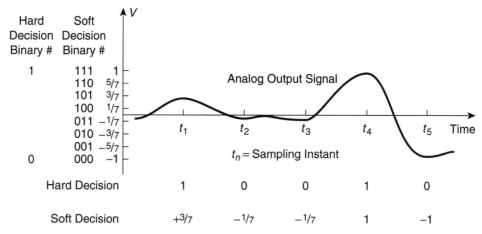

(b) Demodulated Analog Output Signal Resulting from Reception of Signal in (a)
Plus Noise and Resulting Hard and Soft Decisions

FIGURE 6.10   Example of hard and soft decision decoding.

process, instead of being the 01 hard decision shown in Fig. 6.9(a) (0, 7 on the soft decision grid), is now 2, 6. Note that to avoid confusion the octal symbol pairs are shown separated by a comma. Figure 6.11(c) shows the position of this symbol pair on the 8×8
grid. Figure 6.11(d) shows the first section of the encoder trellis originally presented in
Fig. 6.8. From Fig. 6.11(c) it is clear that the distance squared of the received symbol
pair 2, 6 from the upper branch, of branch word 0, 0, is given by $(2 - 0)^2 + (6 - 0)^2 = 40$.
It is also clear that its distance squared from the lower branch, of branch word 0, 7, is
given by $(2 - 0)^2 + (6 - 7)^2 = 5$. Thus, with soft decision decoding, the first section of
the decoding trellis is now as shown in Fig. 6.11(e). The rest of the decoding process
(i.e., discarding unlikely paths, etc.) proceeds in the same fashion as for hard decision
decoding. Clearly, the increase in computation complexity relative to hard decision decoding is minimal. However, when it comes to discarding the unlikely path of two paths
merging at a single node, soft decision decoding significantly minimizes the chance that
these paths have the same cumulative distance metric, thus minimizing the need to arbitrarily discard a path.

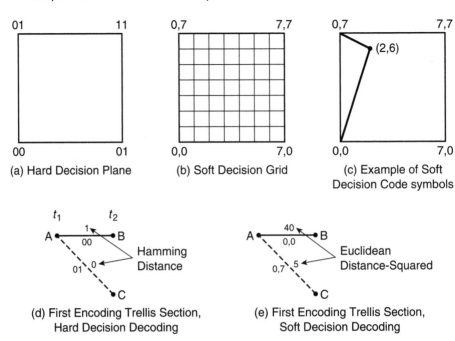

FIGURE 6.11   Hard versus soft decision Viterbi decoding.

The complexity of Viterbi decoding increases exponentially with k, the number of bits inputted at each shift instant. Thus, the simplest realizations of convolution coding where Viterbi decoding is used are for the cases where $k = 1$. The encoder shown in Fig. 6.6 is such a case. Since $k = 1$ leads to a code rate of $1/n$, then the highest rate achievable is $1/2$. It is possible, however, to retain the realization advantages $1/n$ rate coding while increasing the code rate above $1/2$ by employing a technique called *puncturing*. Codes created using this technique are called *punctured codes*.[2] Such codes are created by periodically deleting some of the bits generated by the encoder so that fewer code bits are outputted per data bit inputted. How this is done is most easily explained via an example. Consider the case where we want to create a rate $3/4$ code using the rate $1/2$ encoder shown in Fig. 6.6. We see from the figure that, for the 3-bit input sequence $i_1 i_2 i_3$, the output sequence is the 6-bit one, $l_1 m_1 l_2 m_2 l_3 m_3$. To create a rate $3/4$ code we simply delete, for every 3-bit input sequence $i_1 i_2 i_3$, two of the resulting output bits (l_1 and m_2, for example), creating a 4-bit sequence. Thus, for every three input bits we now have four output bits (i.e., a code rate of $3/4$). Note, however, that the simple "commutator" output structure shown in Fig. 6.6 must now be replaced with one that outputs the four remaining bits in a contiguous sequence in a time period equal to that of the 3-bit input sequence. In general, a rate x/y punctured convolution code can be created from a rate $1/n$ convolution code by deleting $nx - y$ code bits from every nx encoded bits. The performance of the code is dependent on which bits are deleted, so much research has been done to determine optimum codes.

A punctured rate x/y convolution code can be decoded using the same Viterbi decoder as required for the original unpunctured rate $1/n$ code. Thus, punctured code decoding with such a decoder affords the advantage of minimized decoding complexity inherent in $k = 1$ decoders. In order to use such a decoder, however, the punctured code data must be transformed back into a rate $1/n$ structure prior to being fed to the decoder. This is done by inserting dummy symbols, often referred to as erasures, in the positions where bits were deleted when puncturing the code. For soft decision decoding, where $(0, 1)$ is transmitted as $(-1$ V, $+1$ V$)$, the inserted dummy symbols are assigned the value 0 V. Since we don't know the value of the deleted bit, this inserted value is clearly our best guess. Not surprisingly, however, the rate $1/n$ code correcting capability is impaired. This is because puncturing reduces the free distance, d_f, of the code. The good news is that, for optimum codes, it reduces it to a distance that is in nearly all cases the same as that achieved using the standard nonpunctured rate approach for a code employing similar constraint length, K. Only rarely is the free distance reduced to a value less than that achieved with the standard approach. Consider, for example, the case of a rate 2/3 code constructed by puncturing a rate 1/2, $K = 2$, $f_d = 5$ code. With optimum puncturing, the free distance is reduced from 5 to 3. However, 3 is the largest free distance that can be achieved with any rate 2/3, $K = 2$ code. Since the coding gain of a code is a function of d_f, the preservation of the standard approach d_f leads to an approximate preservation of the standard approach coding gain. With punctured codes we thus get the benefit of rate $1/n$ minimized decoding complexity while in most cases sacrificing little or nothing in coding gain.

An additional advantage to the use of puncturing is that it allows the dynamic selection of codes of many different rates using just one basic encoding/decoding structure. Such dynamic selection may be appropriate in situations where the channel condition varies significantly and a variable throughput rate is acceptable. Under a good channel condition a high rate option is selected, but if the channel condition deteriorates to the point where BER approaches an unacceptable level, the coding rate is reduced, thus improving the BER to an acceptable level.

Convolution codes with Viterbi decoding do not perform well in channels subject to error bursts but tend to be superior to RS codes of similar complexity for random error correction. In general, convolution encoding is not used in the standard stand-alone sense as described above in fixed broadband PTP wireless systems. However, it is used in PMP systems and is indeed applied in PTP systems, but in such a way that it becomes a part of the modulation structure. Such a structure is referred to as *trellis coded modulation (TCM)* and will be reviewed next.

6.2.4 Trellis Coded Modulation (TCM)

When block or convolution codes as described previously are used with wireless systems to improve error rate performance, the coding is separate from the modulation. The resulting redundancy thus increases the bit rate and hence the required transmission bandwidth. In trellis coded modulation, however, modulation and convolution coding are combined in such a way that coding gain is achieved with no increase in bandwidth. To grasp the concept of redundant coding without an increase in bandwidth, consider a QPSK system. Here incoming bits are segmented into groups of two, and the four (2^2) possible states of each group are mapped into four phase states on the carrier, resulting in a theoretical spectral efficiency of 2 bits/s/Hz. Consider next an 8-PSK. Here three bits of the incoming data stream are used to create a carrier state. Thus 8 (2^3) carrier states are possible, and the spectral efficiency is 3 bits/s/Hz. Imagine now an encoder of rate 2/3 that therefore outputs three coded bits for every two input bits. If these bits were used to modulate the QPSK system, a 50% larger bandwidth would be required compared to the case where no encoder was used. But what if these bits were used to modulate the 8-PSK system? Yes, we still have a bit rate higher by 50%, but we now have a spectral efficiency that's better by 50%. The net result is that we now have a coding system and an 8-PSK modulation system in tandem that has the same spectral efficiency as the uncoded QPSK modulation system only. If you are a skeptic, you will no doubt say "this is all very well and good, but didn't we give up BER versus E_b / N_0 in going from 4- to 8-PSK, so isn't this likely to off-set any coding gain?" With the coding and modulation functions designed independently, the answer is likely to be yes. However, the good news is that by combining these functions in a highly prescribed fashion, TCM results in a net positive coding gain. The first TCM schemes were proposed by Ungerboeck and Csajka in 1976.[6] In 1982 Ungerboeck published their basic principles,[7] and in 1987 published an excellent two-part tutorial on the subject.[8,9] Ungerboeck called the proposed schemes trellis coded modulation because they can be described by trellis diagrams similar to the trellis diagrams of binary convolution codes, but with the branches labeled with modulation signals, not binary code bits. He showed that, in the presence of Gaussian noise, TCM schemes can, using simple four-state encoders, result in asymptotic coding gain of about 3 dB, and that with more complex higher state encoders the coding gain can reach 6 dB or more.

In TCM a convolution encoder is used to govern the selection of modulation signals that generate coded modulated signal sequences. Since the output of the encoder includes redundancies, these redundancies result in a signal constellation with more points than would be required without coding. Typically, as in the preceding encoder/8-PSK example, encoders are used that produce one more output bit for a given number of input bits. From a code rate perspective, their code rates k/n equal $k/(k+1)$. This leads to a doubling of the combination of bits possible and hence a doubling of the number of modulation states (constellation points) required as compared to the uncoded situation. The coding and modulated signal mapping functions are designed jointly so as to maximize directly the *free Euclidean distance*, also referred to as the *minimum Euclidean distance*, between coded modulated signal sequences (as opposed to maximizing the free Hamming distance of the convolution encoder output sequences). Like the free Hamming distance, the free Euclidean distance is the smallest Euclidean distance between any pair of code sequences beginning and ending on the same state. Also, like the free Hamming distance, we need only compare pairs of sequence that begin and end on the zero state. The maximization of free Euclidean distance is achieved by allowing only certain sequences of successive constellation points. For the same average power, the smallest Euclidean distance between signal points in the enlarged constellation is less than that in the original constellation, as shown in Fig. 6.12 for the case of QPSK enlarged to 8-PSK. However, despite this, TCM is such as to result in an increase in the free Euclidean distance between transmitted sequences compared to the original constellation, thereby improving error performance. In the receiver, the noisy signals are decoded by a soft-decision Viterbi decoder. Euclidean distances of the received signal points from the possible ideal constellation points are calculated, and the allowed sequence closest in Euclidean distance to the received sequence is found, much like the process used in decoding a conventional convolution coded signal. As a result, an estimate of the original bits encoded is made and outputted.

Figure 6.12(b) shows the Euclidean distances between a received signal point of an 8-PSK system and the ideal points.

In summary, demodulation and decoding are a joint process with the Euclidean distance as the decoding metric. As a result, we have with TCM a scheme where Euclidean distance is the key metric throughout, being used in both the encoding-modulation joint process as well as the demodulation-decoding joint process. This process of designing a TCM scheme consists of three important steps. The first is *set partitioning* of the signal constellation. Next the partitioned constellation points are mapped to code trellis transition branches via a set of heuristic rules that assure coding gain. Finally, an encoder is found whose trellis is identical to that mapped.

The first step, set partitioning,[7] takes into account the Euclidean distance of pairs of points in the constellation and provides a relationship between the Euclidean distance between points and the binary labels of those points. It accomplishes this by partitioning the original constellation of points in amoeba-like fashion into a series of subsets of diminishing size and increasing minimum distances between the signal points within each of these

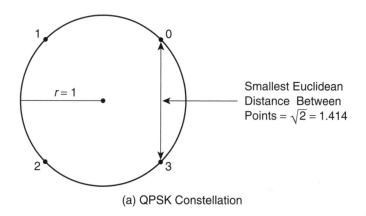

(a) QPSK Constellation

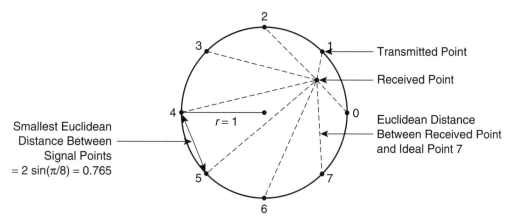

(b) Euclidean Distances Between a Received Point and Ideal 8-PSK Constellation Points

FIGURE 6.12   QPSK and 8-PSK constellations.

subsets. Binary labels can then be assigned in a logical fashion to constellation points according to the subsets in which they appear, and therefore according to their Euclidean distance. The following example demonstrates this procedure.

EXAMPLE 6.5   SET PARTITIONING OF AN 8-PSK CONSTELLATION

The set partitioning and labeling of an 8-PSK constellation is shown in Fig. 6.13. The original set A_0 is first split into sets B_0 and B_1, then B_0 is split into C_0 and C_2, and B_1 split into C_1 and C_3. Finally, each C is split into two Ds, creating eight subsets, each containing a single point. Note that, as the sets split, the minimum distance between two points in a multipoint set increases. Specifically, making the normalizing assumption that the constellation radius is unity, the minimum distances in subsets A, B, and C—namely, Δ_0, Δ_1, and Δ_2—are $2\sin(\pi/8)$, $\sqrt{2}$, and 2, respectively.

Let's designate the binary label of each constellation point as X, Y, and Z, where X is the *most significant bit (MBS)*, Y is the *center bit (CB)*, and Z is the *least significant bit (LSB)*. As indicated in Fig. 6.13, bit Z, the LSB, is assigned according to which B subset the point is in. Bit Y, the CB, is assigned according to which C subset the point is in, and bit X, the MSB, is assigned according to which D subset the point is in. As a result of this bit assignment process, the full binary label assigned to each constellation point is as indicated in the figure.

The next step in the process, the mapping of the partitioned constellation points to code trellis transition branches in order to assure coding gain, is carried out according to rules originally set out by Ungerboeck.[7] These heuristic rules are as follows:

1. All constellation points should be used with equal frequency and with a fair amount of regularity and symmetry.
2. All trellis branches starting from the same node (state) assigned constellation points from either subset B_0 or B_1, but never both simultaneously.
3. All trellis branches terminating in the same node assigned constellation points from either subset B_0 or B_1, but never both simultaneously.
4. All parallel branches are assigned constellation points from either subset C_0, C_1, C_2, or C_3, but never a mixture between them. (Parallel branches are pairs of branches that begin on the same node and end on the same node. Parallel paths are required whenever the number of states is less than the number of constellation points.)

In developing a trellis, these rules must all be applied with the exception that when there are only two states rules 2 and 3 cannot be simultaneously applied. These rules are such that only certain sequences of subsets are allowed, which lead to only certain sequences of transmitted symbols being allowed. This restriction is such as to result in a free Euclidean distance that exceeds the minimum distance between signaling points of the uncoded reference modulation. The application of these rules is best illustrated with a specific example.

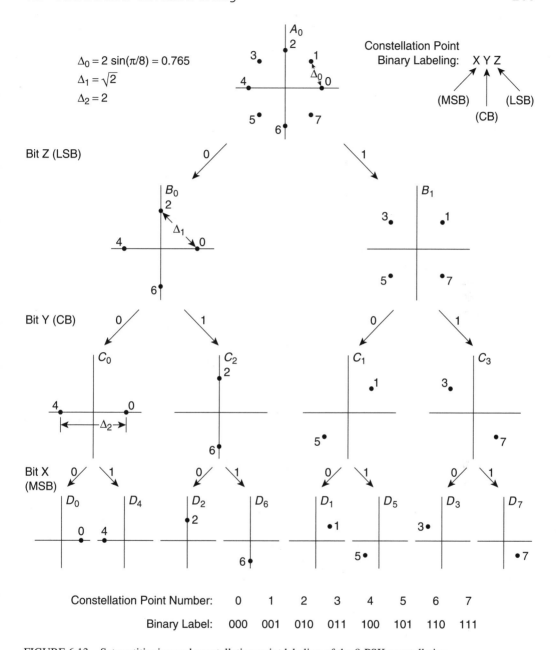

FIGURE 6.13   Set partitioning and constellation point labeling of the 8-PSK constellation.

EXAMPLE 6.6   THE MAPPING OF CONSTELLATION POINTS ON TO THE BRANCHES OF A TRELLIS

Figure 6.14 shows a possible mapping of the 8-PSK constellation points, partitioned as in Example 6.5, onto the branches of a four-state trellis with parallel paths. (Ignore the bold branches at this time. We will return to them later when we explore the coding gain associated with this structure.) The branch assignments are made by examining the subsets in Fig. 6.13 and then applying constellation points to the branches in accordance with the rules listed. An examination of Fig. 6.14 will show that all four rules listed have been observed. Note that each pair of parallel branches in the trellis corresponds to one of the four *C* subsets. Note also that, because the branch labels are constellation points, not binary codewords, the states of the trellis have been designated with a combination of constellation points. A particular state is identified with respect to the constellation points that appear on the branches leaving nodes on that state, with the upper branch labels being listed first. As the labeling (codewords or constellation points) on branches leaving a node is unique to each state, this alternate state labeling serves as a good proxy.

Now that the mapping of the partitioned constellation points to code trellis transition branches to create a code trellis has been explained and demonstrated, we turn to the final task. This task is the use of the created trellis to, in effect, reverse engineer an encoder that exhibits the properties of the trellis. As before, this step is demonstrated via example.

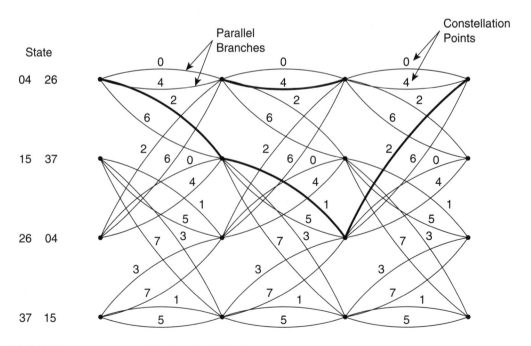

FIGURE 6.14   Four-state trellis for coded 8-PSK.

EXAMPLE 6.7   USING A CODED TRELLIS TO FIND AN ENCODER WITH AN IDENTICAL TRELLIS

In this example, we use the coded 8-PSK trellis in Fig. 6.14 to find a rate 2/3 encoder that has an identical trellis. Since the actual output of a real encoder is a binary label, the first thing we do is to redraw the trellis (one section of the fully developed trellis will do), but with the constellation points replaced with their binary labels per the relationship show in Fig. 6.13. This redrawn trellis is shown in Fig. 6.15(a). Note that we can now indicate the states of the trellis in binary form. The uppermost state has to be 00 since this is the only state that, every time a successive 0 bit is inputted to the encoder, remains unchanged and the encoding results in a 000 output in the upper branch. The designation of the rest of the states follows from the tree diagram construction rules when applied to a two-state encoder.

Close observation of the trellis in Fig. 6.15(a) reveals an important feature. This is that the MSB, and only the MSB, differs between the branches in every parallel branch pair. For example, in the uppermost parallel branch pair, labeled 000, 100, only the MSB (the first bit in each binary label) varies. Hence, of the two encoder input bits used to create three coded output bits, one can be used directly as the MSB and hence the bit that chooses between branches in a parallel pair. This choice represents the selection of one of the points in a C subset. Note, therefore, that there is no need for this bit to be fed into the physical encoder, and as a result is uncoded. The physical encoder is thus reduced from a rate 2/3 one to a rate 1/2 one with trellis labeling as shown in Fig. 6.15(b). The new composite encoding structure is now as

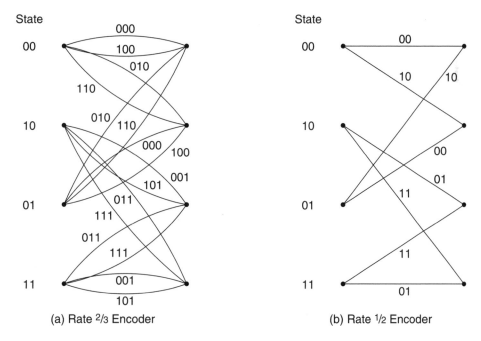

(a) Rate 2/3 Encoder   (b) Rate 1/2 Encoder

FIGURE 6.15   Binary labeled trellises for coded 8-PSK.

shown in Fig. 6.16(a). The second input data bit is fed into the rate 1/2 encoder and the two resulting output bits are used for the LSB and the CB, these bits representing the selection of one of the four C subsets.

The last step in our final task is now, therefore, that of finding a rate 1/2 encoder with a trellis structure that matches that in Fig. 6.15(b). This can be done by drawing, for each of the output bits (i.e., the CB and the LSB), a truth table, referred to as a Karnaugh map, that shows the relationship between that output bit, on one hand, and the encoder state and input data bit, on the other. Table 6.4(a) shows the map for the LSB, and Table 6.4(b) shows that for the CB. Referring to the Table 6.4(a), we see that the LSB is independent of the input data bit and thus a function only of the state of the encoder. Specifically, we see that it is equal to S1, the value in the first "state" register. We recall from Section 6.3 that this is the second register in the encoder register line up, the first register containing the input bit as a result of the convention being used in this text. Thus the output of this register can be used to give us the LSB. Referring to Table 6.4(b), we see that the CB is dependent on both the input data bit and the state of the encoder. Specifically, with a little study, we see that it is equal to the modulo-2 addition of the input data bit and S2, the value in the second "state" register. Thus, the CB can be created by carrying out this addition. The encoder resulting from the generation of the LSB and CB as suggested above is shown in Fig. 6.16(b). It is one of the encoder realizations developed by Ungerboeck.[7]

An 8-PSK coded system utilizing the encoder structure shown in Fig. 6.16(b), and soft decision Viterbi decoding should, having been designed using Ungerboeck's principles, result in a coding gain relative to an uncoded QPSK system. This will now be demonstrated and in so doing provide further insight into how TCM achieves its objectives. Since uncoded QPSK is our reference, a review of its "coding" properties is in order. Such a system can be viewed as having a fictitious one-state encoder where the output equals the input and its trellis diagram, in terms of constellation points as labeled in Fig. 6.12(a), is as

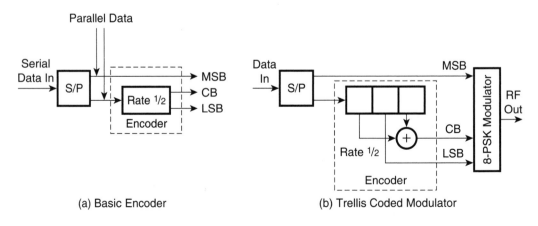

(a) Basic Encoder   (b) Trellis Coded Modulator

FIGURE 6.16   Encoder structure for coded 8-PSK.

TABLE 6.4   Karnaugh Maps for Fig. 6.15(b) Trellis Structure

State S1 S2	Input data bit 0	1
0 0	0	0
1 0	1	1
0 1	0	0
1 1	1	1

(a) LSB

State S1 S2	Input data bit 0	1
0 0	0	1
1 0	0	1
0 1	1	0
1 1	1	0

(b) CB

shown in Fig. 6.17. It is seen to consist of "parallel" transitions that do not restrict the sequences of constellation points that can be transmitted. In other words, as expected, there is no sequence coding. Hence the decoder makes independent, nearest-neighbor, final decisions for each noisy constellation point received. As shown in Fig. 6.12(a), the smallest normalized Euclidean distance between constellation points is $\sqrt{2}$. This distance is in effect the "free distance," d_{ref} say, of the uncoded QPSK reference modulation.

Having established the reference free Euclidean distance, we now seek to determine the free Euclidean distance of the coded 8-PSK system, d_{fe} say, and hopefully find that it is larger than d_{ref}, since, as we may recall, this was the goal in the first place. Free Euclidean distance of a TCM scheme is determined in much the same fashion as that used to determine the free Hamming distance of a convolution code set out in Section 6.2.3. We simply take one reference path and determine the Euclidean distances of all other paths that begin and end on the reference path. The smallest such distance is the free Euclidean distance of the scheme. We choose as our reference path the all-zero (uppermost) path in Fig. 6.14. Two comparison paths are highlighted in bold lines. One is the single parallel branch labeled 4. The other is the sequence of branches labeled 2, 1, 2. For the single parallel branch, the Euclidean distance, taken from the partitioning diagram of Fig. 6.13, is $\Delta_2 = 2$. For the 2, 1, 2 sequence, the individual Euclidean distances from the all-zero branches are

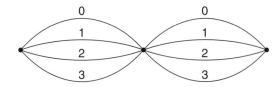

FIGURE 6.17   Uncoded QPSK's one-state trellis diagram.

Δ_1, Δ_0, and Δ_1, respectively. The Euclidean distance of a sequence is the root of the sum of its squares. Thus the Euclidean distance of the sequence 2, 1, 2, Δ_{212} say, is given by

$$\Delta_{212} = \sqrt{\Delta_1^2 + \Delta_0^2 + \Delta_1^2} = \sqrt{2 + (0.765)^2 + 2} = 2.14 \tag{6.8}$$

Thus, the distance of the single parallel branch (i.e., 2) is less than that of the 2, 1, 2 sequence. In fact, analyses of addition paths would show that the single parallel branch has the smallest distance from the all-zero path. Thus the free Euclidean distance of the coded 8-PSK scheme, d_{fe}, is 2, which, as hoped (and predicted by Unger-boeck) is greater than d_{ref}, which equals $\sqrt{2}$.

Let's stand back and review what's happened here. In the case of the 4-PSK system, d_{ref} was set by the Euclidean distance of a parallel branch. In the coded 8-PSK system it was also set by a parallel branch and hence single transitions that get no "coding" protection. However, the coding scheme was chosen such that the uncoded parallel transition had the largest possible Euclidean distance, 2, this distance being larger than that of the reference system. This is why Ungerboeck insisted via mapping rule 4 that, if there are parallel transitions, they be assigned from the same C subset, as points in such subsets have the largest separation possible. Now TCM trellises don't always have parallel transitions. If fact, if coded 8-PSK uses an eight-state trellis, then parallel transitions are un-called for. The result remains unchanged, however. With a trellis designed with Ungerboeck's rules, the free Euclidean distance, which now results from a sequence of signal state transitions, not just one transition, is always larger than that of the uncoded reference.

How much does TCM improve error performance as measured by asymptotic coding gain, G_{as} say? Not surprisingly, it's related to the relationship between d_{fe} and d_{ref}. Specifically, G_{as} is given by[1]

$$G_{as}(\text{dB}) = 20\log_{10}\left(\frac{d_{fe}}{d_{ref}}\right) \tag{6.9}$$

Thus, for the coded 8-PSK system described previously, G_{as} is given by

$$G_{as} = 20\log_{10}\left(\frac{2}{\sqrt{2}}\right) = 3 \text{ dB} \tag{6.10}$$

Though it has been convenient to demonstrate many of the features of TCM via coded 8-PSK, such a scheme is rarely used in modern broadband fixed wireless systems. TCM, when applied, is normally in conjunction with QAM schemes. A simplified 16-QAM modulator is shown in Fig. 6.18(a) and its 32-QAM TCM equivalent shown in Fig. 6.18(b). In TCM, as was seen in the 8-PSK encoding structure in Fig. 6.16, the en-coder need not code all incoming bits. Thus, in Fig. 6.18(b), only two of every four in-coming bits are coded with an 8-state, rate 2/3 encoder. Ungerboeck[9] has provided asymptotic coding gains for coded 32-QAM versus 16-QAM and coded 64-QAM versus 32-QAM. Some of these gains are shown in Table 6.5.

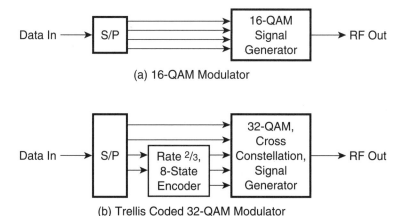

FIGURE 6.18   Simplified 16-QAM and trellis coded 32-QAM modulators.

The entire discussion of TCM has implicitly assumed the transmission of information bits in a two-dimensional (2-D) constellation, the two dimensions being the I axis and the Q axis. Naturally, the concepts covered can be applied equally to transmission in one-dimensional (1-D) constellations such those resulting from DSBSC-PAM systems (Section 4.3.1). In addition, however, it is possible to apply TCM concepts to the transmission of information bits in more than two dimensions. When this is the case the modulation is referred to as *multidimensional trellis coded modulation*.[9,10] How can this be possible? Surely in a given signaling state interval we can't simply manufacture more axes beyond the I and Q ones. Indeed we can't, but the way around this limitation is to create multidimensional signals as a sequence of constituent 1-D or 2-D signals. For example, a four-dimensional (4-D) signal can be transmitted as a sequence of two constituent 2-D signals. The most popular multidimensional TCM schemes for wireless communications are 2K-D schemes, where K is an integer of value ≥ 2 and where m bits are transmitted per constituent 2-D signal and hence mK bits per 2K-D signal. The best way to communicate the basic idea behind the construction of a 2K-D signal is via an example. For our example we assume K $= 2$ (i.e., 4-D) and as the constituent 2-D signal a 128-QAM one. A pair of uncoded 128-QAM signals transmitted

TABLE 6.5   **Asymptotic Coding Gains for Trellis Coded QAM (Values from Ref. 9.)**

Number of States	G_{as} of Coded 32-QAM versus Uncoded 16-QAM	G_{as} of Coded 64-QAM versus Uncoded 32-QAM
4	3.01	2.80
8	3.98	3.77
16	4.77	4.56
32	4.77	4.56
64	5.44	5.23
128	6.02	5.81

serially can be viewed as one comprised of $128 \times 128 = 16{,}384$ states in a 4-D space. Since a space with 16,384 states can carry 14 bits per state ($16{,}384 = 2^{14}$), a rate 13/14 TCM encoder is used. Thus, 13 information bits are converted to 14 coded bits by the encoder. The 14 coded bits are then mapped to a contiguous pair of 128-QAM signals. The result is that one redundant bit is added in every two signaling intervals.

Yes, multidimensional TCM codes can be created, but why? There are two main reasons. One is that for the same information rate spectral efficiency as 2-D TCM, they afford either a better coding gain for comparable complexity or the same coding gain for lower complexity. To get an intuitive reason as to why this is so, recall that with 2-D schemes a redundant bit is added in every signaling interval and hence the size of the TCM constellation is double that of the uncoded constellation. The cost of this doubling is an increase in BER versus E_b / N_0 of about 3 dB. Admittedly this performance loss is more than made up for by the gain of the encoding operation, but nonetheless it's real. With multidimensional TCM the cost incurred as a result of an expanded signal constellation is less than the approximate 3 dB with 2-D TCM because fewer redundant bits are added for each constituent 2-D signaling interval. With 4-D TCM the cost is reduced to about 1.5 dB and with 8-D TCM to about 0.75 dB.[9] Thus, for encoding complexity, and hence encoding operation gain, comparable to that of a 2-D scheme, the net coding gain is higher. Alternatively, for lower encoding complexity and hence lower encoding operation gain, the same net coding gain can be achieved. The second main reason for the creation of multi-dimensional TCM codes is that they afford better immunity to phase ambiguities.

6.2.5 Code Interleaving

Code interleaving[1,11] is a technique applied to mitigate the impact of error bursts. A simple example will illustrates how *block interleaving* addresses error bursts in a simple block code.

EXAMPLE 6.8 HOW BLOCK INTERLEAVING IMPACTS THE DECODING OF SIGNALS CORRUPTED WITH ERROR BURSTS

In this example, an encoder creates the original four bit codewords shown in Fig. 6.19(a). These codewords are fed to an interleaver that creates the interleaved words shown in Fig. 6.19(b). Assume these interleaved words are then transmitted over a wireless noisy channel. As a result, a burst of five contiguous errors appears on the demodulated interleaved words, as indicated in Fig.6.19(b). Note, however, that when de-interleaved as shown in Fig. 6.19(c), the five errors are now spread over the five original codewords. Assuming a decoder that can correct just one bit per codeword, it can, nonetheless, decode the original five codewords without error. Without interleaving, a burst of five contiguous errors would have caused errors in two code words, which would have been beyond the capability of the decoder to eliminate.

The number of coded blocks (codewords) involved in the interleaving process is referred to as the *interleaving depth*. Thus, our simple interleaver in Example 6.8 has an interleaving depth of five. The larger the interleaving depth, the longer the burst of contiguous errors that can be corrected but the greater the delay introduced. A block interleaver consists

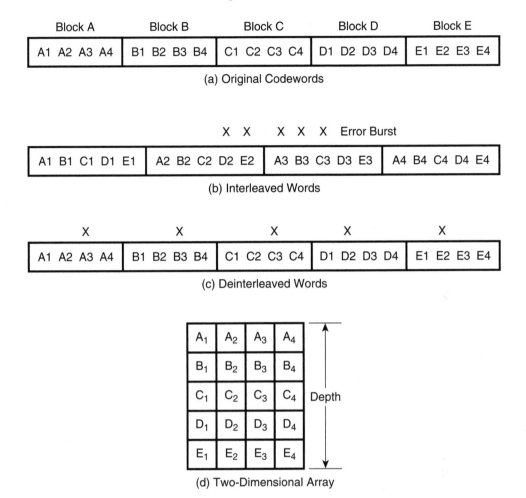

FIGURE 6.19   Block interleaving.

of a structure that supports a two-dimensional array, of width equal to the codeword length and depth equal to the interleaving depth. For the interleaver of Example 6.8, the data is fed in row by row until the array is full, as shown in Fig. 6.19(d), then read out column by column, resulting in a permutation of the order of the data. At the receive end of data transmission, the original data sequence is restored by a corresponding de-interleaver. A common designation for an interleaver is π, and that for its corresponding de-interleaver π^{-1}.

6.2.6 Concatenated Codes

6.2.6.1 Serial Concatenated Codes

A concatenated code is one that combines the performance of two or more separate coding schemes in series. A primary reason for using such a code is to achieve a low error rate with individual components of low complexity relative to that required by a single, more complex code. A very popular concatenated coding scheme is one involving two coding components, where the outer encoder (the first encoder encountered by the data stream) is an RS type, and the inner encoder is a convolution type. On the decoding side, the inner encoder (the first decoder encountered) is a Viterbi soft decision convolution decoder follower by an RS decoder. With Viterbi decoding, the output errors tend to occur in bursts. As noted in Section 6.2.2.2, RS codes are good at handling error bursts up to a certain length. To guard against being presented with bursts that exceed this length an interleaver is normally inserted between the encoders and a de-interleaver between the decoders. Fig. 6.20 shows a block diagram of a RS/convolution concatenated coding scheme with interleaving. In wireless broadband PTP systems, this scheme is not normally used. Rather, what is common is a slight variant, where the outer code continues to be an RS one but the inner code is TCM. Figure 6.21[12] shows the performance for such a scheme. The uncoded system for reference purposes is the 32-QAM one. The coded system is an Ungerboeck 16-state TCM 64-QAM inner code with a (255, 239) RS outer code. The spectral efficiency of the TCM 64-QAM system (without the impact of the RS encoder) is the same as the uncoded 32-QAM system (i.e., 5 bits/s/Hz). The RS encoder reduces this figure by a factor of 239/255. Thus the spectral efficiency of the fully coded system is 4.69 bits/s/Hz. The figure shows results for two different interleavers, one with a depth I of 2 and the other with a depth I of 8. (Also shown is the result for a code labeled TPC. This will be discussed in Section 6.2.6.2.) With the depth 8 interleaver the performance is near optimum and, as expected, better than that achieved by the depth 2 interleaver. From the figure we note that, for a BER of 10^{-4}, the coding gain of the depth 8 interleaver version relative to the uncoded 32-QAM system is 4 dB, with the value increasing rapidly as the BER decreases. Note, however, that this comparison is not entirely fair, as the 32-QAM uncoded system has a slightly better spectral efficiency.

6.2.6.2 Iterative Decoding of Concatenated Codes (Turbo Codes)

Serial concatenated codes, though highly effective, are less than optimum in a very important respect. Only their innermost decoder is normally based on soft decision decoding. The outermost decoders operate on hard decisions, and thus useful confidence information is not incorporated into their decision making. To address this weakness a process called *iterative decoding* is utilized. When the concatenated coding structure is such that

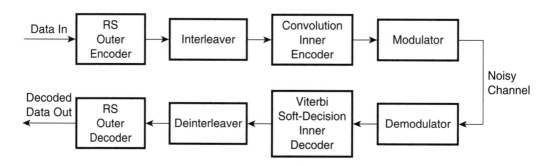

FIGURE 6.20 RS/convoution concatenated coding scheme.

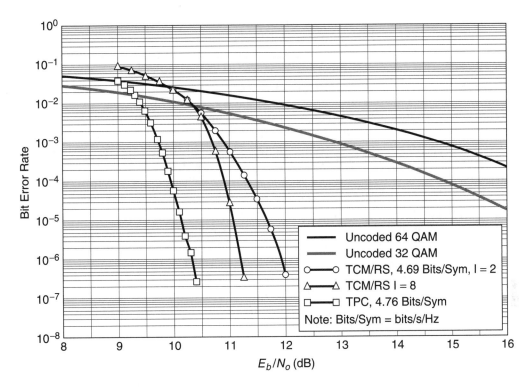

FIGURE 6.21   Performance of TCM/RS and TPC coding. (By permission from Ref. 12, © 1999 IEEE.)

iterative decoding is used, the coding is referred to as *turbo coding*. Two classes of turbo codes have received much attention: *turbo convolution codes (TCCs)*, which, as the name implies is based on concatenated convolution codes, and *Turbo product codes (TPCs)*, also called *block turbo codes*, which are based on concatenated block codes.

Turbo convolution codes made a dramatic entry into the coding arena in 1993 via a paper, now regarded as a classic, by Claude Berrou, Alain Glavieux, and Punya Thitimaj-shima.[13] In 1948 Claude Shannon published what is now called the Shannon-Hartley capacity theorem.[14] This theorem, expressed in terms of a digital transmission system perturbed by additive white Gaussian noise, shows that for a given spectral efficiency there exists a value of E_b / N_0 below which there can be no error-free communication, this value being given by

$$\frac{E_b}{N_0} \geq \frac{2^\eta - 1}{\eta} \tag{6.11}$$

where η is the spectral efficiency in bits/s/Hz.

Equation (6.11) shows that as η decreases, E_b / N_0 for error-free transmission also decreases, until it approaches a limit, which can be computed to −1.6 dB. Up to the early 1990s, the best coding schemes allowed E_b / N_0 to approach the Shannon predicted limit no closer than about 3 dB. Then along came Berrou et al., who described a concatenated convolution code using iterative decoding that was claimed, after 18 iterations, to approach within 0.7 dB of the Shannon bound for a BER of 10^{-5}. The codes they presented form the foundation of what is now called turbo convolution codes. In one version of the TCC encoding scheme, the information bits are processed through identical convolution encoders to generate parity bits and the outputs of both encoders along with the information bits are transmitted over the channel. However, the outputs of the encoders are made to be different by feeding the information bits, unmodified, to the first encoder while passing them through an interleaver prior to feeding them to the second encoder. On the receive end there are two decoders, each fed with the information bits and the appropriate parity bits. Decoding is iterative, with the decoders sharing information with each other after each iteration. After several iterations the output tends to converge on the optimum. These codes perform exceptionally well for BERs greater than about 10^{-5} However, as the BER decreases below this level their performance weakens so that an error floor is established, typically at a BER level of less than about 10^{-10}. In fact, studies have shown[15] that the code originally unveiled by Berrou et al. has an error floor of about 10^{-7}. Further, TCCs not only suffer from considerable latency (coding delay) but are difficult to implement at the high code rates required to conserve spectral efficiency in wireless systems. In fact, the typical TCC code rate is $\leq 1/2$. Thus, unless ways can be found to minimize these limitations, such codes do not appear suitable for those fixed wireless broadband systems that are required to have a residual error rate of not more than about 10^{-12}. Some researchers have studied TCM systems where, instead of standard convolution coding, turbo convolution coding is applied. Such arrangements are thus referred to as *turbo trellis coded modulation (T-TCM)*. Though interesting, T-TCM is likely to have the same error floor and latency disadvantages as TCC in series with standard modulation when considered for application to low error floor fixed wireless broadband systems. The value of Berrou et al's work, however, as it relates to such systems, is that, in demonstrating that iterative decoding was the key, it spurred researchers into looking for other types of codes utilizing this decoding technique that could approach the Shannon limit but without the limitations outlined above. From this work came turbo product codes.

Turbo product codes (block turbo codes) are based on relatively simple systematic linear block codes in a concatenated structure called *product codes* and are well described by Burr.[2,11] Product codes (also called *array codes*) are bit oriented and are constructed in multiple dimensions. The construction is most easily demonstrated by considering a two dimensional code (i.e., one where two codes are concatenated). Fig. 6.22 will aid in this discussion. Assume that the outer code C_1 is a (n_1, k_1) code and the inner code C_2 is a (n_2, k_2) code. Encoding is done by first passing the incoming bits into the outer encoder. The outputted codewords, of length n_1, are then fed into a block interleaver that has n_1 columns and k_2 rows. Once the array is full, the data is read out by columns, each column containing k_2 bits. Each grouping of k_2 bits is the encoded by the inner encoder to create a codeword of length n_2; thus the full contents of the interleaver creates n_1 codewords, each of length n_2 (i.e., a new $n_1 \times n_2$ array). The resultant final array codeword embodied in this array is a member of the $(n_1 \times n_2, k_1 \times k_2)$ product code, P say. Note that the array of product code P includes in its bottom right hand corner a $((n_1 - k_1) (n_2 - k_2))$ array of redundant bits created by the encoding of previously created redundant bits. These bits are referred to as "checks on checks." Because the individual codes are linear and the same type of coding is used for all columns and rows, it turns out that the same set of checks on checks are created if the

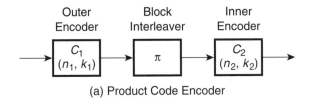

(a) Product Code Encoder

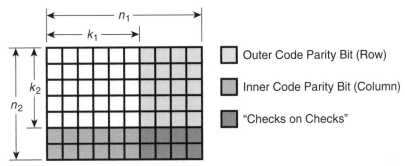

(b) Two-Dimensional Product Code Structure

FIGURE 6.22   Product code encoder and code structure.

k_1 columns are encoded first and then the resulting n_2 rows next. Hence, not only are all the columns of product code P codewords of C_2, but all the rows of product code P are codewords of C_1. Thus, in decoding, either the rows or the columns can be decoded first. Product codes are not new. What's new is decoding them with iterative decoding. It's when this is done that the code is referred to as a TPC.

In iterative decoding, the individual decoders are *soft-input, soft-output (SISO)* decoders. Such decoders are fed with soft decisions and output soft decisions. For SISO decoders, the soft information is derived from the *log-likelihood ratio*[2,11] for each data bit. Simply put, this ratio is the logarithm of the ratio of the probability that a given bit is of value 1 to the probability that it is of value 0. Its sign indicates the most likely hard decision on the bit under consideration, and its magnitude is a measure of the certainty of the likely hard decision. In essence, the log-likelihood ratio is a measure of the total information available regarding a particular bit. The log-likelihood ratio of a data bit at the output of a decoder embodies information both from the received data bit as well as the redundant bits associated with that decoder. That information from the received bit is called the *intrinsic information.* That from the redundancy is called the *extrinsic information* and is derived by subtracting the log-likelihood ratio of the received data bit from the total log-likelihood ratio of that bit at the decoder output. For the purpose of iteration, it is the extrinsic information that is passed between the decoders as the intrinsic information is already available. An example of a turbo product code iterative decoder is shown in Fig. 6.23. Note that each decoder has two inputs, one for intrinsic information, which is directly from the received data, and the other for extrinsic information, which is from the output of the other encoder. Note also the de-interleavers and re-interleavers are required to have the appropriate data fed into each decoder. At the start of the decoding process the inner decoder processes all n_1 rows of the received $n_1 \times n_2$ array and outputs n_1 rows $\times$ n_2 columns of extrinsic information to the de-interleaver. The de-interleaver outputs n_2 rows $\times$ n_1 columns of extrinsic information that is fed to the outer decoder along with a de-interleaved version of the received array. The resulting extrinsic information from the outer decoder is then re-interleaved, fed back to the inner decoder along with a de-interleaved/re-interleaved version of the original received array, and the entire process is iterated once more. As the number of iterations increase, the decoder converges on the correct answer. The composite decoder operates at a rate such that several iterations can be carried out in the time between the arrival of received $n_1 \times n_2$ data blocks. Usually a fixed number of iterations is used, varying from about 4 to about 10, depending on the type and length of code. However, with some realizations the option to have as many as about 30 iterations is provided. With a large number of iterations the decoder performance increases but at the expense of increased latency. At the end of the last iteration the combination of extrinsic information from the outer encoder and the de-interleaved received information is used to generate and output hard decision decoded data.

The advantages of TPCs over TCCs are many with regard to fixed wireless applications. First, and most important, TPCs can be designed to have no error floor. Second, they perform better in the presence of error bursts. This is because bursts that exceed the error-correction capability of a single row of the code are highly likely to be corrected when the columns containing those errors are decoded, as the errors are now spread over several columns. Third, by utilizing individual high rate block codes, TPCs can more easily be created with code rates close to unity. Fourth, because their decoding process is less complex than that for an equivalent TCC, hardware implementation is simpler and higher throughput is possible. TPCs operating at rates of hundreds of Mb/s are now available. For an example of the performance of TPCs relative the TCM/RS we return to Fig. 6.21. The TPC performance curve shown is for a $(64, 57) \times (64, 57)$ code, Gray code mapped to the 64-QAM constellation. Thus, unlike the TCM/RS approach, the coding function and

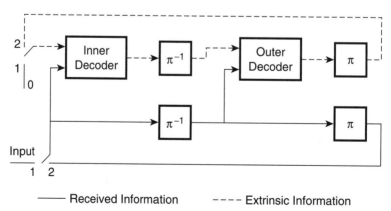

Note: The switches are in position 1 for the first iteration and in position 2 for subsequent iterations.

FIGURE 6.23   Composite turbo product code iterative decoder. *(By permission from Ref. 11.)*

the modulation function are independent, as are the demodulation and decoding functions. This arrangement results in a spectral efficiency of $6 \times (57/64) \times (57/64) = 4.76$ bits/s/Hz, slightly better than that of the TCM/RS approach, which is 4.69. From the figure we see that at a BER of 10^{-4}, the TPC approach has about a 1-dB coding gain advantage over the TCM/RS approach with interleaving of depth 8. Further, the TPC manufacturer claims that the TPC approach provides a latency (delay) that is almost five times less than that created by the TCM/RS approach with interleaving of depth 8. When the TCM/RS interleaving depth is 2, both the TPC and TCM/RS approaches have about the same latency, but now the TPC coding advantage increases to almost 1.5 dB at a BER of 10^{-4}.

6.3   ADAPTIVE EQUALIZATION

6.3.1   Introduction

As was indicated in Chapter 5, frequency selective fading distorts the amplitude and phase characteristics of the transmitted signal, resulting in intersymbol interference (ISI). Further, in all practical systems, there is always some level of ISI due to nonlinearities and the imperfect (non-Nyquist) nature of the combined transmitter and receiver filtering. ISI degrades BER performance and hence, if significant enough, some means must be employed to minimize it. One approach, which can be viewed as operating in the frequency domain and which addresses the effect of frequency selective fading, is to introduce in the receiver at IF a frequency selective network that has a transfer characteristic that is the inverse of that of the fading channel. However, as the transfer characteristic of the fading channel is time varying, it is necessary that the equalizing circuit also vary in time, continuously adapting its characteristic in response to the channel. Thus, such a circuit is known as an *adaptive equalizer*. Adaptive frequency domain equalizers were the first type employed in digital radios. It is also possible to compensate adaptively for ISI in the time domain at baseband. Such compensation has the advantage of addressing all ISI acquired prior to decoding. Time domain equalization employing linear circuitry only is also possible at IF, but as the circuitry required is more complex than that required at baseband, it is usually carried out at baseband in broadband wireless systems.

6.3.2   Frequency Domain Equalization

As indicated in the introduction, *frequency domain equalizers* (*FDEs*) operate at IF and attempt to correct any amplitude distortion in the received signal as a result of frequency selective fading. A simplified block diagram of such an equalizer, which is from Chamberlain et al.,[16] is shown in Fig. 6.24. Such equalizers operate by monitoring the IF power spectrum via narrow-band filters at two or three in-band frequencies and comparing the measured powers with each other or with their undistorted relative levels in a control logic circuit. The logic circuit computes the degree of equalization required and sends the appropriate control signals to the frequency domain filter that acts to compensate for the distortion.

The simplest form of an FDE is an amplitude slope equalizer, which corrects for amplitude slope only. Figure 6.24 demonstrates spectrum correction with such an equalizer. Amplitude slope is normally accompanied by only mild group delay distortion that contributes only minimally to system performance degradation. As a result, amplitude slope equalizers are normally designed with flat group delay characteristics.

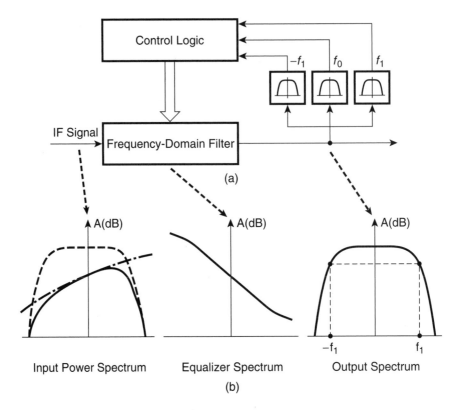

FIGURE 6.24   Frequency domain equalization. (a) Block diagram. (b) Example of spectrum correction with a slope equalizer. *(By permission from Ref. 16, ©1986 IEEE.)*

Unfortunately, amplitude slope is only one component of frequency selective fading. In such fading, the distorting amplitude function is similar to that shown in Fig. 5.27. Any segment within that function, of width equal to the receiver bandwidth, is equally likely to occur. Because of the wide spacing of notches, amplitude slope is more likely to occur than notches. However, notches are still a reality, and when they occur they result in the largest negative impact on ISI performance. As a result, amplitude slope equalizers, used on their own, are not very effective.

To overcome the limitations of slope equalizers, some early digital receivers employed notch-equalizing circuitry in series with a slope equalizer. Such circuitry usually took the form of a resonator filter, the sharpness of which was controlled to be close to the inverse of the fading induced notch. In some equalizers this equalizing notch was fixed at the channel center frequency, but in more sophisticated versions, it adaptively followed the fade notch. The group delay properties of an equalizing notch filter are such that it naturally cancels the group delay associated with minimum phase fading. However, a minimum phase fade and its amplitude equivalent nonminimum phase fade have delay shapes of opposite polarity. As a result, in notches created by nonminimum phase fades, group delay, instead of being equalized by notch-equalizing circuitry, is approximately doubled, leading to performance that is worse than that for minimum phase, being about the same as that achieved with a slope equalizer only.

Because of the limitation in amplitude equalization resulting from the approximate nature of the technique, coupled with the limitations in addressing group delay, some residual distortion is unavoidable with the use of FDEs. *Time domain equalizers* (*TDEs*), which will be discussed next, do not suffer from these limitations and so, in general, provide better performance. Nonetheless, amplitude slope FDEs are still useful, either (a) as stand-alone devices in systems that are only mildly impacted by frequency selective fading, such as systems with low complexity modulation (for example, QPSK and 8-PSK); or (b) in combination with TDEs where the result is better than using a TDE alone. Outage improvement factors of between 1.5 and 5 have been reported for channels operated with adaptive FDEs only. Further, and perhaps more important, the more sophisticated ones extend the operation of carrier and clock recovery circuits in deep fades. A baseband TDE alone is not able to do this, as it requires the carrier and clock recovery circuitry to be operational in order to perform its task.

6.3.3 Time Domain Equalization

6.3.3.1 Introduction

In frequency domain equalization the strategy is to restore as best as possible the modulated signal to its undistorted state. However, the specific cause of BER degradation due to fading is ISI, which is a time domain effect. Equalization in the time domain is therefore the approach that addresses the issue most directly and as a result is the most effective. For a raised cosine filtered received pulse to produce zero ISI it must have an amplitude versus time characteristic shown in Fig. 4.4, where the amplitude is zero at all sampling instances other than at the one coinciding with peak pulse amplitude. The effect of frequency selective fading is to introduce nonzero sampling responses to the right and left of the pulse. Those responses to the right result in ISI in future pulses, which is referred to as *postcursor* ISI (from past symbols). This ISI dominates during minimum phase fading. Those responses to the left result in ISI in past pulses, which is referred to as *precursor* ISI (from future symbols). This ISI dominates during nonminimum phase fading. The impact of frequency selective fading on ISI is well described by Siller.[17] A highly simplified explanation of how this ISI comes about is as follows: In minimum phase fades, the refracted ray is weaker than the direct ray and is delayed in time relative to it. Intuitively, then, one expects the refracted ray to distort the pulse associated with the direct ray at time periods after its peak. This is indeed the case, resulting in postcursor ISI. In nonminimum phase fading, the refracted ray, when weaker than the direct ray, arrives at the receiver ahead of the direct ray. As a result, one now intuitively expects it to distort the pulse associated with the direct ray at time periods before its peak. This also is the case, resulting now in precursor ISI. Nonminimum phase fading also occurs when the refracted ray is delayed relative to the direct ray but has an amplitude that's greater than it. In such a situation, the receiver treats the stronger refracted ray as the direct ray; hence the true direct ray arrives ahead of the direct ray, leading again to precursor ISI. From the preceding, it is clear that effective time domain equalization, of signals subject to frequency selective fading, requires elimination of postcursor and precursor ISI.

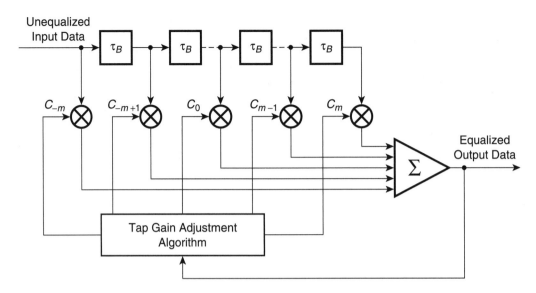

FIGURE 6.25   Linear transversal equalizer with $2m + 1$ taps.

6.3.3.2 Adaptive Baseband Equalization Fundamentals

The adaptive form of the baseband *transversal equalizer* (*TVE*), also called the *tapped delay-line equalizer*, is the most common form of *time domain equalizer* (*TDE*) used on fixed wireless systems. It can be configured in many forms, but before considering some of these, a review of its basic principles is in order. Figure 6.25 shows a block diagram of a *linear nonrecursive* or *feedforward* transversal equalizer in its simplest form. It consists of a delay line with $2m + 1$ taps, tapped at intervals of τ_B, where τ_B is the symbol interval of the data stream being equalized. The signal on each tap is weighted by a variable gain factor c, and the weighted outputs are added and sampled at the symbol interval to create the output. The equalizer works by adjusting the tap gains to appropriately weighted versions of preceding and following pulse amplitudes at the prescribed sampling instances, thus canceling interference by them. A simple example is presented to convey the fundamentals of TVE operation.

EXAMPLE 6.9   ILLUSTRATION OF THE OPERATION OF A BASEBAND TRANSVERSAL EQUALIZER

Consider the pulse shown in Fig. 6.26. As a result of transmission impairments, it has a maximum amplitude of 1 V, an amplitude of 0.1 V one sampling interval earlier than the peak, an amplitude of 0.2 V one sampling period later than the peak, but zero amplitude at all other sampling intervals. This pulse, with no nearby adjacent pulses, is applied to a three-tap TVE. When the peak of the pulse is at the first tap of the equalizer, at time t_{-1} say, the output voltage, v_{-1}, is given by

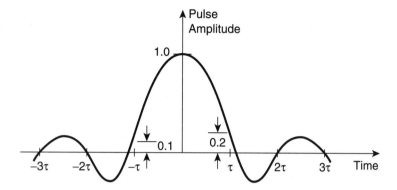

FIGURE 6.26 Equalizer input pulse waveform example.

$$v_{-1} = 1c_{-1} + 0.1c + 0c_{+1} \tag{6.12}$$

One sampling instant later, at time t_0 say, the peak of the pulse is at the second tap, and the output voltage, v_0, is given by

$$v_0 = 0.2c_{-1} + 1c_0 + 0.1c_{+1} \tag{6.13}$$

and one interval later, at time t_{+1} say, the peak of the pulse is now at the third tap, and the output voltage, v_{+1}, is given by

$$v_{+1} = 0c_{-1} + 0.2c_0 + 1c_{+1} \tag{6.14}$$

Now it's time to stand back and ask just how are we trying to accomplish our goal of zero ISI with this equalizer. A good way to visualize the function of the equalizer is to consider it to be a filter that, in conjunction with the preceding filters, restores the Nyquist criteria for zero ISI. If it succeeds in doing this, the pulse output should have a nominal value of 1 at the sampling instant when it's at its peak, and zero at all other sampling instances, so as not to interfere with other pulses. For our simple equalizer, the desired output pulse sample is v_0 taken at time t_0. Thus, for the equalizer to be 100% successful in removing ISI, v_0 would have to be of value 1 and the outputs at all other sampling instants would have to be zero, assuming that only our one pulse was transmitted. We can prevent it from not interfering with an immediately preceding and immediately following pulse by requiring that $v_{-1} = v_{+1} = 0$. When this is the case, then by Eqs. (6.12) and (6.14) we get

$$c_{-1} = -0.1c_0 \tag{6.15}$$

$$c_{+1} = -0.2c_0 \tag{6.16}$$

and, by substituting Eqs. (6.15) and (6.16) into Eq. (6.13), a peak output voltage v_0 of $0.96\,c_0$. Hence, for v_0 to be equal to 1, we have

$$c_0 = 1/0.96 \tag{6.17}$$

We have shown that with the tapped signals weighted with the factors found, a single pulse passing through the equalizer should not create ISI in adjacent pulses. To convince ourselves that there is in fact no ISI when there are adjacent pulses, consider the case where a stream of three identical pulses passes through the equalizer. Let t_0 be the time that the center pulse is at the center tap. If the pulses adjacent to it cause it no ISI, then the value of the center pulse, v_0, at time t_0 should be 1.0, just as it was when unaccompanied. To determine v_0, we need to know the sum of the three pulse amplitudes at each tap at time t_0. Labeling the first pulse in as P_1, the second as P_2, and the third as P_3, then at the first tap at time t_0, the amplitudes of P_1, P_2, and P_3, respectively, are 0, 0.2, and 1.0, for a total of 1.2. It can similarly be shown that at the second tap the combined amplitude is 1.3, and at the third 1.1. Applying these amplitudes and the determined values of c_{-1}, c_0, and c_{+1} to Eq. (6.13), we get, as required,

$$v_0 = (1.2 \times -0.1c_0) + 1.3c_0 + (1.1 \times -0.2c_0) = 0.96c_0 = 1.0 \tag{6.18}$$

Returning now to the case of one pulse passing through the equalizer, we note that because we have no more variables to control the output we cannot prevent the pulse from causing ISI at sampling instances other than immediately adjacent to itself. Thus, two sampling intervals before the peak of the pulse arrives at the center tap, at time t_{-2} say, the resulting output voltage is

$$v_{-2} = 0.1c_{-1} + 0c_0 + 0c_{+1} = -0.01c_0 \neq 0 \tag{6.19}$$

and two sampling intervals after the peak of the pulse leaves the center tap, at time t_2 say, the resulting output voltage is

$$v_{+2} = 0c_{-1} + 0c_0 + 0.2c_{+1} = -0.04c_0 \neq 0 \tag{6.20}$$

Clearly, in Example 6.9, we could have adjusted v_{-2} and v_{+2} to zero had we had two more taps, one on each side of the existing three. Obviously, then, we can further and further decrease the residual ISI by adding more and more taps. In order to force m zero outputs at the sampling instants before the desired pulse sampling instant and thus minimize precursor ISI, and m zero outputs after it, thus minimizing postcursor ISI, the TVE requires $2m+1$ tap points. Because this equalizer forces the output associated with a given pulse to zero at $2m$ sampling times, it is called a *zero-forcing* (ZF) equalizer. Determining the $2m+1$ weighted values c requires the solution to $2m + 1$ linear simultaneous equations. The more dispersed the pulse signal, the more the number of taps that are required to force the ISI to a negligible level. Fortunately, because the time dispersion of pulse signals caused by multipath fading is only over a relatively small number of symbol cycles, equalizers with a manageable number of taps, usually no more than about 20, are very effective in reducing ISI to insignificant levels.

The zero-forcing approach neglects the effect of noise. Further, such an approach is only guaranteed to minimize the worst ISI case if the peak pulse distortion before equalization is less than 100% (i.e., if the data eye pattern at the equalizer input is open). Better performance can be achieved in the presence of noise if the tap gains are adjusted to minimize the mean square error at the output of the equalizer, this error being the sum of squares of all the ISI terms, including that at the desired pulse, plus the noise power. An equalizer based on this approach is said to employ a *stochastic gradient algorithm* (SGA) and is called a *least mean square* (*LMS*) equalizer or a *minimum mean square error* (*MMSE*) equalizer. Such an equalizer maximizes the signal to distortion ratio at its output within the constraints of the number of taps and delay.

The fundamentals of adaptive equalization with a transversal equalizer are well presented by Lucky et al.[18] for the ZF version and by Sklar[1] for the LMS version. It's a relatively complex affair, and only a cursory overview will be presented here. In wireless systems, for a feedforward equalizer, it is normally achieved by continually adjusting the tap gains during data transmission via an algorithm that is based on the error, e_k say, between the signal at the output of the equalizer and the estimate of the transmitted signal made by converting the equalizer output to hard decisions. For the ZF equalizer, this error is correlated with the hard decision data stream to compute estimates of the required tap gains. For the LMS equalizer, the error is correlated with the signal values present in the equalizer delay line. A highly simplified block diagram of a three-tap ZF adaptive equalizer based on this approach is shown in Fig. 6.27, and the equivalent LMS version is shown in Fig. 6.28. The equalization process is an iterative one, which works to minimizing the magnitude of the error signal. Because the equalizer learns by employing its own decisions, the process is called *decision directed*. For this equalizer to function properly the input symbol sequence must be random; therefore scrambling is required at the transmitter. As this is always done in fixed wireless systems for other reasons, it does not present a problem.

A more complex but much more effective version of the adaptive TVE is a nonlinear one, which consists of a feedforward section plus a decision feedback section. Such an equalizer is normally referred to as a *decision feedback equalizer* (*DFE*), with the feedforward component being understood. Note that decision feedback is not to be confused with decision directed. They have different meanings as will become obvious later. A simplified block diagram of the DFE equalizer is shown in Fig. 6.29. The feedforward section is as discussed previously. The key feature of this equalizer is that, via the feedback section, it uses previously detected symbols (decisions) to eliminate the ISI on pulses that have just been equalized by the feedforward section. The ISI being removed

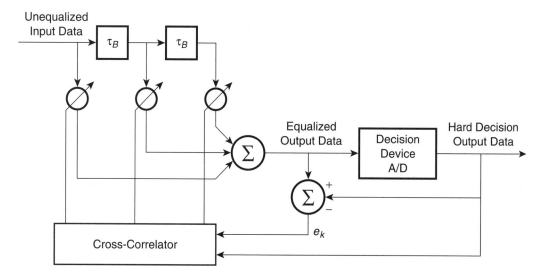

FIGURE 6.27   Three-tap decision directed ZF adaptive transversal equalizer.

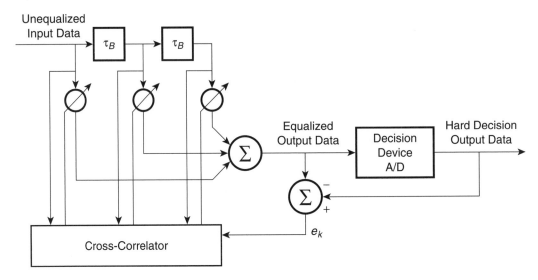

FIGURE 6.28   Three-tap decision directed LMS adaptive transversal equalizer.

was caused by the trailing edges (in time) of previous pulses. Thus, the distortion on the pulses exiting the feedforward section that was caused by previous pulses (i.e., postcursor interference) is subtracted. Since the feedback section eliminates postcursor ISI, the feedforward section is configured to attempt to compensate only for precursor ISI. Adaptive equalization can be achieved via either the ZF algorithm or the LMS algorithm, with

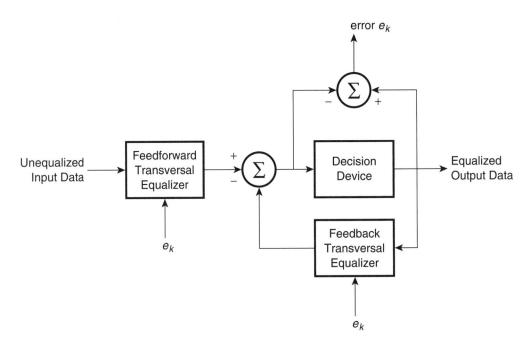

FIGURE 6.29   Decision feedback equalizer.

the error sample, e_k, as shown in Fig. 6.29, being inputted to both the feedforward and feedback cross correlators.

Recall that postcursor ISI dominates during minimum phase fading. Because postcursor ISI is canceled in the feedback section, and since this section operates on noiseless quantized levels, its output is free of channel noise. As a result, effective equalization is possible even for minimum phase fades that produce an infinite in-band notch. In nonminimum phase fading, however, where precursor ISI dominates, the behavior of the filter is primarily controlled by the feedforwad section. As this section is susceptible to noise, equalization effectiveness is degraded by large in-band notch depth.

Though effective on linear modulation systems, it should noted that the equalizers just described are ineffective in reducing ISI resulting from frequency selective fading on systems employing nonlinear modulation and demodulation processes such as multilevel FM. This is because, in such systems, the component of the demodulated signal due to the receipt of a delayed (or advanced) reflected signals (i.e., the component that results in ISI) is a nonlinear function of the component due to the direct signal (i.e., the component consisting of the desired symbols).

6.3.3.3 QAM Adaptive Baseband Equalization

In an ideal QAM demodulator, as described in Chapter 4, the in-phase data stream at the low pass filter output is independent of the modulation on the quadrature phase and vice versa. However, when the transmitted signal is subjected to frequency selective fading, this is no longer the case, with each output rail being contaminated with information from the other.[17] This effect is called *quadrature crosstalk*. To understand this effect, consider a QAM transmission system that is ideal in every respect, except that it is experiencing frequency selective fading. During this fading assume a direct ray and a refracted ray, with the refracted ray having a received amplitude b relative to the direct ray and delayed in time by τ seconds relative to the direct ray. Assume that the direct ray, $r_d(t)$ at the demodulator input, is given by

$$r_d(t) = m_i(t)\cos 2\pi f_c t + m_q(t)\sin 2\pi f_c t \qquad (6.21)$$

where $m_i(t)$ is the transmitter baseband in-phase signal modulating signal
          $m_q(t)$ is the transmitter baseband quadrature signal modulating signal
and f_c is the carrier frequency

Then the refracted ray, $r_r(t)$, at the demodulator input, is given by

$$r_r(t) = br_d(t-\tau)$$
$$= bm_i(t-\tau)\cos 2\pi f_c(t-\tau) + bm_q(t-\tau)\sin 2\pi f_c(t \qquad (6.22)$$

and the total signal at the demodulator input, $r_t(t)$, is

$$r_t(t) = r_d(t) + r_r(t) \qquad (6.23)$$

On the in-phase side of the demodulator, $r_t(t)$ is multiplied by $\cos 2\pi f_c t$, resulting in an output, $r_i(t)$, say. Substituting Eqs. (6.21) and (6.22) into Eq. (6.23) and carrying out this multiplication, using standard trigonometric identities and neglecting the components of the output centered at $2f_c$, we get

$$r_i(t) = \frac{1}{2} m_i(t) + \frac{1}{2} b \cdot m_i(t-\tau)\cos 2\pi f_c \tau - \frac{1}{2} b \cdot m_q(t-\tau)\sin 2\pi f_c \tau \qquad (6.24)$$

On the quadrature side of the demodulator, $r_i(t)$ is multiplied by $\sin 2\pi f_c t$, resulting in an output, $r_q(t)$ say. Computing $r_q(t)$ via a similar exercise to that carried out to determine $r_i(t)$ leads to

$$r_q(t) = \frac{1}{2} m_q(t) + \frac{1}{2} b \cdot m_q(t-\tau)\cos 2\pi f_c \tau + \frac{1}{2} b \cdot m_i(t-\tau)\sin 2\pi f_c \tau \qquad (6.25)$$

In Eqs. (6.24), the first term on the right side is clearly the desired output. The second term is a response due to the delay of the in-phase modulated signal and is referred to as an "in-rail" response. The third term is a "crosstalk" response, being due to the delay of the quadrature modulated signal, and is referred to as a "cross-rail response." Similar responses are noted in Eq. (6.25).

As a result of quadrature crosstalk, adaptive transversal equalization of the I and Q data streams of a QAM system cannot be done independently of each other. Figure 6.30 shows the block diagram of a DFE for a QAM system. As can be seen, each input stream drives two feedforward equalizers, one for in-rail equalization, the other for cross-rail equalization. Similarly, each output stream drives two feedback equalizers for in-rail and cross-rail equalization. One may well ask why this doubling of equalizers is necessary. On the I rail, say, wouldn't one feedforward equalizer largely remove all precursor ISI, and one feedback equalizer largely remove all postcursor ISI, regardless of its source? The an-

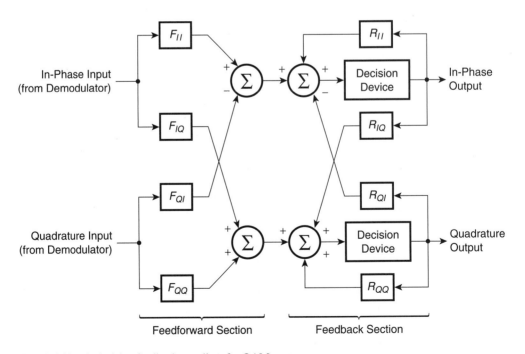

FIGURE 6.30   A decision feedback equalizer for QAM systems.

swer is unfortunately no. The equalizer algorithm only removes ISI resulting from imperfections of a linear nature created within that I channel. As a result, separate equalizers are required to eliminate cross-rail ISI. In Fig. 6.30, note that on the upper signal summers, which handle the in-phase equalization, the inputs from equalizers F_{QI} and R_{QI}, which are driven by quadrature signals, are of negative value, as dictated by the third right-hand term of Eq. (6.24). Similarly, on the lower signal summers, which handle the quadrature equalization, the inputs from equalizers F_{IQ} and R_{IQ}, which are driven by in-phase signals, are of positive value, as dictated by the third right-hand term of Eq. (6.25).

6.3.3.4 Initialization Methods

In the equalization structures just described it has been tacitly assumed that at the start of operation the signal at the decision device output is a fairly accurate replica of the original transmitted signal. This condition allows the generation of a mostly valid error signal and hence initial adjustments of the equalizer tap coefficients. This in turn leads to convergence on an ISI free output. In practice, as a result of signal distortion, this is not always the case. It is necessary, therefore, to take action to assure that this condition is quickly met. To accomplish this, one of the two following broad methods is normally applied. In the first method, prior to the transmission of information data, a finite *training sequence*, known to the receiver, is transmitted. In the receiver, a synchronized version of this sequence is generated and used in place of the decision device output signal for error generation. At the end of the sequence the equalizer switches to the decision device output for error generation. The sequence is made long enough so that the equalizer tap values are adjusted to the point that when regular information data transmission commences the equalizer is fully operative. Such a scheme, though reliable, is not normally used with fixed broadband systems. Because these systems are subject to fading that can interrupt service, then, after every interruption, a message would have to be sent to the transmitter to inject a training sequence. This clearly increases the complexity of the system, is wasteful of throughput capacity, and, in the case of point-to-multipoint systems, is a real stop and start nightmare.

A second method of initial equalizer alignment, referred to as *blind equalization*,[19] is therefore normally used with fixed wireless systems. With this method no training sequence is transmitted. Instead, during the initial acquisition phase, the equalizer adjusts tap values in response to sample statistics instead of in response to sample decisions, as is the case in the normal decision directed phase. Blind equalization typically employs one of three major algorithms: the *reduced constellation algorithm (RCA)*,[20] the *constant-modulus algorithm (CMA)*,[21] and the *multi-modulus algorithm (MMA)*.[22] A detailed description of these algorithms and equalizers employing them is beyond the scope of this text. However, as the constant-modulus algorithm is probably the most widely used for initial equalization of QAM signals, a review of its salient features is in order.

Consider a QAM equalizer where the output of the in-phase adaptive filter for the nth symbol is I_n and the output of the quadrature adaptive filter of the nth symbol is Q_n. Then the equalizer outputs can be represented as one complex output, Z_n, where $Z_n = I_n + jQ_n$. The CMA equalizes the QAM signal constellation by finding the set of tap values that minimizes the cost function $D^{(p)}$, where $D^{(p)}$ is the expectation (average value) of the square of the difference between the modulus of the equalizer complex output, $|Z_n|$, raised to the power p, and a positive real constant, R_p, that is a function of the constellation structure. Thus

$$D^{(p)} = E(|Z_n|^p - R_p)^2$$

(6.26)

For the case of $p = 2$ the algorithm is well documented and relatively easy to implement. As can be seen from Eq. (6.26), the minimization $D^{(p)}$ is in effect an effort to fit the complex equalizer output to a ring of constant modulus (i.e., magnitude), hence the nomenclature of the algorithm. The function $D^{(p)}$ characterizes the amount of ISI at the equalizer output[21]; thus its minimization leads to the minimization of ISI. For signals possessing a constant modulus (e.g, 8-PSK) excessive dispersive fading, noise, or interference will significantly vary the modulus of the modulated states, which show up at the equalizer output as a significant variation of Z_n. Thus, it seems intuitively that minimization of $D^{(p)}$, which forces Z_n back to its original modulus, leads to the reduction of ISI and hence to partial equalization. What is remarkable about the CMA-based equalizer, and certainly not intuitive, however, is that it also partially equalizes signals not possessing a constant modulus, such as QAM signals.

A key feature of the CMA approach is that, again as can be seen from Eq. (2.26), $D^{(p)}$ is independent of the carrier phase. Thus its operation is phase invariant and can operate independently of the receiver carrier-tracking loop. However, before switching from the blind mode to the decision directed mode, separate constellation phase recovery is necessary in order to properly align the constellation.

6.4 CROSS-POLARIZATION INTERFERENCE CANCELLATION (XPIC)

As was mentioned in Chapter 5, co-channel cross-polarization transmission is a method of doubling channel transmission capacity. However, its effectiveness is dependent on maintaining a minimum XPD for the given modulation method. This is accomplished by the use of an adaptive *cross-polarization interference canceler* (*XPIC*). The principle of operation of such an equalizer may be explained with the aid of Fig. 6.31. T_v and T_H are the transmitted signals, R_V and R_H are received signals, and O_v and O_H the output signals of

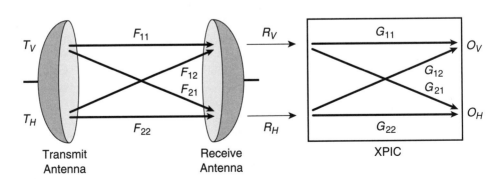

FIGURE 6.31   Cross-polarization interference and cancellation model.

the canceler. F_{ij} and G_{ij} are the transfer functions of the antenna system/space and the canceler, respectively. From the figure, it follows that

$$R_V(f) = F_{11}T_V(f) + F_{12}T_H(f) \tag{6.27}$$

$$R_H(f) = F_{22}T_H(f) + F_{21}T_V(f) \tag{6.28}$$

and

$$O_V(f) = G_{11}R_V(f) + G_{12}R_H(f) \tag{6.29}$$

$$O_H(f) = G_{22}R_H(f) + G_{21}R_V(f) \tag{6.30}$$

Substituting Eqs. (6.27) and (6.28) into Eqs. (6.29) and (6.30), we get

$$O_V(f) = (G_{11}F_{11} + G_{12}F_{21})T_V(f) + (G_{11}F_{12} + G_{12}F_{22})T_H(f) \tag{6.31}$$

$$O_H(f) = (G_{21}F_{11} + G_{22}F_{21})T_V(f) + (G_{22}F_{22} + G_{21}F_{12})T_H(f) \tag{6.32}$$

The adaptive canceler works by choosing values of G_{ij} so that O_V and O_H are, at all times, good approximations of the transmitted signals T_V and T_H, respectively. Clearly, to accomplish this, it must force the factors $(G_{11}F_{12} + G_{12}F_{22})$ and $(G_{21}F_{11} + G_{22}F_{21})$ toward zero. Practical XPICs can be realized as transversal filters and can be implemented at RF, IF, or baseband. The circuitry of such XPICs closely resembles that of adaptive transversal equalizers used for ISI reduction on quadrature channels. As with ISI equalizers, the baseband version is the easiest to realize and hence the most common.

6.5   AUTOMATIC TRANSMITTER POWER CONTROL

Automatic transmitter power control (ATPC) is a design feature of a digital wireless link whereby the transmitter output power is automatically adjusted as a function of the associated far-end receiver signal level. Under normal (i.e., nonfaded) path conditions the transmitter operates at its *coordinated transmit power*. At this power level, the system operates comfortably within its desired nonfaded BER requirements. However, during severe fading conditions, transmitter output power is increased, typically by up about 10 dB, to its *maximum transmit power* so as to increase the link fade margin and so assure the desired availability.

The primary value of ATPC is that, when the transmitter output is at its coordinated power, interference is reduced to microwave systems nearby. Further, although its interference property increases when, as a result of fading on its own path, it operates at maximum power, this only occurs for a very small percentage of the time. As a result, it's highly unlikely that the other path would be experiencing a fade simultaneously and thus be susceptible to interference. As a result, many licensing authorities allow links to be installed with the coordinated transmit power being used to coordinate adjacent system interference analysis, easing the coordination process.

Other potential advantages of ATPC are as follows:

- Elimination of receiver front-end overloading in higher frequency links designed with rain fade margins so high that their nominal (unfaded) received signal levels, when the transmitters operate at full power, overload the front ends.
- Elimination of receiver front-end overloading in shorter, lower-frequency links operating in frequency congested areas. Such links require large antennas to provide the necessary frequency discrimination. Thus, when the transmitters operate at full power, the nominal received signal levels overload the front ends.
- Reduced DC power consumption and improved reliability of the transmitter power amplifier. The former will be the case if the amplifier DC biasing is varied with the RF output level so as to result in minimum DC consumption, consistent with the desired linearity. The latter results from the former as lower DC consumption results in lower RF power transistor junction temperatures and hence higher transistor reliability.
- Improved system gain. This is achieved by setting the transmitter maximum output level higher than it would be without ATPC. At this higher level, amplifier distortion results in a dribble error rate that's higher than the desired nonfaded BER requirement. However, these errors only occur during deep fade conditions and are insignificant when compared to those being created by the fade. Typically the increased output power, and hence improved system gain afforded by this approach, is on the order of 1 dB or less for systems with FEC and adaptive equalization.

Systems that are licensed based on the coordinated transmitter power must take special precautions to guard against failure that results in the transmitter operating at full power continuously. Typically, full-power operation is monitored and if it exceeds a predetermined time, normally no more than minutes, an alarm condition is declared and control circuitry attempts to return the power to its coordinated level.

REFERENCES

1. Sklar, B., *Digital Communications: Fundamentals and Applications*, Prentice Hall PTR, 2001.
2. Burr, A., *Modulation and Coding for Wireless Communications*, Pearson Education, 2001.
3. Wilson, S. G., *Digital Modulation and Coding*, Prentice Hall, 1996.
4. Gallager, R.G., *Information Theory and Reliable Communication*, John Wiley and Sons, 1968.
5. AHA Application Note, *Primer: Reed-Solomon Error Correction Codes (FEC)*, Advanced Hardware Architectures, Inc., Pullman, Washington, 1995.
6. Ungerboeck, G., and Csajka, I., *On Improving Data-Link Performance by Increasing the Channel Alphabet and Introducing Sequence Coding*, 1976 Int. Symp. Inform. Theory, Ronneby, Sweden, June 1976.
7. Ungerboeck, G., "Channel Coding with Multilevel/Phase Signals," *IEEE Trans. Information Theory*, Vol. IT-28, January 1982, pp. 55–67.

8. Ungerboeck, G., "Trellis-Coded Modulation with Redundant Signal Sets, Part I: Introduction," *IEEE Communications Magazine*, Vol. 25, No. 2., February 1987, pp. 5–11.

9. Ungerboeck, G., "Trellis-Coded Modulation with Redundant Signal Sets, Part II: State of the Art," *IEEE Communications Magazine*, Vol. 25, No. 2., February 1987, pp. 12–21.

10. Wei, L., "Trellis-Coded Modulation with Multidimensional Constellations," *IEEE Trans. Inform. Theory*, Vol. IT-33, No. 4, July 1987, pp. 483–501.

11. Burr, A., "Turbo-Codes: The Ultimate Error Control Codes?", *Electronics and Communications Engineering Journal*, August 2001, pp. 155–165.

12. Hewitt, E, *Turbo Product Codes for LMDS*, 1999 IEEE Radio and Wireless Conference (RAWCON), Denver, Colorado, Aug. 1999.

13. Berrou, C., Glavieux, A., and Thitimajshima, P., "Near Shannon Limit Error-Correcting Coding and Decoding: Turbo-Codes," *IEEE Proceedings of International Conference on Communications*, Geneva, Switzerland, 1993, pp. 1064–1070.

14. Shannon, C. E., "A Mathematical Theory of Communications," *Bell Sys. Tech. J.*, Vol. 27, July and October 1948, pp. 379–473, 623–656.

15. Valenti, M., "Inserting Turbo Code Technology into the DVB Satellite Broadcasting System," *Proc. IEEE Military Commun. Conf. (MILCOM)*, Los Angeles, Calif., Oct. 2000, pp. 650–654.

16. Chamberlain, J. K., Clayton, F. M., Sari, H., and Vandamme, P., "Receiver Techniques for Microwave Digital Radio," *IEEE Communications Magazine*, Vol. 24, No. 11, Nov. 1986, pp. 43–54.

17. Siller, C. A., Jr., "Multipath Propagation: Its Associated Countermeasures in Digital Microwave Radio," *IEEE Communications Magazine*, Vol. 22, No. 2, February 1984.

18. Lucky, R.W., Salz, J., and Weldon, E. J., *Principles of Data Communication*, McGraw-Hill, 1968.

19. Garth, L., Yang, J. J., and Werner, J., *An Introduction to Blind Equalization*, TD-7 of ETSI/STS TM6, Madrid, Spain, January 1998.

20. Godard, D. N., and Thirion, P. E., *Method and Device for Training an Adaptive Equalizer by Means of an Unknown Data Signal in a QAM Transmission System*, U.S. Pat. 4 227 152, Oct. 7, 1980.

21. Godard, D. N., "Self-Recovering Equalization and Carrier Tracking in Two-Dimensional Data Communication Systems," *IEEE Trans. Commun.*, Vol. 28, No. 11, November 1980.

22. Yang, J., Werner, J. J., and Dumont, G. A., *The Multimodulus Blind Equalization Algorithm*, Proc. Thirteenth International Conf. On Digital Signal Processing, Santorini, Greece, July 1997.

CHAPTER 7

POINT-TO-POINT SYSTEMS IN LICENSED FREQUENCY BANDS

7.1 INTRODUCTION

Previous chapters have dealt largely with theoretical aspects of fixed point-to-point wireless systems. In this chapter we turn our attention to practical aspects of such systems operating in *licensed bands*. The term *licensed band* is used to signify an operating block of frequencies where the user has to be granted a license by the appropriate government authority to operate on specific frequencies within it. Further, such frequencies have to be coordinated with others in the nearby environment to assure acceptable levels of interference. Point-to-point licensed systems form the oldest and largest category of fixed wireless systems. Broadband versions thereof operate primarily in frequencies from 2 to 60 GHz with capacities from a few Mb/s to hundreds of Mb/s. In the sections that follow, the regulatory environment for such systems is first reviewed. This is followed by an overview of common single- and multichannel configurations, and, finally, by an examination of key aspects of a number of commercial systems.

7.2 SPECTRUM MANAGEMENT AUTHORITIES

The electromagnetic spectrum is a finite natural resource, and as such its use is regulated by government agencies worldwide. Because frequencies operating near borders know no political boundaries, governments tend to try to manage frequencies in harmony with neighboring countries. The global organization that manages the big picture of frequency spectrum allocation and utilization on behalf of governments worldwide is the *International Telecommunications Union,* or *ITU*. The ITU is a specialized agency of the United Nations and oversees the creation and maintenance of myriad standards and recommendations relative to telecommunications networks. Wireless activities within the ITU are handled by its Radiocommunication Sector (*ITU-R*), which, in addition to playing a vital role in the management of the radio frequency spectrum, is charged with defining the technical characteristics, and creating the operational procedures, for a large range of wireless services. The set of rules governing the use of the radio spectrum that is developed and adopted by the ITU-R is called the *Radio Regulations*. In spectrum management, the portion of the radio frequency spectrum suitable for communications is divided into a large number of frequency bands. Bands are allocated to services on an exclusive or shared basis, and their sizes vary depending on the service or services they are allocated to. The ITU-R maintains a full list of frequency bands allocated in the different regions of the world in its *Table of Frequency Allocations*, which is part of its Radio Regulations. Specific channel arrangements within bands and radio equipment requirements are published in the ITU-R F, Fixed Service Recommendations. Though many countries contribute to the ITU standardization- and recommendation-making process, significant input comes from Europe, North America, and Japan.

Input to the ITU-R from most European countries is via the *European Telecommunications Standards Institute* (*ETSI*), a nonprofit organization whose mission is to produce telecommunications standards for use throughout Europe and beyond. ETSI was created by the *European Conference of Postal and Telecommunications Administrations* (*CEPT*) in 1988. Since some standards now administered by ETSI were created under CEPT (for example, the plesiochronous digital hierarchy), such standards are referred to in some instances as CEPT standards, and in other instances as ETSI standards. CEPT recommendations regarding frequency spectrum use are prepared by *the Electronics Communications Committee* (*ECC*). However, prior to October 2001, this work was performed by the *European Radiocommunications Committee* (*ERC*). Thus, several current recommendations are ERC recommendations.

In the United States, regulatory responsibility for the radio spectrum is handled by the *Federal Communications Commission* (*FCC*) and the *National Telecommunications and Information Administration* (*NTIA*). The FCC manages the spectrum for private sector and state and local government use, while the NTIA manages the spectrum for use by the federal government. The FCC and NTIA are responsible for preparing U.S. input to the ITU-R, this input is coordinated with and delivered by the Department of State. FCC rules and regulations are codified in Title 47 (Telecommunications) of the *Code of Federal Regulations*. Fixed microwave services in licensed bands are addressed under Part 101 of Title 47. Though the FCC issues spectrum regulations for non-federal-government users, its current practice is to have these users develop a consensus as to their requirements in a given area of interest, and then, to a large extent, incorporate this consensus into its regulations covering this area. A key facilitating organization in the development of such consensus is the *American National Standards Institute* (*ANSI*). ANSI is a private nonprofit membership organization that publishes standards and works closely with the FCC; some of its standards are incorporated into FCC rule making. ANSI does not itself develop standards but rather facilitates development by establishing consensus among qualified accredited groups. A prominent ANSI accredited group regarding telecommunication matters including spectrum regulation is the *Telecommunications Industry Association* (*TIA*). Formed in 1988, the TIA is a leading nonprofit trade association in the communications and information technology industry and represents the communications sector of the *Electronics Industry Association* (*EIA*).

In Canada, frequency use regulation is administered by the *Industry Canada* department of the federal government. Technical requirements for fixed wireless system use in licensed bands, including spectrum utilization issues, are contained in the department's *Standard Radio System Plans* (*SRSPs*).

7.3 FREQUENCY BAND CHANNEL ARRANGEMENTS AND THEIR MULTIHOP APPLICATION

Channels within a frequency band are arranged in a number of ways depending on the service(s) that they support.[1] The arrangement, in general, establishes the center frequency and bandwidth of each RF channel, the polarization of the signal, and the preferred growth pattern. In the most common arrangements for point-to-point broadband communications, a band of total width W is divided, as shown in Fig. 7.1, into three segments: two subbands, one for each direction of transmission, and a guard band, located between the subbands. Each subband is in turn divided into N regularly spaced channels. Different arrangements are created by the way in which the channels are arranged in the subbands. The most common arrangement, developed originally to accommodate analog systems, though now shared with digital ones, is shown in Fig. 7.2. Because of its origin, it is referred to here as the *analog channel plan*. In such a plan, channels lie side by side with no overlap and with alternate channels utilizing alternate polarization. This arrangement facilitates excellent adjacent channel discrimination via filter discrimination as well as cross-polarization discrimination. Transmit/receive frequency pairing is such as to maintain constant separation. An alternate arrangement is the *co-channel plan* shown in Fig. 7.3. Here, each channel can be used twice, by employing simultaneously transmission in both the horizontal and vertical polarization planes. The obvious advantage of this scheme is that it doubles the capacity of the band. However, because of the possibility of cross-polarization interference, it is only well suited to digital systems, preferably with XPIC equipped receivers (Section 6.4). Another arrangement, again more suitable to digital transmission, is the *interleaved channel plan* shown in Fig. 7.4. It is similar to the analog plan, but with alternate channel partial overlap, resulting in three channels in the space occupied by two equally wide channels in the analog plan. The capacity increase of this plan relative to the analog plan is achieved at the expense of adjacent channel discrimination, which is reduced as a result of diminished filter discrimination.

Frequency band channel arrangements are accompanied by requirements that equipment must meet in order to operate in the band. Two key such requirements are channel emission limits and minimum spectral efficiency.

Channel emission limits are important in that they put limits on the interfering capability of the signal. Meeting specified limits normally requires stringent filtering and spurious emission control. Channel limits can be thought of as a mask within which the transmitted spectrum must remain. In the United States, the emission limitations for systems

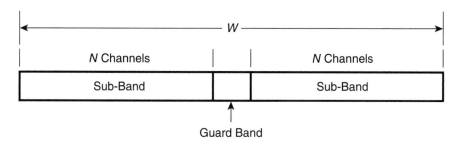

FIGURE 7.1   Standard frequency band structure.

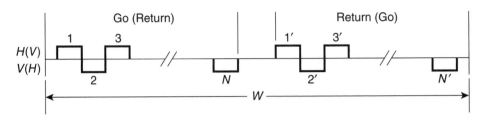

FIGURE 7.2   Analog channel plan.

operating under Part 101 of Title 47 are contained in Section 101.111. As an example of these limitations, for most digital transmission in authorized frequency bands below 15 GHz, the mean output power in any 4-kHz band of emission must be attenuated below the mean total wideband output power of the transmitter in accordance with the following mask:

$$A(f) = 0 \qquad\qquad \text{for } 0 < P < 50\% \qquad\qquad (7.1a)$$

$$= 35 + 0.8(P\text{--}50) + 10(\log_{10} B), \text{ or } 50, \text{ whichever is greater, with}$$
attenuation greater than 80 not required

$$\text{for } 50\% < P < 250\% \qquad\qquad (7.1b)$$

$$= 43 + 10\log_{10}(\text{mean output power in Watt}), \text{ or } 80, \text{ whichever is the}$$
lesser $\qquad\qquad$ for $P > 250\%$ $\qquad\qquad (7.1c)$

where $A(f)$ = attenuation (dB) below the mean output power level
$\qquad\quad f$ = frequency (MHz) at which the attenuation specification is being evaluated
$\qquad\quad B$ = authorized bandwidth (MHz)
$\qquad\quad P = (|f - f_c| / B) \times 100\%$
$\qquad\quad f_c$ = modulated signal center frequency (MHz)

For a 30-MHz bandwidth channel, Eq. (7.1) results in the mask shown in Fig. 7.5. Note that the mask does in fact permit transmission outside the assigned channel, albeit at greatly attenuated levels. As can be seen, this out-of-channel emission is largest in the adjacent channels, and therefore they are the ones most at risk of interference. In Canada emission limits are stated in the relevant SRSP and the ITU states emission limits in the relevant ITU-R F Recommendation.

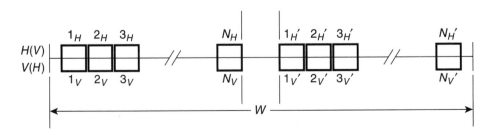

FIGURE 7.3   Co-channel plan.

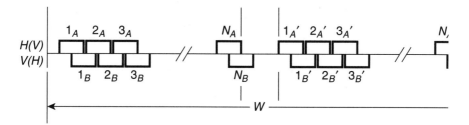

FIGURE 7.4 Interleaved channel plan.

Minimum spectral efficiency (spectral efficiency being information bits per second per Hertz of transmission bandwidth [bits/s/Hz]) is specified in order to guarantee efficient utilization of the spectrum. Required spectral efficiency varies depending on the licensing authority and the frequency band. Normally, the value specified varies from a low

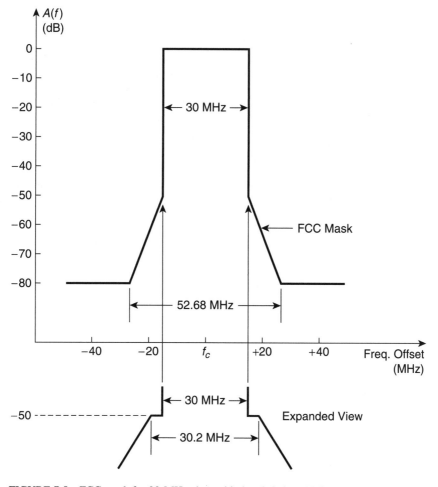

FIGURE 7.5 FCC mask for 30-MHz channel in bands below 15 GHz.

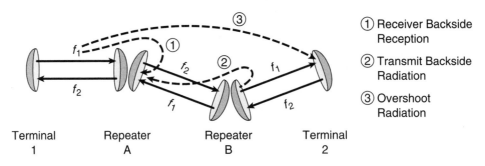

FIGURE 7.6   Three-hop system employing two-frequency plan.

of close to 1 bit/s/Hz, typically in bands above about 12 GHz, to as high as approximately 4.5 bits/s/Hz, typically in bands below about 12 GHz.

We now turn to the multihop application of frequency plans. There are two common methods of doing this. One method is called the *two-frequency plan*, the other the *four-frequency plan*, and each method is described next.

The two-frequency plan on a three-hop system is shown in Fig. 7.6. In this plan, the same frequency pair per two-way channel is used in adjacent hops. Thus, at a repeater site, both receivers facing opposite directions operate on the same frequency. The same is the case for both transmitters. The system shown is susceptible to three types of intrasystem co-channel interference, one example of each being indicated in the figure. In example 1, the receiver at repeater A facing repeater B is interfered with by the transmitted signal from terminal 1, with the only protection being the receiving antenna's discrimination to the interfering signal. This type of interference is called *receive backside reception* and impacts all repeater receivers. In example 2, the same receiver at repeater A as in example 1 is interfered with by the signal from repeater B transmitted toward terminal 2. In this case, the only protection against interference is the attenuation of the interfering signal's transmitting antenna to transmission in the direction of repeater *A*. This type of interference is called *transmit backside radiation* and impacts all receivers except those at the repeaters facing the terminals. Finally, in example 3, the receiver at terminal 2 is interfered with by the transmitted signal from terminal 1, with the attenuating factor to the signal level at the receiving antenna being the distance and topography between the terminals. Because the distance is long (three hop lengths) and obstruction blockage is likely due to earth bulge and terrain obstruction, the probability that this type of interference will be at a level to degrade receiver performance is low. This type of interference is called *overshoot (overreach) radiation*, and its

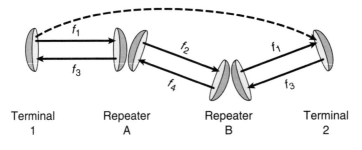

FIGURE 7.7   Three-hop system employing four-frequency plan.

impact on the receiver can be further attenuated by employing a highly directive receiving antenna, assuming that the interfering signal arrives at more than a few degrees off the receiving antenna's boresight. In the case of interference of the types shown in examples 1 and 2, the only way to minimize it is to employ antennas at the repeaters with high front-to-back ratios, typically shrouded antennas. The advantage of the two-frequency plan is that it allows maximum usage of the frequency band capacity by reusing frequencies on every hop, but, as seen previously, this advantage comes at the expense of significant exposure to co-channel interference, the minimization of which requires expensive antennas.

The four-frequency plan on a three-hop system is shown in Fig. 7.7. The advantage of this plan over the two-frequency plan is that the receivers operating on the same frequency are two hops apart. As a result, the only type of co-channel interference is overshoot radiation, which, as discussed previously, has a low probability of degrading receiver performance. The obvious disadvantage to this plan is that, with its application, the band can be used to only one-half of its capacity.

No mention of polarization has been made so far regarding either the two- or four-frequency plan. Polarization can be used to minimize intrasystem co-channel interference. For example, if dual polarized transmission is not being used on each channel, then, if the polarization is changed after every two hops, the co-channel overshoot interference in both frequency plans is significantly reduced as a result of the cross-polarization discrimination of the receiving antenna. Further, in the case of the two-frequency plan, interference due to receiver backside reception and transmitter backside radiation between adjacent hops with differing polarization is minimized.

7.4 POPULAR FREQUENCY BANDS

A large number of bands have been assigned by the ITU for broadband fixed point-to-point wireless applications. For most of these bands, several channel arrangements with several bandwidth options are specified. However, some of these bands are more popular than others, and for each such band, some channel arrangements and bandwidths are more popular than others. Table 7.1 provides a listing of the more popular non-U.S. ITU-R bands, bandwidths, and associated minimum spectral efficiency requirement, and Table 7.2 provides the equivalent U.S. data.

TABLE 7.1 Popular ITU-R (Non-U.S.) Frequency Bands and Channel Bandwidths

Band (GHz)	Freq. Range (GHz)	Ch. Bandwidths (MHz)	ITU Rec. (F.)
L6	5.925 – 6.425	29.65	383–7
U6	6.425 – 7.11	40, 80	384–7
7	7.11 – 7.9	7, 14, 28	385–7
8	7.725 – 8.5	7, 14, 29.65	386–6
11	10.7 – 11.7	40	387–9
13	12.75 – 13.25	7, 14, 28	497–6
15	14.4 – 15.35	7, 14, 28	636–3
18	17.7 – 19.7	13.75, 27.5, 55	595–7
23	21.2 – 23.6	7, 14, 28,56	637–3
26	24.5 – 26.5	7, 14, 28	748–4
38	37 – 39.5	7, 14 ,28 , 56	749–2

TABLE 7.2   Popular U.S. FCC Part 101 Frequency Bands, Channel Bandwidths, and Minimum Spectral Efficiency

Band (GHz)	Freq. Range (GHz)	Ch. Bandwidths (MHz)	Min. Spectral Efficiency (bits/s/Hz)
L6	5.925 – 6.425	3.75/5/10/30	3.3/3.7/4.5/4.5
U6	6.525 – 6.875	3.75/5/10	3.3/3.7/4.5
10	10.55 – 10.68	3.75/5	3.3/3.7
11	10.7 – 11.7	3.75/5/10/30/40	3.3/3.7/4.5/4.5/3.6
18	17.7 – 19.7	5/10/20/40	1
23	21.2 – 23.6	5.0 – 50	N.A.
39	38.6 – 40	5.0 – 50	N.A.

7.5   SINGLE- AND MULTICHANNEL SYSTEM ARCHITECTURES

For broadband point-to-point terminals, which are always two way, system architectures are normally such as to minimize the antenna system hardware as much as possible. The most effective approach is for transmit and receive frequencies on the same polarization to share a common antenna and antenna transmission line. This is accomplished by using an *antenna coupler,* which is also called an *antenna combiner.*

For a single channel (two-way) terminal, antenna coupling takes the form of an *antenna duplexer*, as shown in Fig. 7.8. A key component of the duplexer shown is the antenna circulator. A basic circulator is a three-port device, constructed from ferrite material, with behavior such that an input signal to any port circulates unidirectionally and exits at the next port on the unidirectional path. Thus, in the figure, the transmit signal of frequency f_1 that enters port 1 of the antenna circulator exits port 2 and proceeds via transmission line to the antenna. Likewise, the signal received by the antenna of frequency f_2 enters port 2, exits port 3, and proceeds to the receiver. Because the transmit

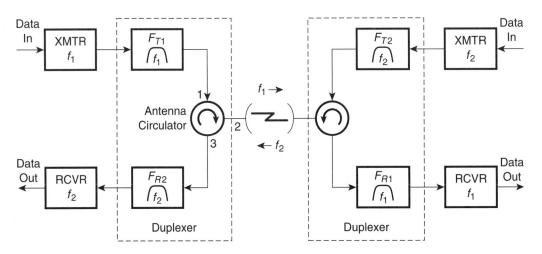

FIGURE 7.8   Single-channel antenna duplexing.

signal can never be totally absorbed by the antenna system, a small fraction of it returns to port 2 and ends up at the receiver input. The duplexer filter, F_{T1}, in the transmit leg, limits the level of noise and spurious emission that falls within the receiver passband. It is designed to do this to a degree that any such unwanted input that appears at the receiver front end is at a level low enough as to not degrade receiver BER performance. The filter F_{R2} in the receive leg of the duplexer helps ensure that the level of the transmitted signal reaching the receiver front end does not overload it, resulting in nonlinear behavior that degrades the BER performance. The loss experienced by a transmit or receive signal through a duplexer varies depending on the design but typically runs between about 1 and 2 dB.

Often single channel systems are operated in an equipment protection mode. A common version of such protection is shown in Fig. 7.9 and is referred to as *monitored hot standby protection*. In such a scheme, two fully operational transmitters and receivers are employed at each terminal. On the transmit side both transmitters are modulated with the input data, but only one, transmitter A in the figure, is connected to the antenna duplexer, this transmitter being referred to as the working transmitter. The other transmitter, which is referred to as the standby transmitter (transmitter B in the figure),

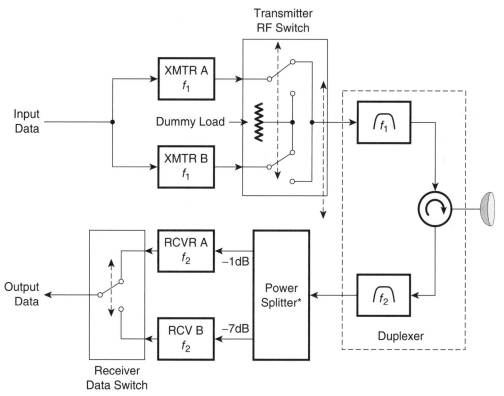

FIGURE 7.9   Single-channel hot stand-by terminal.

is connected to a dummy load. The operation of the transmitters is continually monitored and, if the working transmitter fails, the transmitter RF switch switches the standby transmitter to the antenna duplexer, thus restoring transmission almost instantly. On the receive side, the RF input signal is normally unevenly split by a power splitter. Typically, the splitter imparts a loss of about 1 dB in the path of the working receiver (receiver A in the figure) (i.e., the receiver normally connected to the data output port), and a loss of about 7 dB in the path of the standby receiver. Should the working receiver fail, then the receiver data switch switches to the standby receiver, receiver B in the figure, restoring transmission almost instantly, as in the case of a transmitter failure. When operating on receiver B, the received signal is attenuated by 6 dB relative to when operated on receiver A. However, as operation on receiver B is rare, this penalty is deemed acceptable, versus the use of an equal loss splitter that would impart about 3.5 dB loss to each receiver, and thus an additional 2.5 dB loss to path A relative to the 1/7 dB splitter arrangement.

For terminals supporting more than one channel on the same polarization, the antenna coupler is an expanded version of the duplexer and is structured as shown in Fig. 7.10. By the use of additional circulators and filters, branching networks are created on each side of the antenna circulator. The signal from transmitter 1, S_{T1} say, passes through filter F_{T1}, into port 1 of branching circulator C_{t1}, out of its port 2 and to the input of filter F_{t2}. The input of F_{t2} is reflective to signals outside its passband. As a result, S_{T1} is reflected back to port 2 of, C_{t1} reenters it, exits port 3, and continues on in a similar fashion until it exits the antenna circulator, C_a. On the receive side, receive signals behave similarly to transmit signals on the transmit side and are being reflected off filters whose passbands are removed from the signals' occupied bands. Note that for a single channel with frequency diversity protection as discussed in Chapter 5, only one transmit and one receive branching circulator is required.

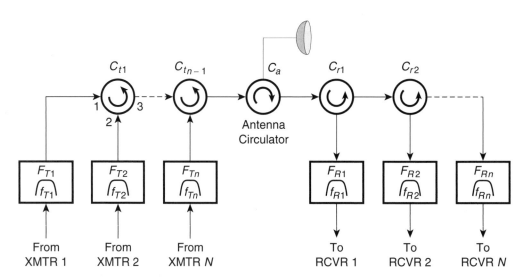

FIGURE 7.10   N-channel antenna coupler.

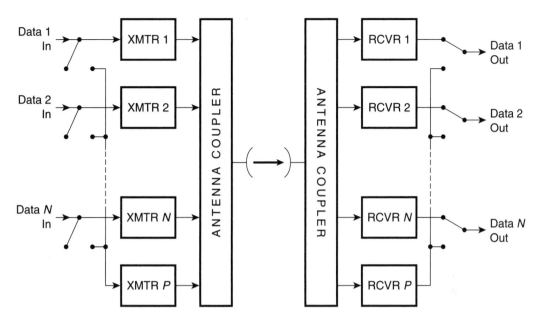

FIGURE 7.11   1 for *N* protection system (one-way, for simplicity).

A common protection scheme for multiple channels is one where they are protected with an additional channel (or channels) operating on an additional frequency (or frequencies). Such a scheme for protecting N working channels with one protection channel, referred to as a 1 for N protection system, is shown in Fig. 7.11. In such a system, the antenna coupler supports $N + 1$ channels. Excessive BER on any of the working receiver outputs triggers, via a reverse channel to the far-end transmitters, a bridging of the input data to the associated far-end transmitter to the protection transmitter, and a switch of the receiver output data port from the failed receiver to the protection receiver. Such a system protects against both equipment failure and path failure. When acceptable BER performance is restored on the failed channel, traffic is routed back through it, making the protection channel available for the next working channel that fails.

7.6   TYPICAL RADIO TERMINALS

In this section three products that operate in licensed bands will be reviewed in order to provide in insight into practical systems utilizing the modulation and other realizations techniques discussed in previous chapters. The products were chosen to allow the review to encompass a range of capacities, operating frequencies, and modulation methods.

The first product reviewed is a 16 x E1, 18 GHz, 4-CPFSK type, manufactured by Witcom Ltd. The second is a 3 x DS3, 6 GHz, 4-D 128-QAM trellis coded, offered by Harris Corporation. The third is a 2 x OC-3, 23 GHz, 256-QAM type, provided by Stratex Networks. Key parameters of these products are summarized in Table 7.3.

TABLE 7.3   Key Parameters of Radio Terminals Reviewed

Parameter	Witcom WitLink-2000	Harris Constellation™	Stratex Networks Altium MX 311
Operating frequency range (GHz)	17.7–19.7	5.925–6.425	21.2–23.6
Channel bandwidth (MHz)	27.5	30	50
Modulation	4-CPFSK	4-D 128-QAM TCM	256-QAM
Primary data	16 E1	$3 \times$ DS3	$2 \times$ STM-1
Primary data rate (Mb/s)	16×2.048	3×44.736	2×155.52
Primary data line code	AMI or HDB3	B3ZS	Optical
Auxilary data rate (kb/s)	64	1544 (DS1) + 102 max.	4×64
Information bit rate (Mb/s)	32.83	155.52	311.29
Composite bit rate (Mb/s)	40.5	160.23	335.16
Spectral efficiency at info. bit rate (bits/s/Hz)	1.19	5.2	6.2
Spectral efficiency at comp. bit rate (bits/s/Hz)	1.47	5.3	6.7
Xmtr. output power, typical (dBm)	27	29	14.5
ATPC range, if equipped (dB)	30 in 1 dB steps	10	7 or 10
Rcvr. noise figure (dB)	7	4	7
Rcvr. threshold at 10^{-6} BER, typical (dBm)	−76	−72	−61.5
Rcvr. threshold at 10^{-3} BER, typical (dBm)	−78	−73.5	−63.5
System gain at 10^{-6} BER rcvr. threshold (dB)	103	101	76
System gain at 10^{-3} BER rcvr. threshold (dB)	105	102.5	78
Residual BER	10^{-12}	10^{-12}	10^{-13}
Rcvr. overload level at 10^{-6} BER (dBm)	−10	−17	−23
Dispersive fade margin at 10^{-3} BER (dB)	N.A.	50	>38
Co-channel T/I (dB)	<23	31	<37
Adjacent channel T/I (dB)	<1	−14	<3
FEC type (n, k)	RS (255, 215)	RS (238, 232)	RS (255, 239)
FEC overhead (%)	18.6	2.59	6.69
Adaptive freq. domain equalizer type	N.E.	IF slope	N.E.
Adaptive time domain equalizer type	N.E.	11 feedforward taps	20 feedforward taps

Note: N.A. = not available. N.E = not equipped.

7.6.1   Witcom WitLink-2000, 16 x E1, 18 GHz, 4-CPFSK Radio Terminal

The Witcom terminal is divided into two units, the indoor unit or IDU and the outdoor unit or ODU; these units are interconnected by a single coaxial cable. The ODU incorporates an antenna, with the option for two sizes provided—namely, 1 foot or 2 feet in diameter. A photo of this terminal along with a laptop computer for network management purposes is shown in Fig. 7.12. A simplified block diagram of the IDU is shown in Fig. 7.13(a) and that of the ODU in Fig. 7.13(b). These block diagrams indicate major components and the signal flow from transmit data in to receive data out. A few key points to note in this realization are as follows:

- It uses quadrature modulation, as described in Section 4.4, to generate a 4-CPFSK modulated signal centered at 400 MHz.
- It uses differential detection, as described in Section 4.4, to demodulate a 4-CPFSK modulated signal centered at 30.357 MHz.
- The transmitter portion uses dual upconversion, both stages of which take part in the ODU. First the 400-MHz modulated IF from the IDU is upconverted to 2910 MHz; then this signal is in turn upconverted to the desired output frequency.
- The receiver portion uses three stages of downconversion, two in the ODU, one in the IDU. First, in the ODU, the incoming RF signal is downconverted to 1900 MHz, followed by a second downconversion to 140 MHz. This signal passes from the ODU to the IDU, where it is downconverted for a final time to the "near baseband" frequency of 30.375 MHz.
- It uses one level of FEC coding, this being Reed-Solomon (255, 215) coding.

As indicated in Table 7.3, the information bit rate, defined here as the sum of all the input data rates to the multiplexer, is 32.83 Mb/s. The output bit rate of the asynchronous multiplexer, as a result of bit stuffing, is 34.15 Mb/s. As a result of encoding in the RS (255, 215) encoder, the data rate increases to $34.15 \times (255/215) = 40.5$ Mb/s, referred to in Table 7.3 as the composite bit rate. Since the channel bandwidth is 27.5 Mhz, the spectral efficiency at the information bit rate is given by $32.83/27.5 = 1.2$ bits/s/Hz and at the composite bit rate given by $40.5/27.5 = 1.5$ bits/s/Hz.

FIGURE 7.12   The Witcom Witlink-2000 terminal and network management computer. *(Courtesy of Witcom Wireless Communication.)*

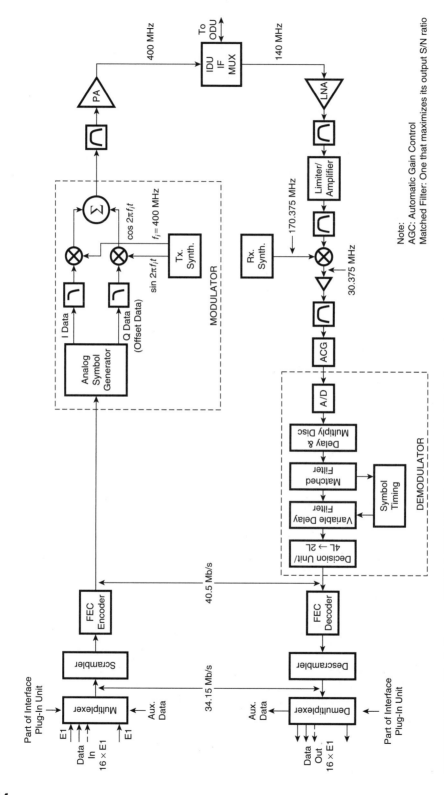

FIGURE 7.13(a) Block diagram of Witlink-2000 16 x E1, 4-CPFSK, 18-GHz indoor unit. (*Courtesy of Witcom Wireless Telecomunication.*)

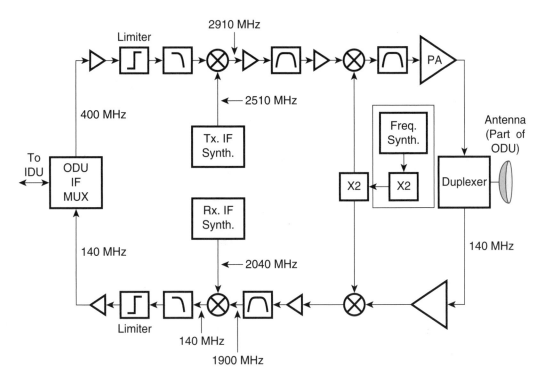

FIGURE 7.13(b)   Block diagram of Witlink-2000 16 x E1, 4-CPFSK, 18-GHz outdoor unit. *(Courtesy of Witcom Wireless Telecommunication.)*

An interesting comparison is the BER performance of this real system compared to an ideal uncoded one, of bit rate the same as the pre-encoded bit rate of this real one. For this real system, the modulation index, m, is 0.75. Figure 4.44 shows that, for an ideal 4-FSK uncoded system, with $m = 0.75$, the E_b / N_0 for a BER of 10^{-6} is 15.3 dB. By Eq. (4.189), the received signal level, P_{Si}, that results in a given E_b / N_0 is

$$P_{Si}(dBm) = \frac{E_b}{N_0}(dB) - 174 + 10\log_{10} f_b + F(dB) \tag{7.2}$$

For the real system, the pre-encoder bit rate, f_b, is 34.15 Mb/s and the noise figure, F, is 7 dB. Applying these values to Eq. (7.2), we get, for the ideal system, $P_{Si} = -76.4$ dBm. From Table 7.3 we note that for the real encoded system, $P_{Si} = -76$ dBm for a BER of 10^{-6}. Thus, the real performance, as a result of Reed-Solomon coding gain, is essentially the same as the theoretical uncoded performance, despite implementation losses.

Figure 7.14 shows the eye diagram of the four-level amplitude modulated signal at the matched filter output of the Witcom receiver under a high received signal level condition. The slight closure of the eyes is due to implementation imperfections.

Figure 7.15 shows the transmitted spectrum and the appropriate ETSI mask, with the spectrum comfortably within the mask. By Eq. (4.178), which was derived from Carson's rule (Section 4.4), the minimum transmission bandwidth for a 4-FSK system

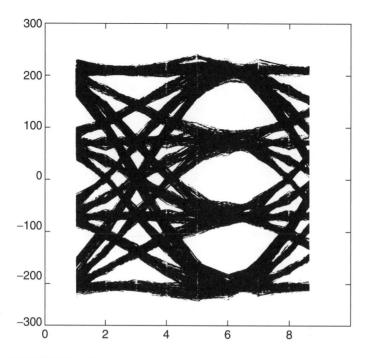

FIGURE 7.14   Eye diagram of 4-level PAM signal at matched filter output of Witlink-2000 receiver. *(Courtesy of Witcom Wireless Telecommunication.)*

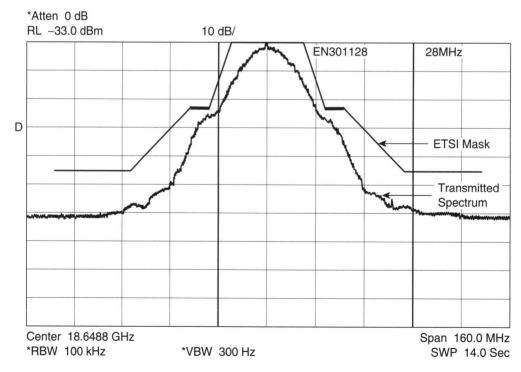

FIGURE 7.15   Transmitted spectrum and ETSI mask of Witlink-2000 18-GHz, 16 x E1 transmitter. *(Courtesy of Witcom Wireless Telecommunication.)*

with $m = 0.75$ and $f_b = 40.5$ MHz is 35.4 MHz. However, this system operates in an authorized bandwidth of only 27.5 MHz. The discrepancy is likely due to the following. First, the mask width isn't constant but increases at lower spectral density levels, making a fit easier. Second, the real system used CPFSK that, because it eliminates phase discontinuities, reduced the radiated bandwidth. Finally, Eq. (4.178) is only a rule of thumb.

The system just described can be converted to one that transports 2 10/100 Base-T Ethernet signals plus 4 E1 streams. This is accomplished by changing the front loadable interface plug-in unit on the IDU, where multiplexing and de-multiplexing takes place, to the wireless bridge option. On the transmit side, this plug-in unit feeds the bursty Ethernet traffic into local buffer memory. It then clocks it out and multiplexes this new stream with the 4 E1 streams to create a stream of data that after RS FEC encoding has a rate of 40.5 Mb/s. In this mode, the user can trade error correction capability for throughput by selecting, via software, one of three options of the RS code (n, k) parameters. These options are (255, 215), (255, 235) and (255, 247). On the receive side of this plug-in unit, the signal processing is reversed.

7.6.2 Harris Constellation™ 3 x DS3, 6 GHz, 4-D 128-QAM Trellis Coded Radio Terminal

The Harris Constellation™ terminal is an all indoor unit, as shown in Fig. 7.16. Its simplified block diagram is shown in Fig. 7.17, which indicates major components and the signal flow from transmit data in to receive data out. Some key features to note in this realization are as follows:

- It uses 4-D 128-QAM TCM as described in Section 6.2.4.
- There are two levels of coding, the outer code being the shortened RS (238, 232) code and the inner code being the modulation trellis code. Since the system employs 4-D 128-QAM TCM, the TCM overall encoder is a rate 13/14 one (Section 6.2.4), and 13 input bits to it result in $2^{14} = 16,384$ 4-D constellation points that are mapped into two successive 128-QAM constellations. The overall encoder is constructed with a standard rate 1/2 convolution encoder as the basic encoding unit. For every 13 input bits, 10 are passed through uncoded and 3 are routed to the convolution encoder single input via a parallel to serial converter. The convolution encoder output is punctured (Section 6.2.3) to create a rate 3/4 code. Its four coded output streams along with the 10 uncoded ones make up the overall encoder output.
- There are two levels of upconversion in the transmitter assembly and two levels of downconversion in the receiver assembly.
- Automatic transmitter power control (ATPC) is provided. When the received signal level at the far-end receiver drops below a predetermined threshold, a request for increased transmitter power is sent upstream to the transmitter. For a maximum of 5 minutes, the transmitter increases the output power by a range of up to 10 dB to bring it to its full level.
- The service channel unit supports a one-voice, one-data channel system. The voice channel information is transmitted over the system at a 64 kb/s rate. Data can be handled in the half-duplex or full-duplex mode. This data can be asynchronous,

FIGURE 7.16   Harris Constellation™
3xDS3 terminal. *(Courtesy of Harris
Corporation.)*

at a bit rate of 9.6, 19.2, or 38.4 kb/s, or sampled, of rate ranging from DC to
4.8 kb/s.

- The transmitter section of the high-level mux multiplexes the 3 DS3 input signal
  into an STS-3 Synchronous Payload Envelope. The 1 DS1 wayside channel, the
  data from the service channel unit, and system internal data are fed into unused
  STS-3 overhead bit locations. The output of the high-level mux is an STS-3 signal.

The information bit rate of this system referred to in Table 7.3 is the output bit
rate of the high-level mux (i.e., 155.52 Mb/s). A signal at this rate is fed via the scrambler into the Reed-Solomon FEC encoder. This encoder outputs 238 bytes for every
232 input bytes. However, the 232 input bytes consist of 231 information bytes and 1
overhead byte used for framing purposes. Thus the encoder output signal has a rate of
$155.52 \times (238/231) = 160.23$ Mb/s, referred to in the Table 7.3 as the composite bit rate.
This signal is fed, via the interleaver, to the TCM encoder, which therefore outputs a
signal of rate $160.23 \times (14/13) = 172.56$ Mb/s.

The theoretical optimum spectral efficiency of uncoded 128-QAM is 7 bits/s/Hz and
thus that of 4-D 128-QAM TCM is $(13/14) \times 7 = 6.5$ bits/s/Hz. However, as pointed out
in Chapter 4, this is not achievable in practice, as it would require unrealizable brick wall
filters. This Constellation™ system uses Nyquist baseband filtering with an excess bandwidth of 20%; this filtering is split equally between the transmitter and receiver. Its transmit portion leads to a spectral efficiency at the composite bit rate into the TCM modulator
of 5.3 bits/s/Hz. This excess bandwidth is at the low end of what is normally

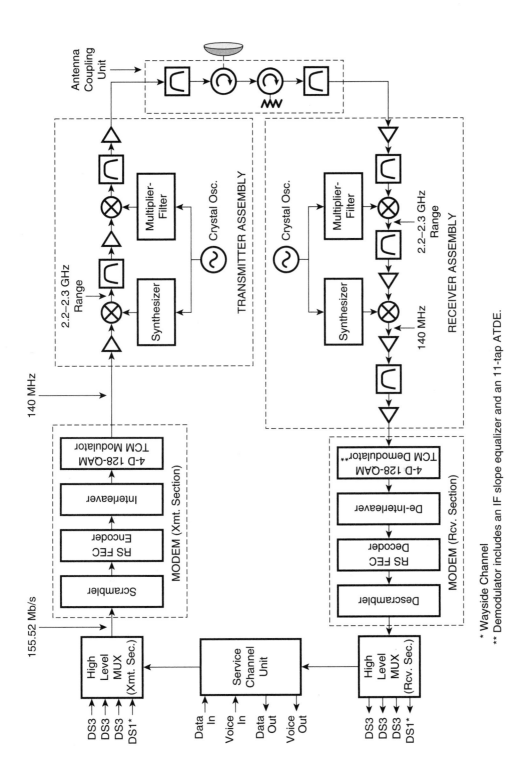

FIGURE 7.17   Block diagram of Harris Constellation™ 3xDS3, 4-D, 128-QAM TCM, 6-GHz terminal. (*Courtesy of Harris Corporation.*)

* Wayside Channel
** Demodulator includes an IF slope equalizer and an 11-tap ATDE.

implemented, as going much lower is difficult to realize without incurring significant implementation loss.

Figure 7.18 shows a typical BER versus C/N ratio for this system. As was done with the Witcom system, let's compare actual BER performance against a theoretical uncoded equivalent. Such an equivalent system must transmit 155.52 Mb/s in a 30-MHz bandwidth. This implies a spectral efficiency of 5.2 bits/s/Hz. This can be achieved with 64-QAM and an excess bandwidth of about 15%. Such an excess bandwidth is possible but, as indicated previously, difficult to realize without significant implementation loss. Nonetheless, as we are only doing a theoretical comparison, let's use it anyway. Figure 4.20 shows that, for an uncoded 64-QAM system, the theoretical E_b/N_0 for a BER of 10^{-6} is 19 dB. Substituting this value of, E_b/N_0 a 4-dB noise figure, and a 155.52 Mb/s bit rate into Eq. (7.2), results in a received signal level, P_{Si}, of –69.1 dBm. From Table 7.3 we note that for the encoded system, $P_{Si} = -72$ dBm for a BER of 10^{-6}. Thus, the performance of the real system, as a result of Reed-Solomon coding and TCM, is better than that of the theoretical uncoded one by approximately 3 dB, despite implementation losses.

Figure 7.19 shows a photograph of the system's 128-point cross constellation.

It should be noted that, by simply replacing the high-level mux (HLM) plug-in card with a different HLM card, the capacity of the terminal described can be changed to OC-3 plus 1 DS1 wayside channel. Also, by replacing the HLM card with a different HLM card and adding a simple connector panel, the capacity can be changed to STM-1/STS-3 plus 1 DS1 wayside channel.

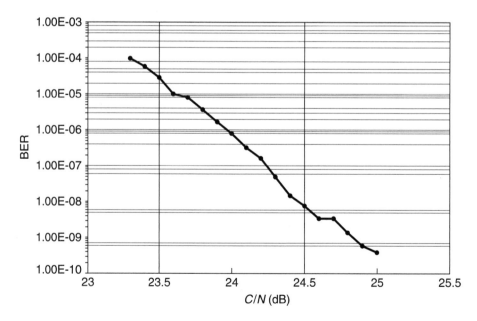

FIGURE 7.18   Typical BER versus C/N(dB) curve for Harris Constellation™ 3xDS3 receiver. *(Courtesy of Harris Corporation.)*

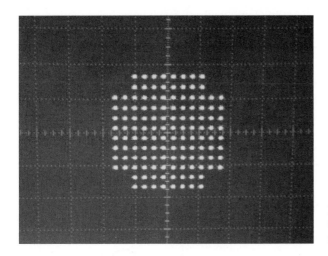

FIGURE 7.19   The Harris Constellation™ 3xDS3 128 point cross constellation. *(Courtesy of Harris Corporation.)*

7.6.3   Stratex Networks Altium MX 311, 2 x OC-3, 23-GHz, 256-QAM Radio Terminal

A photo of the Altium MX 311 terminal is shown in Fig. 7.20. Like the Witcom terminal, it has both an indoor unit and an outdoor unit. A highly simplified block diagram of its 256-QAM version configured in the nonprotected mode is shown in Fig. 7.21. However, it can also be configured to operate in the monitored hot-standby protected mode. It will be observed from the Fig. 7.21 that it utilizes dual ups and downconversion, all of which takes place in the ODU. Note also that the "data" interface to the external world is optical.

FIGURE 7.20   The Stratex Networks Altium MX 311 terminal. *(Courtesy of Stratex Networks.)*

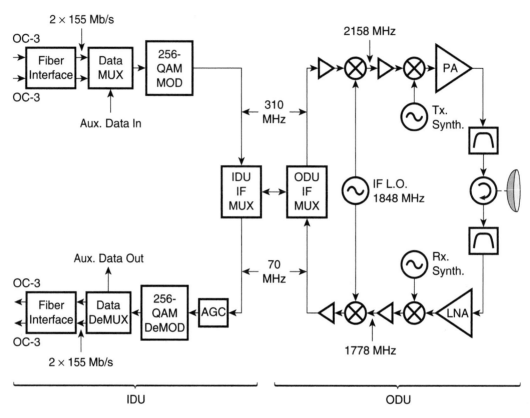

FIGURE 7.21   Block diagram of Altium MX 311, 2xOC-3 256-QAM, 23-GHz terminal. *(Courtesy of Stratex Networks.)*

256-QAM has been used with this version of this product because of the need to fit its extremely high data rate, at least by wireless standards, into 50 MHz of authorized bandwidth. Figure 7.22 shows a plot of the radiated spectrum of an Altium MX 311, 256-QAM system operating at 22.566 GHz and the appropriate FCC prescribed mask. The theoretical optimum spectral efficiency of 256-QAM is 8 bits/s/Hz. Like the Harris terminal, this system employs Nyquist baseband filtering split equally between the transmitter and receiver. Here the excess bandwidth is approximately 25%. The transmitter portion of this filtering leads to a spectral efficiency at the composite bit rate of 6.7 bits/s/Hz. It uses one level of FEC coding, this being Reed-Solomon (255, 239) coding.

The receiver employs a highly advanced decision-directed (linear) 20-tap time domain equalizer that is set adaptively to either a carrier phase acquisition mode or a tracking mode. Both modes of operation differ only in the way the error term is computed for updating coefficients. During phase acquisition, the equalizer is switched into what's referred to as a *blind mode* (Section 6.3.3.4). In such a mode, the equalizer is able to acquire lock even with a rotating and/or heavily disturbed input signal. After acquisition, the equalizer is switched into a LMS mode (Section 6.3.3.2) to enable a high degree of channel equalization.

Atten 10 dB
RL −10.0 dBm   10 dB/

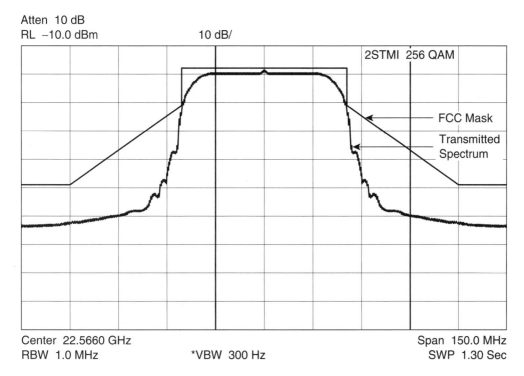

Center 22.5660 GHz   Span 150.0 MHz
RBW 1.0 MHz   *VBW 300 Hz   SWP 1.30 Sec

FIGURE 7.22   The Altium MX 311 256-QAM transmitted spectrum and FCC mask. *(Courtesy of Stratex Networks.)*

Figure 7.23 shows BER versus E_b / N_0 as well as received signal level for both the coded and uncoded system. The dramatic improvement as a result of coding is immediately obvious. Once more, let's compare actual BER performance, in this case both coded and uncoded against the theoretical uncoded equivalent. Since the composite bit rate, which is the RS (255, 239) encoder output rate, is 335.16 Mb/s, then the input rate to the encoder is $335.16 \times (239/255) = 314.13$ Mb/s. It is the theoretical uncoded system at this rate, with a noise figure of 7 dB, which we will compare the real uncoded and coded systems against. Figure 4.20 shows BER versus E_b / N_0 for a theoretical 256-QAM uncoded system. Substituting values of E_b / N_0 and the noise figure and bit rate of the equivalent uncoded system into Eq. (7.2), one is able to create the BER versus received signal level curve shown in Fig. 7.23 for the theoretical uncoded system. We note that for a BER of 10^{-6}, its received signal level, P_{Si}, equals −58.5 dBm. However, also from Fig. 7.23, we see that for the real uncoded system, the received signal level is −56.6 dBm for a BER of 10^{-6}. Thus, at the 10^{-6} BER level, there is an implementation performance loss of 1.9 dB. For the real encoded system, on the other hand, the received signal level for a BER of 10^{-6} is, from Fig. 7.23, −61.9 dBm. Thus, for a BER of 10^{-6}, the performance of the real encoded system, as a result of the Reed-Solomon coding gain, is 3.4 dB better than that of the theoretical uncoded one, despite an implementation loss of 1.9 dB.

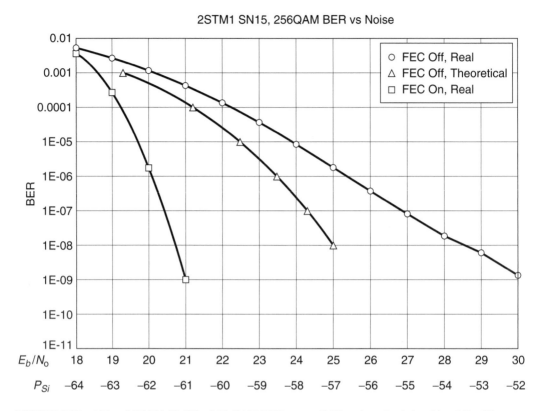

FIGURE 7.23   Altium MX 311 23-GHz, 256-QAM BER versus E_b/N_o and received signal level P_{Si}. *(Courtesy of Stratex Networks.)*

REFERENCE

1. Ivanek, F., editor, *Terrestrial Digital Microwave Communications,* Artech House, 1989.

CHAPTER 8

POINT-TO-POINT SYSTEMS IN UNLICENSED FREQUENCY BANDS

8.1 INTRODUCTION

In this chapter we study point-to-point wireless systems that operate in unlicensed bands, such bands being alternatively referred to as license-free bands. In such bands, an operator does not require a license from the spectrum management authority in order to install and operate a system. Additionally, no prior interference coordination with other systems operating in the band is required, though such coordination is certainly advisable. The equipment used must, however, meet certain stringent requirements. Sounds like a recipe for chaos! And indeed it is. Fortunately, because of the equipment restrictions, it is controlled chaos. Because of the possibility of frequency sharing, uncontrolled interference immediately springs to mind. In such an environment, successful operation is clearly facilitated by the use of equipment that is unusually resistant to interference while imparting minimum interference to others. A class of systems that exhibits such properties is *spread spectrum* systems, and such systems are permitted to operate as point-to-point links in unlicensed bands. Point-to-point systems employing standard digital modulation techniques such as discussed in Chapter 4 are also permitted in some unlicensed bands. In general, in unlicensed bands, transmitted power levels and spectral power densities are, by regulation, severely limited in order to minimize interference. In the specific case of point-to-point links, high-directivity/high-gain antennas are usually deployed in order to further minimize interference from and to other systems while enhancing path performance. We begin with a review of relevant spread spectrum theory. Next, key spectrum management requirements and allocated bands are discussed, and finally key features of a commercial system employing spread spectrum modulation are presented.

8.2 SPREAD SPECTRUM TECHNIQUES

In a spread spectrum system, for a given information data input signal, the modulated carrier normally (but not necessarily, as will be seen in Section 8.2.2) occupies a much larger spectrum (i.e., is much more spread out), than that occupied by a carrier modulated with a conventional technique such as one of those reviewed in Chapter 4. As bandwidth is a limited resource, this spectrum spreading would seem to be an inefficient use of spectrum. And were there only to be one carrier accommodated in this widened spectrum this would indeed be the case. However, the key feature of spread spectrum is that it permits many users simultaneously to use the same bandwidth without significantly interfering with each other. As a result, in a multiple user environment, spread spectrum systems can be very spectrally efficient. In such systems the spreading of the spectrum is achieved by adding a second modulation process to a conventional one, the second process being normally (but again, not necessarily, as will be seen in Section 8.2.2) independent of the information data. Two spread spectrum techniques lend themselves as candidates for application in fixed point-to-point broadband links—namely, *direct sequence spread spectrum* (*DSSS*) and *frequency hopping spread spectrum* (*FHSS*).

8.2.1   Direct Sequence Spread Spectrum (DSSS)

8.2.1.1   The Basic Principles

To understand the basic principles of standard DSSS it is helpful to first review BPSK. In BPSK a binary modulating signal $b(t)$, of amplitude ± 1, bit duration τ_b, and bit rate $f_b = 1/\tau_b$, is multiplied with a carrier $\sqrt{2P_S}\cos 2\pi f_c t$ to create a BPSK modulated signal $s(t)$ of average signal power P_S and given by

$$s(t) = \sqrt{2P_S}\,b(t)\cos 2\pi f_c t \tag{8.1}$$

The power spectral density $G_{BPSK}(f)$ of this signal, previously stated in Eq. (4.90), has the familiar $(\sin x/x)^2$ format and is given by

$$G_{BPSK}(f) = P_S \tau_b \left[\frac{\sin \pi (f - f_c)\tau_b}{\pi (f - f_c)\tau_b} \right]^2 \tag{8.2}$$

Note that, for a given P_S, the spectral density at any frequency is proportional to τ_b and hence inversely proportional to f_b. Recall also from Fig. 4.16 that the width of the main lobe is equal to $2/\tau_b = 2f_b$. Thus, the larger f_b, the wider the spectrum. Imagine now that the modulating signal is not the information signal, $b(t)$, but a pseudorandom binary sequence (also referred to as a *pseudonoise* [*PN*] sequence), $g(t)$, of amplitude ± 1, bit duration τ_C, and bit rate $f_C = 1/\tau_C$, where $f_C = nf_b \gg f_b$ (note that f_C is not to be confused with f_c, the carrier frequency). Now clearly, for the same P_S, the signal modulated with $g(t)$ will have a spectrum with density at f_c that's $1/n$ the equivalent density of the signal modulated with $b(t)$ and a main lobe width that's n times as wide. Standard DSSS works by, in effect, double modulating a carrier, once with an information signal, a second time with a non-information-carrying pseudorandom signal of bit rate much higher than the information signal.

A conceptual diagram of DSSS transmitters employing BPSK as the basic modulation is shown in Fig. 8.1. The spread spectrum signal $v(t)$ can be created by multiplying the BPSK signal $s(t)$ by a pseudorandom signal $g(t)$ and is thus given by

$$v(t) = g(t)s(t) \tag{8.3a}$$

$$= \sqrt{2P_S}\, g(t)b(t)\cos 2\pi f_c t \qquad (8.3b)$$

Figure 8.1(a) shows a modulator that creates a DSSS signal in this fashion. Because the multiplication processes involved in creating of $v(t)$ are linear, we note that $v(t)$ can equally be created by first multiplying $b(t)$ by $g(t)$, then multiplying the result by the carrier $\sqrt{2P_S}\cos 2\pi f_c t$. A modulator for creating a DSSS signal in this latter fashion is shown in Fig. 8.1(b). That this double modulation process spreads the spectrum is more easily seen via this second approach. Fig. 8.2 shows the waveforms and the spectral

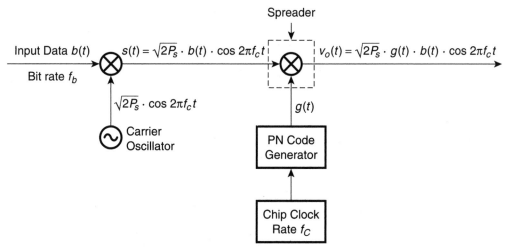

(a) BPSK Modulation followed by Spreading

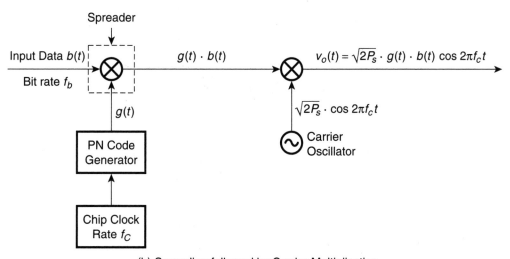

(b) Spreading followed by Carrier Multiplication

FIGURE 8.1   Direct sequence spread spectrum transmitter realizations.

densities of the data sequence $b(t)$, the pseudorandom sequence $g(t)$, and the product of these two sequences, $b(t)g(t)$. Note that, as is standard practice, the edges of $b(t)$ and $g(t)$ are aligned, and this alignment necessitates that n be an integer. Note also that f_C, the bit rate of $g(t)$, is called the *chip rate*, since $g(t)$ can be regarded as chopping the bits of information data into chips. From Fig. 8.2 the product sequence $b(t)g(t)$ is seen to have the same bit rate as $g(t)$ and is in fact a pseudorandom signal itself. Thus, the spectrum of $v(t)$ appears identical to the spectrum created by modulating the carrier by $g(t)$ only (i.e., it is spread out by a factor of n relative to a BPSK only modulated signal, and its density at f_c is $1/n$ times that of the unspread BPSK modulated signal.) Figure 8.3 shows an unspread BPSK spectrum and the direct sequence spread spectrum resulting from spreading the unspread BPSK spectrum with a PN sequence rate f_C that's 10 times greater than that of the information rate f_b.

The DSSS received signal can be demodulated with a receiver as shown in Fig. 8.4. We assume for simplicity that the path results in no attenuation to the transmitted signal, as this assumption has no impact on our current theoretical analysis. Thus, the received signal is $v(t)$. First $v(t)$ is despread by multiplying it with a synchronized copy of the pseudorandom sequence, $g(t)$, created by a local PN generator. As $g(t) \times g(t) = 1$ ($1^2 = 1$, $-1^2 = 1$), the output of this first multiplier is $s(t)$, the prespread BPSK modulated signal. This signal is then multiplied by the recovered carrier, $\cos 2\pi f_c t$, and the multiplier output

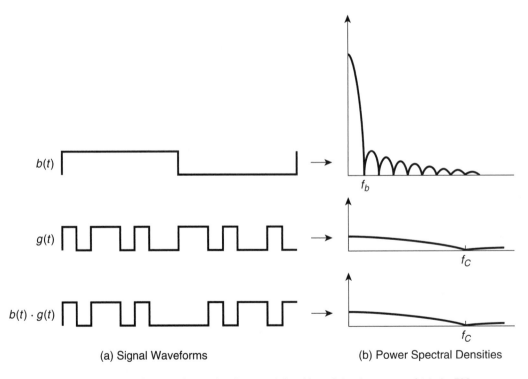

(a) Signal Waveforms   (b) Power Spectral Densities

FIGURE 8.2   Typical waveforms and associated spectral densities of the data stream $b(t)$, the PN sequence $g(t)$, and the product sequence $b(t)\cdot g(t)$.

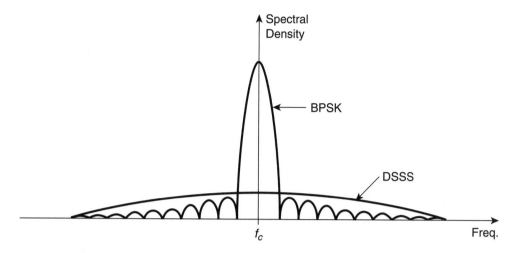

FIGURE 8.3   BPSK and DSSS spectra, where $f_C = 10f_b$.

is filtered and digitized to create $\hat{b}(t)$, the receiver's estimate of $b(t)$. Note that, because of the linear nature of the multiplication operations, the order of multiplication can, as is the case in modulation, be reversed.

Let us now turn our attention to the impact of Gaussian noise, $n(t)$, at the input to the receiver shown in Fig. 8.4. This noise is first "chipped" as a result of being multiplied by the PN sequence, $g(t)$. The impact of chipping is that, at nominally random times dictated by the PN sequence, the polarity of the noise waveform is reversed. This reversal has no effect on the power spectral density or the probability density function of the noise, the noise being a random variable to start with. Hence, at the output of the despreader, the BPSK signal and the statistical properties of its accompanying noise are unaffected by the spreading/despreading processes. Thus the probability of bit error performance is the same as for a standard BPSK system and hence given by Eq. (4.89); that is,

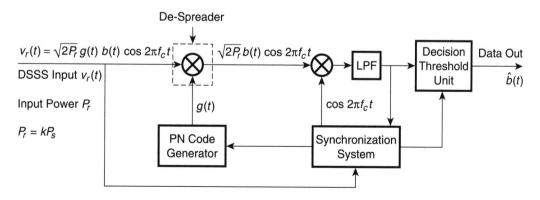

FIGURE 8.4   Direct sequence spread spectrum receiver.

$$P_{be} = Q\left[\left(2\frac{E_b}{N_0}\right)^{\frac{1}{2}}\right] \tag{8.4}$$

8.2.1.2   Interference

We have seen that, as a result of spectrum spreading, the spectral density of the BPSK-derived DSSS system is significantly reduced in any given narrowband and thus its effect on any narrowband receiver in its vicinity is greatly reduced. Further, this is achieved at no penalty to BER performance. It's certainly good not to interfere with one's neighbor. But what about the potential for being interfered with by one's neighbor? Here also, DSSS imparts significant advantage.

Consider the case where the DSSS signal $v_r(t)$ shown in Fig. 8.4 is interfered with by a sinusoidal signal, $v_J(t)$, of normalized power P_J, of frequency equal to the carrier frequency f_c, and with a phase difference relative to the carrier of θ. The input to the receiver, $v_{in}(t)$ is then given by

$$v_{in}(t) = v_r(t) + v_J(t) \tag{8.5a}$$

$$= \sqrt{2P_r}\, g(t)b(t)\cos 2\pi f_c t + \sqrt{2P_J}\, \cos(2\pi f_c t + \theta) \tag{8.5b}$$

Thus, given that $g^2(t) = 1$, the output of the despreader, $v_o(t)$, is given by

$$v_o(t) = g(t)v_{in}(t) \tag{8.6a}$$

$$= \sqrt{2P_r}\, b(t)\cos 2\pi f_c t + \sqrt{2P_J}\, g(t)\cos(2\pi f_c t + \theta) \tag{8.6b}$$

The first term in Eq. (8.6b) is the expected BPSK modulated signal. The second term is due to the single frequency interfering signal, but note its reincarnation! Its now a BPSK signal, where its modulating signal is $g(t)$, and thus has the same spectrum width of the DSSS signal. The despreading process has spread the spectrum of the interferer! Figure 8.5 shows the dramatic spectral role reversal between $v(t)$ and $v_J(t)$ resulting from despreading. As a consequence of filtering associated with the BPSK demodulation process, only that spectral portion of the interfering component of $v_o(t)$ that falls within the BPSK bandwidth can impact BER performance. As this component represents a small percentage of the energy in the original interferer, the interferer's ability to degrade BER performance is significantly reduced. Clearly, if the interfered was not of frequency f_c but some other frequency, $f_c + x$ say, then despreading would spread it as before but about the frequency $f_c + x$. For the case where the sinusoidal interferer is of frequency f_c and is of such level relative to the noise level as to control the probability of bit error performance, P_{be}, it can be shown[1] that P_{be} is given by

$$P_{be} = \frac{1}{2}erfc\left[2\left(\frac{P_r}{P_J}\right)\left(\frac{f_C}{f_b}\right)\right]^{\frac{1}{2}} \tag{8.7a}$$

$$= Q\left[4\left(\frac{P_r}{P_J}\right)\left(\frac{f_C}{f_b}\right)\right]^{\frac{1}{2}} \tag{8.7b}$$

What if the interferer was not a single frequency signal but a standard modulated or DSSS modulated signal of bandwidth very much less than that of the desired DSSS signal? Once more the despreader in the receiver would spread the interfering spectrum over a much wider bandwidth than the BPSK bandwidth. The spectrum width of this spread interferer would be at least as large as that of the DSSS signal, and likely larger. To understand this, recall that with a DSSS signal the edges of $b(t)$ and $g(t)$ are aligned and the chip rate is an integer multiple of the information bit rate. These constraints result in the bit rate of the product $b(t)g(t)$ being the same as that of $g(t)$. Let us assume, as is likely to be the case, that the edges of $b_J(t)$, the final modulating sequence of the interfering signal, are not aligned with those of $g(t)$. Then the multiple $b_J(t)g(t)$ at the output of the despreader will consist not only of pulses of width $1/f_C$, but, as shown in Fig. 8.6, of some that are narrower. These narrower pulses will create a $(\sin x/x)^2$ component of the spectrum that is wider than that of the chip rate spectrum, resulting in an overall spectrum that's widened.

Finally, what if the interfering signal is a DSSS signal created by a transmitter with similar bit rate and chip rate to the desired? Assuming, as is likely if the interference is from an independent source, that there is no synchronization between the systems, then again, because of nonaligned signal edges, spreading beyond the chip rate spectrum will result.

The net result, then, of the preceding interference scenarios is that regardless of the spectrum width of the interfering signal, this width after despreading is likely to be at least the width of the chip rate spectrum. As spreading is accompanied by a decrease in spectral density, this spreading results in a decrease in the interfering energy in the BPSK bandwidth centered about f_c and consequently a reduction in the interfering capability of the interferer.

The ratio f_C/f_b is a measure of the extent to which the interfering capability of an unwanted signal is reduced as a result of despreading. Also, it can viewed as a measure of the improvement to the receiver signal-to-noise in the signal bandwidth ratio as a result of despreading. This ratio is called the *processing gain, G_P,* of the system, and the greater it is,

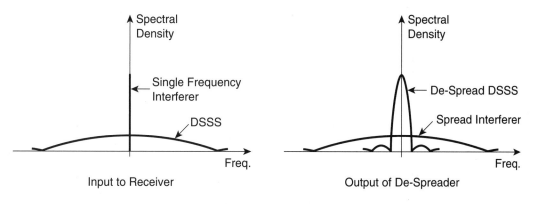

FIGURE 8.5   Impact of despreading on single-frequency interferer.

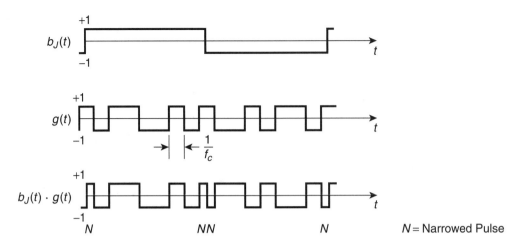

FIGURE 8.6 The effect of multiplying interferer modulating data $b_J(t)$ with the PN sequence $g(t)$.

the greater is the system's ability to suppress the impact of in-band interference. Since BW_{INFO}, the bandwidth of the unspread, information-data-modulated-only signal, is proportional to f_b, and BW_{SS}, the bandwidth of the spread spectrum signal, is proportional to f_C, then processing gain can also be defined in terms of these bandwidths. Thus, G_P is given by

$$G_P = \frac{f_C}{f_b} = \frac{BW_{SS}}{BW_{INFO}} \tag{8.8}$$

Processing gain expresses an improvement factor for the system. It does not, however, indicate the actual level of interference tolerable by the system relative to the desired signal at the receiver input, (J/S) say. This level is referred to as the jamming margin, M_J, which is given by[2]

$$M_J(dB) = \left(\frac{J}{S}\right) = G_P - \left(\frac{S}{N}\right)_{out} - L_{sys} \tag{8.9}$$

where G_P is the processing gain (dB)
$(S/N)_{out}$ is the theoretical ratio (dB) of the signal-to-noise in the Nyquist bandwidth at the de-spreader output (and hence at the PSK demodulator input) required to maintain minimum acceptable BER
L_{sys} is cumulative system loss (dB) to theoretical $(S/N)_{out}$ performance due to filtering, synchronization, etc.

Equation (8.9) can be intuitively understood by considering first the situation where the spreader/despreader is turned off and thus there is no processing gain and where the desired signal level at the demodulator input, S, is sufficiently high that thermal noise impact is negligible. Imagine now that the jamming signal level at the demodulator input, J_{di}, is such as to increase the BER to the minimum acceptable level. If, from a BER point of view, J_{di} acts on the receiver the same as does thermal noise in the receiver Nyquist bandwidth of total power the same as J_{di}, then we would expect (S/J_{di}) to be the same as the actual signal to thermal noise ratio at the demodulator input required to result in minimum acceptable BER. This actual signal to noise ratio is $(S/N)_{out} + L_{sys}$ so we have

$$(S/J_{di}) = (S/N)_{out} + L_{sys} \qquad (8.10)$$

Consider now the situation where the spreader/despreader is turned on. Now the jamming signal level J_{di} is reduced by G_P at the demodulator input thus reducing the BER. Therefore, to maintain the same BER we must increase J, the jamming level at the receiver input, by G_p. In other words, we must have

$$J = J_{di} + G_p \qquad (8.11)$$

Substituting Eq. (8.11) into Eq. (8.10), we get Eq. (8.9).

EXAMPLE 8.1   THE JAMMING MARGIN OF A BPSK-DERIVED DSSS SYSTEM

Determine the jamming margin of a BPSK-derived DSSS system where the processing gain $G_p = 11.8$ dB ($f_C/f_b = 15$) and the cumulative system loss $L_{sys} = 2$ dB.

Solution

Since $S = E_b f_b$, then for all digital systems, the signal-to-noise in the bit rate bandwidth ratio equals $(E_b f_b/N_0 f_b) = (E_b / N_0)$. For a BPSK system, the Nyquist bandwidth at the demodulator input is equal to the bit rate bandwidth; thus the ratio of signal-to-noise in the Nyquist bandwidth at this point equals the (E_b / N_0). From Fig. 4.20, we have, for a BER 10^{-6}, a theoretical (E_b / N_0), and therefore (S/N) at the input to a BPSK demodulator, of 10.5 dB. Thus, by Eq. (8.9), the jamming margin, M_J, is given by

$$M_J = 11.8 - 10.5 - 2 = -0.7 \text{ dB}$$

As a result, the system could operate with interference power in the spread bandwidth of up to -0.7 dB of the desired signal power without increasing BER above 10^{-6}.

8.2.1.3   QPSK-Derived DSSS

So far we have analyzed only BPSK-derived DSSS systems. In fixed point-to-point broadband systems, however, QPSK, or DQPSK, is often the fundamental modulation. This is largely because, for the same information transfer rate and processing gain, the spread spectrum bandwidth with QPSK is half that with BPSK, for a jamming margin, as we shall see, that's only 3 dB lower. Using QPSK as the fundamental modulation frees up adjacent spectrum for other systems in the same geographical area. An analysis of the operation of a typical QPSK derived DSSS system will show how the aforementioned properties are achieved. Figure 8.7(a) shows a typical QPSK-derived DSSS transmitter, and Fig. 8.7(b) shows a typical associated receiver, up- and downconversion being omitted for the purpose of simplification. The input data to the transmitter, $b(t)$, of bit rate f_b, is first divided by the serial to parallel converter into an in-phase stream $b_i(t)$ and a quadrature stream $b_q(t)$, of baud rates f_{bi} and f_{bq}, respectively. Thus

$$f_{bi} = f_{bq} = f_b / 2 \tag{8.12}$$

The in-phase and quadrature streams are spread with PN sequences $g_i(t)$ and $g_q(t)$, respectively, each with the same chip rate f_C, and the transitions of $b_i(t)$ and $b_q(t)$ are aligned with those of $g_i(t)$ and $g_q(t)$. Recognizing that the transmitter is, in effect, two BPSK transmitters combined at the output, we have at the output a signal $v(t)$ given by

$$v(t) = v_i(t) + v_q(t) \tag{8.13a}$$

$$= \sqrt{P_S}\, g_i(t) b_i(t) \cos 2\pi f_c t + \sqrt{P_S}\, g_q(t) b_q(t) \sin 2\pi f_c t \tag{8.13b}$$

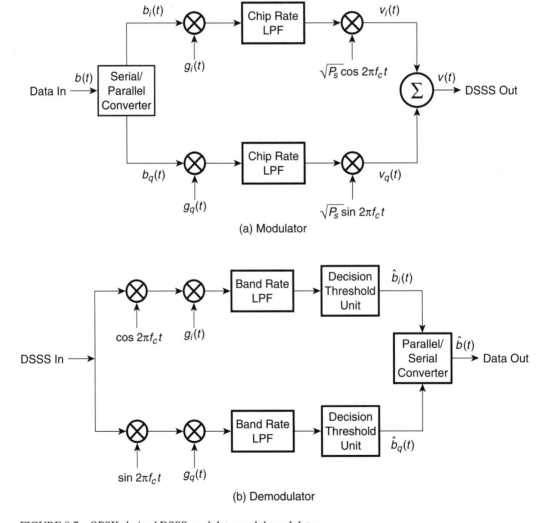

FIGURE 8.7   QPSK-derived DSSS modulator and demodulator.

where P_s is the average power $v(t)$.

At the demodulator input the received signal is split and fed into the in-phase and quadrature arms, the outputs of which are combined with a parallel to serial converter the create $\hat{b}(t)$, the demodulator's estimate of $b(t)$. The processing gain of this receiver is equal to the processing gain of each of its individual BPSK derived receivers and is thus given by

$$G_P = \frac{f_C}{f_{bi}} = \frac{f_C}{f_{bq}} = 2\frac{f_C}{f_b}$$

(8.14)

Hence, for the same processing gain as a (single) BPSK-derived system, the QPSK-derived system requires a chip rate-to-input data rate ratio that is half that required for the BPSK-derived system. Thus, for a given input data rate, the chip rate and hence the DSSS bandwidth required by the QPSK derived system is half that required by a (single) BPSK-derived version.

The Nyquist bandwidth of the QPSK system at the demodulator input equals $f_{bi} = f_{bq} = f_b/2$, half that of a BPSK system of equal bit rate. Hence the ratio of signal to noise in the Nyquist bandwidth is twice that of the ratio in the bit rate bandwidth and thus 3 dB more than (E_b / N_0). Since, for a given BER, QPSK and BPSK have the same (E_b / N_0), then for a given bit rate the signal to noise in the Nyquist bandwidth ratio for a QPSK system is 3 dB larger than for a BPSK system. For example, for a BER 10^{-6}, we have from Fig. 4.20, at the input to a QPSK demodulator, a theoretical (E_b / N_0) of 10.5 dB and therefore a (S/N) of 13.5 dB. As a result of this 3-dB increase in (S/N), the jamming margin for a QPSK-derived system is, by Eq. (8.9), 3 dB less than that for a BPSK-derived one with the same bit rate and processing gain, assuming the same system losses, L_{sys}. In practice, L_{sys} for the QPSK derived system is likely to be slightly more than that for the BPSK-derived system, as the QPSK circuitry is somewhat more complex.

8.2.1.4   Pseudonoise (PN) sequences

Based on the preceding description of DSSS systems, it is clear that the use of pseudorandom binary sequences (*pseudonoise* or *PN* sequences) plays a key role in their operation and performance. A top-level review of such sequences as used in fixed broadband wireless systems is therefore in order. As the name suggests, these sequences are not truly random, rather, they are periodic (i.e., they repeat themselves exactly after a given length). Nonetheless, they have many characteristics that are similar to truly random binary sequences, such as having a nearly equal number of 1s and 0s, a limit on the number of consecutive 1s and 0s (*runs*) in any period, low *crosscorrelation*, and low *autocorrelation* between shifted versions of itself. Correlation is a very important measure of the randomness of pseudorandom sequences. When a sequence is compared to shifted versions of another sequence the outcome is referred to as the crosscorrelation, whereas when compared to shifted versions of itself the outcome is called the autocorrelation. An observation over a long period of time of two truly random binary sequences would lead to the conclusion that there was zero crosscorrelation between them. For a binary PN sequence of normalized chip duration equal to unity, the autocorrelation, $R_a(\tau)$, may be defined as

$$R_a(\tau) = \frac{1}{p} \text{ (number of agreements } - \text{ number of disagreements)}$$

(8.15)

where τ is the discrete cyclic shift between sequences being compared

p is the number of chips in the period

agreements and disagreements are in a comparison over one full period of the sequence

The most common PN sequences employed in single-user spread spectrum systems, such as broadband fixed point-to-point systems, are *maximal length (ML)* sequences. Such sequences are generated by the use of linear feedback shift registers and, for an n-stage shift register, have a sequence length p of $2^n - 1$. Figure 8.8 shows a general implementation of an n-stage linear feedback shift register, which consists of n registers in series and feedback via modulo-2 addition. Maximal length sequences have several key properties, including the following:

1. The difference between the number of 1s and number of 0s in a period is equal to one.
2. A well-defined statistical distribution for the runs of 1s and 0s. Specifically, in each period, one-half of the runs have length 1, one-fourth have length 2, one-eighth have length 3, and so on.
3. An autocorrelation function as shown in Fig. 8.9. Note that, for any relative cyclic shift τ greater than one normalized chip duration, $R_a(\tau)$ is a constant value that equals $-1/p$. Thus the greater the value of p, the less the autocorrelation for τ greater than one normalized chip duration.

8.2.1.5   Performance in a Multipath Environment

DSSS signals are more resistant to multipath fading than their fundamental modulation counterpaths. This is easily understood by considering the received multipath signal in the time domain and, for simplicity, assuming a direct signal and only one reflected signal delayed by τ seconds relative to the direct signal. Assume that the receiver is synchronized to the direct signal. Then, at the receiver input, if the PN sequence component of the direct

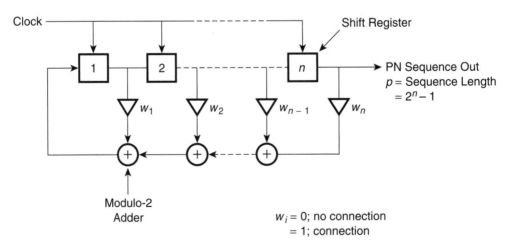

FIGURE 8.8   Generalized feedback shift register.

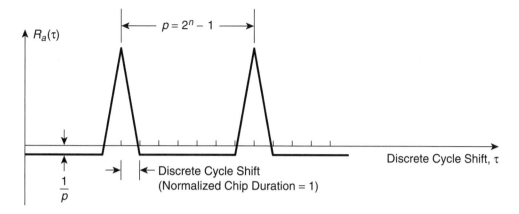

FIGURE 8.9   PN sequence autocorrelation function.

signal is $g(t)$, the PN sequence component of the reflected signal will be $g(t - \tau)$. Thus, when the reflected signal is multiplied in the despreader by $g(t)$, the output will be proportional to $g(t) \times g(t - \tau)$. Assuming $g(t)$ to be a maximal length sequence, then its autocorrelation function will be as in Fig. 8.9. Therefore, the larger the value of τ, the less correlated will be the reflected signal with the direct signal, and if τ approaches or exceeds one chip length, then the correlation will be extremely low and the output of the despreader resulting from the reflected signal will appear more like noise. Consider, for example, an 4XT1 DSSS system, with a modulating data rate after FEC encoding of 7 Mb/s, QPSK as the fundamental modulation, and a processing gain of 15 (11.8 dB). Then, by Eq. (8.14), f_C equals $(15 \times 7)/2 = 52.5$ Mb/s, and hence the chip period equals $1/f_C = 19$ ns. For such a system, if the relative delay τ approaches or exceeds 19 ns, then the output of the despreader resulting from the reflected path will appear more as low-level pseudorandom noise rather than a delayed replica of the desired signal. Note that a relative delay of 19 ns results from a reflected path of length 5.7 meters or 18.7 feet longer than the direct path.

8.2.2   IEEE 802.11b's (Wi-Fi's) Complementary Code Keying (CCK)

Complementary code keying (CCK)[3,4] is a modulation scheme developed by Lucent and Harris Corporation's semiconductor division (now Intersil) and adopted in 1998 by the IEEE 802.11 working group as the modulation scheme for 5.5 and 11 Mb/s data rate wireless local area networks (WLANs) in the 2.4-GHz band. The official IEEE designation of the specification of these WLANs is *802.11b*,[5] but it is also referred to as wireless fidelity, or *Wi-Fi*. In WLANs, the separation between communicating devices is normally measured in hundreds of feet at most. Though our interest is in point-to-point unlicensed systems where the separation between communicating devices is measured in miles, CCK is briefly reviewed here as it now finding application in such systems. As the name implies, CCK utilizes codes referred to as *complementary codes*. Such codes were originally proposed in 1951 by M. Golay for application in infrared spectrometry. However, their properties make them good codes for application in communication systems. These properties include good autocorrelation as well as crosscorrelation. In this context, good autocorrelation means a large correlation only at a zero shift between a sequence and a copy of itself and small or zero correlation otherwise. Good crosscorrelation means small or zero correlation regardless of the shift, including zero, between a pair of non-identical sequences. Codes used in CCK modulation are specifically *polyphase complementary codes*. A subset and simple version of such codes are *binary complementary codes*. We thus start by reviewing the binary version then expand our study to cover the polyphase version.

Complementary codes have been defined by Sivaswamy[6] as comprising "a pair of equal finite length sequences having the property that the number of pairs of like elements with any given separation in one series is equal to the number of pairs of unlike elements with the same separation in the other." (© 1978 IEEE). In binary complementary codes, the elements of the sequences are binary. The complementary properties of such codes are best demonstrated via an example. Consider the pair of binary complementary sequences shown in Fig. 8.10. Sequence 1 has one pair of like elements with a separation of 1 and two pairs of unlike elements with a separation of 1. Sequence 2, on the other hand, has two pairs of like elements with a separation of 1 and one pair of unlike elements with a separation of 1. The numbers of element pairings for separations of 1, 2 and 3 are shown in Table 8.1, where it is seen that complementary properties set out in the preceding definition of complementary codes have indeed been met. In summary, a binary complementary code is a pair of binary sequences having defined complementary properties. Though Golay only discussed pairs of binary complementary sequences, the concept has since been extended to sets (more than two) of such sequences.

We now turn our consideration to polyphase complementary codes. Such codes meet the same complementary code definition given previously. However, here the elements of the sequences are phase parameters. In the case of the codes used in CCK, the elements have four phase parameters, these being 0, $\pi/2$, π, and $-\pi/2$. A four-element section of a sequence using all four possible phases is shown in Fig. 8.11. Mathematically, elements of the four-phase complementary codes used in CCK can be described as the set of complex numbers $(1, j, -1, -j)$. Such codes are thus described as *complex complementary codes*.

CCK is a form of DSSS where complex complementary codes are used as spreading codes. These codes have a code length of 8, and the 8 complex chips (elements) constitute a single symbol. Since there are four possible values of each chip, there are $4^8 = 65,536$ possible codewords. It turns out that from these complex codewords there are sets of 64 complementary codes that are nearly orthogonal. One such set is used with CCK for 11 Mb/s data rate systems, and a subset of 4 from this set is used for 5.5 Mb/s ones.

A CCK modulated signal is, in effect, one where a carrier is modulated with complex eight-element codewords derived from the following formula:

$$C = e^{j(\phi_1+\phi_2+\phi_3+\phi_4)}, e^{j(\phi_1+\phi_3+\phi_4)}, e^{j(\phi_1+\phi_2+\phi_4)}, -e^{j(\phi_1+\phi_4)}, e^{j(\phi_1+\phi_2+\phi_3)}, e^{j(\phi_1+\phi_3)}, -e^{j(\phi_1+\phi_2)}, e^{j\phi_1} \quad (8.16)$$

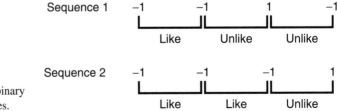

FIGURE 8.10   Pair of binary complementary sequences.

TABLE 8.1   Like and Unlike Element Pairing Results for Sequences 1 and 2 Shown in Fig. 8.1.

Pair Separation	Sequence 1 Like	Sequence 1 Unlike	Sequence 2 Like	Sequence 2 Unlike
1	1	2	2	1
2	1	1	1	1
3	1	0	0	1

The phase parameters ϕ_1 to ϕ_4 determine the phase values of the complex elements of the codeword C.

For the 11-Mb/s option, each codeword represents 8 bits of information. As a result, the input data stream is grouped into 8-bit data words, the bits in each data word labeled as (d7, d6, d5, d4, d3, d2, d1, d0), and these bits are assigned to phase parameters as specified in Table 8.2. The phase parameters then QPSK modulate in a differentially encoded fashion as per Table 8.3. Referring to Eq. (8.16), we note that that the parameter ϕ_1 is contained in all eight elements of the codeword. Thus, it essentially rotates the entire code word to one of four phases. In other words, it QPSK modulates the 8-element codeword. This suggests a clever way to generate C. The last three dibits and hence 6 bits of the 8-bit data word can be used to generate one of the 64 complex complementary spreading codes, which are derived, in effect, from Eq. (8.16) with ϕ_1 set to 0, and the first dibit (d1, d0) used to QPSK modulate the spreading code. A CCK modulator applying this scheme is shown in Fig. 8.12. The 11-Mb/s input data stream is fed to the 1 : 8 serial to parallel converter. The output parallel streams are thus each generated at a bit rate of 1.375 Mb/s. The complex code generator is driven by an 11-Mhz clock and generates 8-chip codes at a rate of 11 Mchips/s. Six bits from the converter output are fed to the complex code generator and used to select one of 64 complex complementary codes. This works since the period of these parallel bits is the same as that of the codewords. The other two bits streams from the converter represent the QPSK modulating I and Q signals. They are spread by the outputs of the code generator and then differentially encoded prior to carrier multiplication and summation. The outputs of the spreader and differential encoder shown in Fig. 8.12

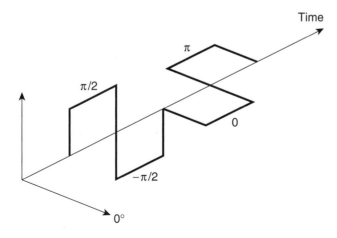

FIGURE 8.11   Four-element section of a complex sequence.

TABLE 8.2   Dibit/Phase Parameter Assignment of 8-Bit Data Word Dibits

Dibit	Phase Parameter
(d_1, d_0)	0_1
(d_3, d_2)	0_2
(d_5, d_4)	0_3
(d_7, d_6)	0_3

TABLE 8.3   Differentially Encoded QPSK Modulation of Phase Parameters

Dibit value	Phase
00	0
01	π
10	$\pi/2$
11	$-\pi/2$

are the real and imaginary codeword elements as per Eq. (8.16). Demodulation is similar to standard QPSK DSSS demodulation except for despreading. Here it is accomplished, not with two correlators, each driven by a PN code, but with a bank of 64 correlators each driven by one of the 64 complex complementary codes followed by a "biggest picker" circuit. This circuit determines which code was transmitted and hence gives 6 bits of the data word. The other 2 bits of the 8-bit data word are determined via QPSK demodulation.

For the 5.5-Mb/s data rate version of CCK, each codeword represents 4 bits of information. The input data stream is therefore divided into 4, with 2 bits used to generate four complex complementary codes of chip rate 11 Mchips/s and the other 2 to QPSK modulate.

With CCK the spreading ratio is only 8 to 1, and thus the processing gain from despreading is only 9 dB. Nonetheless, CCK has an effective processing gain of about 11 dB. How is this possible? The answer lies in two places. First, processing gain as defined by the despreading ratio is really a proxy for S/N improvement as a result of despreading. Second, CCK results in about 2 dB of coding gain relative to BPSK/QPSK and thus, after despreading, the S/N improves by about 11 dB over the S/N in the spread bandwidth. Fine, we say, but now a new question has been raised: Where does the coding gain in

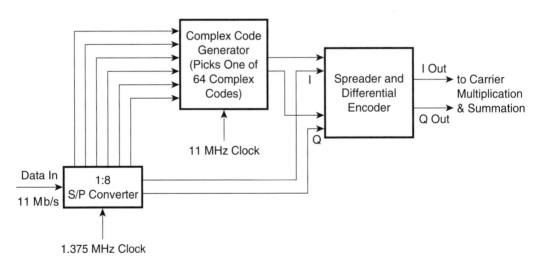

FIGURE 8.12   11-Mb/s CCK modulation process.

CCK come from, as no explicit encoder/decoder is included? The explanation provided by Andren and Webster[4] is as follows: "The modulation basically ties several bits together so that the receiver makes a symbol decision. If a symbol is in error then all of the bits in that symbol are suspect, but not all will necessarily be in error. Thus the symbol error rate and the bit error rate are related. While the SNR required to make a symbol decision correctly is higher than that required to make a one bit decision, it is not as high as required to make all of the bit decisions of a symbol independently and correctly. Thus, some coding gain is embedded in the basic spreading waveform." Like standard QPSK-DSSS, CCK exhibits good tolerance to mutipath fading, in part because the complex complementary codes used possess good autocorrelation properties.

Let's now stand back and ask ourselves just what is accomplished with CCK versus standard QPSK-DSSS. For a start there is a significant saving in bandwidth. For an 11-Mb/s QPSK-DSSS system with approximately 11 dB processing gain, the spreading ratio is 13. The symbol bit rate is 5.5 Mb/s and thus the chip rate is $5.5 \times 13 = 71.5$ Mchips/s. This implies an unfiltered main lobe bandwidth of 143 MHz! This compares with a CCK unfiltered main lobe bandwidth of 22 MHz. Thus, 11-Mb/s CCK is 6.5 times more spectrally efficient. In fact, CCK was specially designed to have the same bandwidth as the 802.11 2-Mb/s specified system. Further, as we have seen, there is about a 2-dB improvement in BER versus E_b / N_0. If left here, it seems like something for nothing. But the seasoned skeptic knows that this is rarely the case. So what has been sacrificed for these benefits? Jamming margin is the obvious suspect, as it is what was shown in Section 8.2.1.3 to be sacrificed in going from BPSK-DSSS to QPSK-DSSS. Let's therefore compute it for both cases and assume (a) that the minimum acceptable BER is 10^{-5} (as this rate is commonly used in WLANs) and (b) that cumulative system implementation losses, L_{sys}, in both cases is 2 dB. Recall also that for both cases $G_p = 11$ dB. The relationship for jamming margin, given in Eq. (8.9) and repeated here for convenience, is

$$M_J(dB) = \left(\frac{J}{S}\right) = G_P - \left(\frac{S}{N}\right)_{out} - L_{sys} \tag{8.17}$$

For the QPSK-DSSS case, given that the Nyquist bandwidth at the demodulator input is $f_b/2$, where f_b is the system bit rate, then the noise in the Nyquist bandwidth is $N_0 f_b/2$. Also, the signal power at the demodulator input can be stated as $E_b \cdot f_b$, where E_b is the energy per bit. Thus

$$\left(\frac{S}{N}\right)_{out}(dB) = \left(\frac{E_b \cdot f_b}{N_0 \cdot f_b / 2}\right)(dB) = \left(\frac{E_b}{N_0}\right)(dB) + 3dB \tag{8.18}$$

[We note that for QPSK there are 2 bits per symbol and thus the energy per symbol, E_s, is equal to $2E_b$ and therefore $(S/N)_{out} = E_s / N_0$].

For QPSK, the E_b / N_0 for a BER of 10^{-5} is found from Fig. 4.20 to be 9.6 dB. Thus $(S/N)_{out} = 9.6 + 3 = 12.6$ dB. Substituting this value and the preceding values of G_p and L_{sys} into Eq. (8.17) gives a jamming margin, M_j, of −3.6 dB.

For the CCK case, how one computes $(S/N)_{out}$ is not intuitively obvious, since some of the information bits are transported in the spreading code, others in the QPSK modulation. For the QPSK-DSSS case it was shown above that $(S/N)_{out} = E_s/N_0$. Andren[7] implies that this relationship holds for CCK, and we assume this to be the case. In the 11-Mb/s option of CCK there are 8 bits per symbol. Thus, $E_s = 8E_b$, and therefore

$$\left(\frac{S}{N}\right)_{out} (dB) = \left(\frac{E_s}{N_0}\right)(dB) = \left(\frac{E_b}{N_0}\right)(dB) + 10\log_{10} 8$$

$$= \left(\frac{E_b}{N_0}\right)dB + 9dB \qquad (8.19)$$

For CCK, the E_b / N_0 for a BER of 10^{-5} is, because of the coding gain, approximately 2 dB better than that for BPSK/QPSK and thus 7.6 dB. As a result, $(S/N)_{out} = 7.6 + 9 = 16.6$ dB. Substituting this value and the preceding values of G_p and L_{sys} into Eq. (8.17) gives a jamming margin, M_j, of –7.6 dB. Measurements of real systems agree closely with this result. Thus, as suspected, the jamming margin of CCK is indeed lower than that of the equivalent QPSK-DSSS system, the difference being 4 dB.

Given that an 11-Mb/s non–spread spectrum BPSK system has the identical RF bandwidth to an 11 Mb/s CCK one, an interesting question is just how much additional jamming margin protection CCK affords relative to BPSK in exchange for its added complexity. For BPSK, the jamming margin is given by Eq. (8.17) with $G_p = 0$. Therefore, for a minimum acceptable BER of 10^{-5}, and assuming L_{sys} to be 2 dB, it computes to –11.6 dB. Thus, CCK buys 4 dB of margin compared to BPSK. Not a large number, but helpful nonetheless. CCK does, however, afford greater than 10 dB processing gain, whereas BPSK affords none. At the time of the adoption of CCK by the IEEE 802.11b working group there was a FCC requirement that G_p be greater than or equal to 10 dB. That requirement, however, was dropped by the FCC in August 2002.

8.2.3   Frequency Hopping Spread Spectrum (FHSS)

In frequency hopping systems, as the name implies, a carrier, modulated with data to be transmitted, is continually changed or "hopped" between a number of predetermined frequencies within a given band so that it spends only a small percentage of its total time at any one frequency. The modulated carrier frequency is pseudorandomly hopped by controlling the carrier frequency synthesizer with a PN sequence generator. At each frequency hop time the PN generator inputs to the synthesizer a frequency word, of length n chips, which dictates one of $N = 2^n$ frequency outputs. At the receiver, the received signal is dehopped by downconverting it to an IF frequency with a synthesizer fed by a PN sequence identical to the one used in the transmitter and in alignment with it as received by the receiver. To avoid the need for, and delay associated with, carrier acquisition in the receiver after every hop, M-FSK modulation is normally used in the transmitter and noncoherent demodulation thereof in the receiver. Figure 8.13 shows the block diagram of a typical FHSS radio terminal.

The bandwidth per hop position, Δf, is called the *instantaneous bandwidth*. This bandwidth is that necessary to include most of the power in the M-FSK modulated carrier

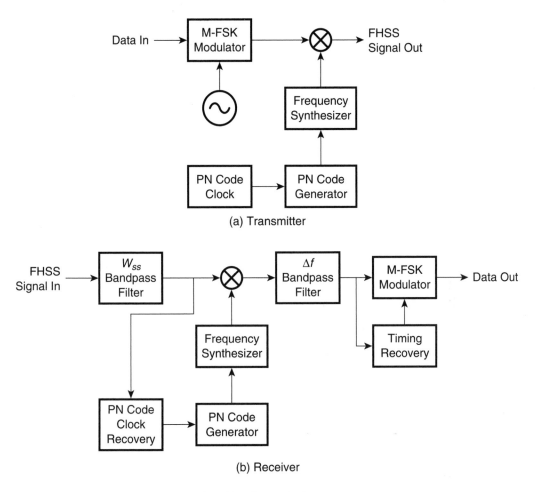

FIGURE 8.13   Block diagram of typical FHSS radio terminal.

burst. The bandwidth of the spectrum within which all the hopping occurs, W_{ss}, is called the *total hopping bandwidth* and is equal to N × Δf. The processing gain, G_p, of a FHSS system is defined by

$$G_P = \frac{W_{ss}}{\Delta f} = N \tag{8.20}$$

If the hopping rate, also know as the chipping rate, is higher than the transmitted data symbol rate, then there are multiple hops per symbol and the system is referred to as a *fast frequency hopping* one. On the other hand, if the hopping rate is lower than the symbol rate, then there are multiple symbols per hop, and the system is referred to as a *slow frequency hopping* one. The hopping rate of a FHSS is limited by the agility of the transmitter and receiver frequency synthesizers used. The data rate is limited by the data throughput possible in the instantaneous bandwidth. It is further limited by the fact that after each hop the receiver frequency synchronizer and clock recovery circuit need a finite amount of time to lock up, during which time no data can be transferred.

Interference occurs in a FHSS system when an undesired signal occupies a particular hopping channel simultaneously with the desired signal. If the undesired signal is from a DSSS system, the desired signal may encounter the undesired signal at all hopping channels, but the undesired signal's spectral density in each narrowband hopping channel is likely to be too low to cause significant performance degradation. If the undesired signal is from another FHSS system, however, a *collision* or *hit* is likely to occur, possibly resulting in serious BER degradation, normally in the form of burst errors.

To gain an intuitive understanding of the performance of FHSS system in a FHSS interference environment, consider the case where several users in the same general geographic area independently but synchronously hop their carrier frequencies. Assume that the desired system uses 4-FSK modulation and incorporates no form of error correction. During periods where there are no hits, the probability of bit error, $P_{be(nh)}$, is given by Eq. (4.164). Whenever a hit with a strong interferer occurs, however, it is reasonable to assume that the probability of error is driven to its maximum value of 1/2. Let us assume that $P_{be(nh)}$ is so low that the average probability of error, P_{be}, is controlled only by those errors generated during hits. Then P_{be} is given by

$$P_{be} = \frac{1}{2} p_h \tag{8.21}$$

where p_h is the probability of a hit.

With N possible hopping channels, the probability that a given interferer will be present in the desired channel (i.e., the probability of a hit), is $1/N$. If there are J interferers, and assuming N is large, then the probability that at least one interferer is present in the desired slot (i.e., the probability of a hit), p_h, is now J/N. Substituting this value of p_h into Eq. (8.21), we get

$$P_{be} = \frac{1}{2} \cdot \frac{J}{N} \tag{8.22}$$

For example, for a receiver with two high-power interfering signals ($J = 2$) and 75 hopping channels available ($N = 75$), then, in an interference environment as described previously, P_{be} would equal 1.33×10^{-2}, an unacceptably high performance for most applications. Recall, however, that the aforementioned environment assumed no error correction. To minimize the impact of hits, FHSS systems do typically employ error correction coding that's good at handling burst errors (for example, Reed-Solomon coding). Another method, often employed if permitted, to minimize the impacts of hits is to incorporate intelligence into the system that allows it to take note of the location of other users in the band and to adapt its hop sets so as to minimize hits with those other users.

8.3   UNLICENSED FREQUENCY BANDS

Unlicensed bands are administered by the same authorities, reviewed in Chapter 7, that administer licensed bands. However, unlike licensed bands, these bands are not subdivided by the authorities into channels. Subdivision, if any, is left up to the manufacturer and the selection of manufacturer determined channels left up to the user.

In the United States, the FCC regulations for unlicensed bands are addressed under Part 15 of Title 47, and there are two classes of such bands. In one class, called the *Industrial, Scientific and Medical (ISM)*, the operational requirements of which are covered in Section 15.247, FHSS and digitally modulated (nonfrequency hopped) systems are allowed. In the other, called the *Unlicensed National Information Infrastructure (U-NII)*, the operational requirements of which are covered in Section 15.407, the types of devices that can be used are not specified and are referred to simply as "Radio Frequency Devices." Because anyone can use these bands, there is no guarantee of noninterference and users operate at their own risk. However, to minimize the likelihood of severe interference, the operating regulations impose several restrictions that include, depending on the band, limitations on transmitter output power, transmitter output power spectral density, and antenna radiated power.

The ISM bands used for broadband outdoor fixed point-to-point communications and some of the key specifications for digitally modulated systems operating therein are given in Table 8.4. Similar data for FHSS systems is given in Table 8.5. Though the 2.4-GHz band is also used by microwave ovens, the pulsed operation of such ovens usually has no impact on well-designed outdoor spread spectrum systems. For digitally modulated point-to-point systems, an important observation is that, in the 2.4-GHz band, maximum peak output power is limited via a "3-for-1" rule when the radiating antenna gain is greater than 6 dBi whereas no such limitation applies in the 5.8-GHz band. For FHSS systems, a specification of particular importance is the maximum hopping channel 20-dB bandwidth. At 5.8 GHz it is limited to 1 MHz. Assuming 4-FSK modulation with a maximum practical spectral efficiency of about 1.5 bits/s/Hz, as suggested in Chapter 4, the implied maximum data rate is about 1.5 Mb/s or 1 T1. However, this doesn't take into account throughput lost as a result of time required for synchronization. The result is that effective T1 transmission is difficult and multiples of T1 just about impossible. Higher levels of FSK modulation could be used to increase throughput, but at the expense of degraded BER performance and increased susceptibility to interference. At 2.4 GHz there

TABLE 8.4   Key Specifications of Digitally Modulated, Fixed, Point-to-Point Systems Operating in the 2.4 and 5.8 GHz ISM Bands, August 20, 2002 Revision

Technical Characteristic	Condition	2.4 GHz ISM Band Specification	5.8 GHz ISM Band Specification
Frequency range (MHz)		2400–2483.5	5725–5850
Width of band (MHz)		83.5	125
Min. system 6 dB bandwidth (kHz)		500	500
Min. processing gain (dB)		0*	0*
Max. peak output power into ant. (dBm)	Radiating antenna gain, G, < or = 6 dBi	30	30
Max. peak output power into ant. (dBm)	Radiating antenna gain, G, > 6 dBi	$30 - ((G-6)/3)$	30
Max. peak power spectral density into ant.		8 dBm in any 3-kHz band	8 dBm in any 3-kHz band

* Note: Processing gain was 10 dB min. prior to 20 August 2002.

TABLE 8.5   Key Specifications of FHSS, Fixed, Point-to-Point Systems Operating in the 2.4 and 5.8 GHz ISM Bands, August 20, 2002 Revision

Technical Characteristic	Condition	2.4 GHz ISM Band Specification	5.8 GHz ISM Band Specification
Frequency range (MHz)		2400-2483.5	5725-5850
Width of band (MHz)		83.5	125
Min. HC carrier freq. separation		Greater of 25 kHz or the 20 dB HC BW	Greater of 25 kHz or the 20 dB HC BW
Min. no. of HCs		15	75
Max. 20 dB BW of HC (MHz)		NA	1
Max. av. time of occupancy on any one channel (s.)	Within period of 0.4 s × no. of HC employed	0.4	NA
Max. av. time of occupancy on any one channel (s.)	Within period of 30 s.	NA	0.4
Max. peak output power into ant. (dBm)	Radiating antenna gain, G, < or = 6 dBi, no. of HCs > or = 75	30	30
Max. peak output power into ant. (dBm)	Radiating antenna gain, G, > 6 dBi, no. of HCs > or = 75	$30 - ((G-6)/3)$	30
Max. peak output power into ant. (dBm)	Radiating antenna gain, G, < or = 6 dBi, no. of HCs < 75	21	NA
Max. peak output power into ant. (dBm)	Radiating antenna gain, G, > 6 dBi, no. of HCs < 75	$21-((G-6)/3)$	NA
Intelligent hopping techniques permitted	Min. of 15 non-over lapping HCs used	Yes	No

Legend: HC = Hopping channel; BW = Bandwidth; NA = Not applicable

is no stated limit on the hopping channel 20-dB bandwidth, but the number of hopping channels has to be at least 15. Taking into account the need to meet out-of-band emission limitations, this results in a maximum hopping channel bandwidth of about 5 MHz. This can support four T1s using 4-FSK. Unfortunately, this increased throughput comes at the expense of radiated power since when the number of hopping channels is less than 75, the radiated power permitted is 9 dB less than when the number of hopping channels is 75 or greater. The net result of these limitations on FHSS throughput capacity and radiated power is that, for outdoor broadband point-to-point systems in the ISM bands, digital modulation (without frequency hopping) is the modulation of choice.

The U-NII bands used for broadband outdoor point-to-point fixed communication and some of the key specifications for systems operating therein are given in Table 8.6. As

TABLE 8.6 Key Specifications of Radio Frequency Devices Used in Fixed, Point-to-Point Communications in the 5.3 and 5.8 GHz U-NII Bands, October 1, 2001 Revision

Technical Characteristic	Condition	5.3 GHz U-NII Band Specification	5.8 GHz U-NII Band Specification
Frequency range (MHz)		5250–5350	5725–5825
Width of band (MHz)		100	100
Max. peak output power into ant. (dBm)	Radiating antenna gain, G, < or = 6 dBi	Lesser of 24 or $11 + 10 \log B$	NA
Max. peak output power into ant. (dBm)	Radiating antenna gain, G, > 6 dBi	Lesser of 24-(G-6) or $11-(G-6)+10 \log B$	NA
Max. peak output power into ant. (dBm)	Radiating antenna gain, G, < or = 23 dBi	NA	Lesser of 30 or $17 + 10 \log B$
Max. peak output power into ant. (dBm)	Radiating antenna gain, G, > 23 dBi	NA	Lesser of 30 – (G-23) or $17 – (G-23)+10 \log B$
Max. peak power spectral density into ant.	Radiating antenna gain, G, < or = 6 dBi	11 dBm in any 1 MHz band	NA
Max. peak power spectral density into ant.	Radiating antenna gain, G, > 6 dBi	11-(G-6) dBm in any 1 MHz band	NA
Max. peak power spectral density into ant.	Radiating antenna gain, G, < or = 23 dBi		17 dBm in any 1 MHz band
Max. peak power spectral density into ant.	Radiating antenna gain, G, > 23 dBi		17-(G-23) dBm in any 1-MHz band

Legend: B = 26 dB system emission bandwidth in MHz.; NA = Not applicable

can be observed from Table 8.6, a major disadvantage of these bands relative to the 5.8-GHz ISM band is the relatively low levels of radiated power permitted.

Availability of unlicensed bands varies around the world. The 5.8-GHz ISM band is available as unlicensed in many other countries and regions, including Canada, China, Brazil, and Southeast Asia, with regulations similar, but not always identical, to those imposed in the United States. The European Union (EU), however, has yet to open this band to point-to-point fixed radio systems. The 2.4-GHz ISM band is also available as unlicensed in many countries around the world, but regulations and the specific frequency range vary. In the EU, the appropriate ETSI standard (EN 300 328) specifies the same frequency range as the FCC in the United States. However, it limits the EIPR (transmitter power + antenna gain relative to isotropic) to only +20 dBm, thus severely limiting the usefulness of the band in an outdoor environment. In general, in Latin America, regulations in this band tend to follow or be close to those of the United States. In Brazil, for example, the maximum peak output power into the antenna is +30 dBm as in the U.S., but the maximum EIRP is limited to +36 dBm. In Canada, only the 2450–2483.5-MHz portion of the band is licensed exempt, and only if the EIRP is limited to +36 dBm. For outdoor systems using the entire 2400–2483.5 MHz, licensing is required. However, while for such licensed systems the maximum transmitter output power is specified at +30 dBm, there is no limit on EIRP.

8.4   TYPICAL RADIO TERMINAL: THE PLESSEY BROADBAND WIRELESS MDR 2400/5800 SR, 4 T1/E1 PLUS 10BASET, 2.4/5.8 GHZ, DSSS CCK UNIT

The Plessey Broadband Wireless MDR 2400/5800 SR terminal operates in the 2.4/5.8-GHz ISM bands. It consists of an indoor unit (IDU) and an outdoor unit (ODU) connected via two cables: a CAT5 cable and a 2 conductor DC cable. Figure 8.14 is a photo of such a terminal and key parameters of both to 2.4- and 5.8-GHz versions are shown in Table 8.7. As can be seen from the table, the data-handling capability of both versions are identical, either 1 to 4 T1s plus 10BaseT, or 1 to 4E1s plus 10BaseT, as well as auxiliary data of up to 115.2 kb/s. The 10BaseT maximum possible throughput rate is a function of the number of T1s/E1s equipped. For example, with no T1s/E1s equipped, this throughput rate is 8 Mbs, whereas with 4 T1s/E1s equipped, it falls to 2/0.15 Mb/s. These designs take advantage of the availability of standard 802.11b WLAN chip sets, which handle date rates of up to 11 Mb/s using CCK modulation. Though developed for LAN coverage, the chip sets are adapted for fixed broadband point-to-point operation over several miles by using them in conjunction with high-gain/high-directivity external antennas. A simplified terminal block diagram is shown in Fig. 8.15. A few key points to note in this realization are as follows:

- On the transmit side of the IDU, the T1/E1 signals, the 10BaseT Ethernet signal, the auxiliary data, and the internal control data are multiplexed in the IDU and then FEC encoded, resulting in a signal of bit rate 11 Mb/s. On the receive side, the process is reversed.
- The 11 Mb/s signal entering the ODU from the IDU is fed to the baseband module, where CCK I and Q modulating streams are created by the transmit baseband processor section of an Intersil HFA3860B chip. These streams are fed to the transmit module, where quadrature modulation takes place via an Intersil HFA3763 chip, creating a CCK-modulated signal, of carrier frequency 374 MHz. This modulated signal is then upconverted to the desired output frequency. The output of the transmit module feeds the power amplifier unit, which provides the final amplification prior to duplexing and transmission via the antenna.
- On the receive side of the ODU, the incoming RF signal from the duplexer is fed to the receive module, where it is low noise amplified and downconverted to an IF frequency of 280 MHz. It is then demodulated to I and Q streams with an Intersil HFA3761 chip and finally restructured to the original 11-Mb/s data stream in the receive baseband processor section of the HFA3860B Intersil chip.

It is interesting to compare the actual BER performance with FEC against the theoretical and actual achievable without FEC. Figure 17 of the HFA3860B data sheet[8] indicates that, for a BER of 10^{-6}, the theoretical E_b / N_0 for 11-Mb/s CCK is 8.7 dB. For both the 2.4- and 5.8-GHz versions of the product under review the typical E_b / N_0, as a result of implementation losses, is 12.5 dB and the typical noise figure, F, is 4.6 dB. Thus, by Eq. (7.2), the theoretical received signal level, P_{Si}, necessary for a BER of 10^{-6} is –90.3 dBm and the expected actual is –86.5 dBm. However, as we see from Table 8.7, the actual

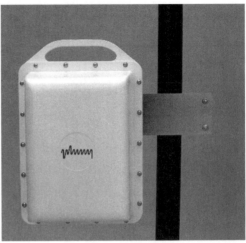

(a) Indoor unit   (b) Outdoor unit

FIGURE 8.14   The Plessey broadband wireless MDR 2400/5800 SR terminal. *(Courtesy of Plessey Broad-band Wireless.)*

TABLE 8.7   Key Parameters of the Plessey Broadband Wireless MDR 2400/5800 SR Terminal

Parameter	Specification
Operating freq. range (MHz): MDR2400/5800	2400–2483.5 / 5725–5850
RF channel bandwidth (MHz)	18
Modulation method	CCK
Primary data	1-4 T1/E1 plus 10BaseT Ethernet
Auxiliary data (kb/s)	115.2 max.
T1 data line code	B8ZS
E1 data line code	HDB3 or AMI
Composite bit rate (Mb/s)	11
Xmtr. output power (dBm)	adj., +24 max
Rcvr. noise figure (dB)	4.6
Rcvr. threshold at 10^{-6} BER, typical (dBm)	−88
Rcvr. threshold at 10^{-3} BER, typical (dBm)	−90
System gain at 10^{-6} BER rcvr. threshold (dB)	112
System gain at 10^{-3} BER rcvr. threshold (dB)	114
Residual BER	10^{-11}
Rcvr. overload level at 10^{-6} BER (dBm)	−30
FEC type, (n,k)	RS (241,221)
FEC overhead (%)	9.05

(Courtesy of Plessey Broadband Wireless.)

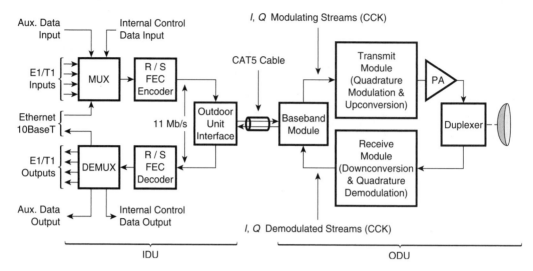

FIGURE 8.15   Simplified block diagram of MDR 2400/5800 SR, 4 T1/E1 plus 10 BaseT, 2.4/5.8 GHz DSSS CCK terminal. *(Courtesy of Plessey Broadband Wireless.)*

typical P_{Si} for a BER of 10^{-6} is −88 dBm. Thus, as result of the use of Reed-Solomon FEC, the actual receiver threshold performance for a BER of 10^{-6} is 2.3 dB worse than the theoretical value, but 1.5 dB better then the practical value, obtainable without FEC.

Note that on October 3, 2003, Stratex Networks, Inc., acquired Plessey Broadband Wireless from Tellumat (Pty) Ltd.

REFERENCES

1. Taub, H., and Schilling, D. L., *Principles of Communication Systems, 2nd ed.,* McGraw-Hill, 1986.
2. Dixon, R. C., *Spread Spectrum Systems*, John Wiley & Sons, 1976.
3. Pearson, R., *Complementary Code Keying Made Simple*, Application Note AN9850.1, Intersil Corp., May 2000.
4. Andren, C., and Webster, M., Intersil Corporation, *CCK Modulation Delivers 11 Mbps for High Rate IEEE 802.11 Extension*, Wireless Symposium/Portable by Design Conference, San Jose, Calif., Spring 1999.
5. IEEE Std. 802.11b-1999, Part II: Wireless LAN Medium Access Control (MAC) and Physical Layer (PHY) specifications: Higher-Speed Physical Layer Extension in the 2.4 GHz Band, The Institute of Electrical and Electronic Engineers, Inc., 20 January, 2000.
6. Sivaswamy, R., "Multiphase Complementary Codes," *IEEE Trans. Inform. Theory,* Vol. IT-24, No. 5, Sept. 1978, pp. 546–552.
7. Andren, C., *Testing for compliance with FCC rules 15-247e*, Intersil Corp., January 11, 2000.
8. Intersil Corporation, *Data Sheet HFA3860B*, July 1999.

CHAPTER 9

FIXED BROADBAND WIRELESS ACCESS SYSTEMS

9.1 INTRODUCTION

In previous chapters, wherever the detailed structure of communication links was addressed, the stated or tacit assumption was that such links had a *point-to-point* (*PTP*) architecture. In this chapter we explore fixed broadband systems where the links form part of either a *point-to-multipoint* (*PMP*) or a *mesh* architecture. Such systems are often referred to as *fixed broadband wireless access* (*FBWA*) systems.

In the simplest PMP architecture, the base station communicates via an omnidirectional antenna with all the remote stations in its coverage area or cell. In a more sophisticated layout the cell is subdivided into sectors and the base station communicates via sector antennas as shown in Fig. 1.3. In broadband communications, such systems are primarily used to provide the last link to subscribers. Such subscribers range from residential or small office/home office (SOHO) ones, where high-speed Internet access and possibly a limited number of standard telephone lines is required, to large business enterprises, where high-speed data, Internet access, and a large number of telephone lines feeding a private branch exchange (PBX) may be the need. Clearly, individual PTP links can be structured to meet such needs. However, unless the individual need is large, such systems tend to be uneconomic. Further, by dedicating a section of bandwidth to one subscriber only who will likely be transmitting information over it at full capacity for only a small percentage of the time, maximum utilization of the available bandwidth is nowhere near possible. PMP systems, on the other hand, if cleverly constructed, can go a long way to overcoming these limitations. By sharing a common hub, for example, per subscriber cost can be significantly decreased, and by sharing the available bandwidth between all subscribers, bandwidth utilization can be greatly increased. All of the basic technologies used to deploy point-to-point systems (for example, digital modulation, up- and downconversion, FEC) are used in PMP systems. However, in addition to these, several other technologies are required. Further, several realizations are possible. The designer is thus faced with a variety of options, and it is not always clear which specific approach is optimum for the specific needs being addressed. There are so many options, in fact, that to address them all in any significant detail would easily fill an entire text. This chapter, therefore, will be limited to a top-level overview of PMP enabling technologies and realization options, with several detailed references provided.

Mesh networks, as indicated in Section 1.5, are in fact comprised of PTP links but configured in such a way that, via software control, traffic can be routed over several different paths. Here we will take a closer look at such networks and contrast them to PMP ones.

9.2   MULTIPLE ACCESS SCHEMES

In PMP broadband systems, transmission in the *downstream* (also referred to as *downlink*) direction (i.e., from the base station to the remotes) can be effected either on a continuous or on a burst basis. In continuous transmission, the carrier can be modulated by a standard TDM signal or by a data signal consisting of packets, where the packets are assigned dynamically to specific remotes. In either case, at each remote, the entire signal is demodulated, and, in the de-multiplexing process, only that data addressed to that remote is extracted. In burst transmission, the framing mechanism can be structured to support either a TDM transmission format or a dynamically assigned packet data format. In the latter case, each burst can be addressed to a specific remote. In the *upstream* (also referred to as *uplink*) direction (i.e., from the remotes to the base station), life is somewhat more complicated. Should all remotes transmit simultaneously on the same frequency, then, without special precautions, massive co-channel interference would occur and the base station receiver would be at a loss as to how to demodulate, much less segregate the incoming data. To allow upstream transmission, a scheme referred to as a *multiple access scheme* must be applied. Three types of multiple access schemes are used in broadband PMP systems—namely *frequency division multiple access* (*FDMA*), *time division multiple access* (*TDMA*), and *code division multiple access* (*CDMA*).

9.2.1   Frequency Division Multiple Access (FDMA)

In FDMA, the total upstream bandwidth available is divided into separate subchannels and each individual subchannel is assigned for transmission between a specific remote station and the base station. Subchannel bandwidths need not be all the same but can be chosen to meet the varying needs of the remote terminals. This subchannel assignment may be either on a permanent basis or on a temporary or so called *on-demand* basis. When assigned on a permanent basis, the scheme is referred to as a *pre-assigned multiple access* (*PAMA*) one, and when assigned on demand, it is called a *demand assigned multiple access* (*DAMA*) one. With a PAMA system, the number of subchannels must equal the number of remote stations. However, with a DAMA system, each frequency can be used on a sequential basis by many remote stations. As all remotes are not likely to be communicating at the same time, more remotes can be served by the system than the number of channels available. Figure 9.1(a) shows an example of subchannel allocation in a DAMA/FDMA system with four subchannels, of frequencies f_1 to f_4, shared between users at eight remote stations. Figure 9.1(b) shows a simplified block diagram of a FDMA base station receiver that processes these subchannels. With FDMA, even though downconversion of all the received signals is usually done as a single process, multiple demodulators are required. This is a disadvantage, in that it limits cost savings, but an advantage, in that it easily allows each upstream link to use the most optimum modulation method. For example, a remote close to the base station could use a high-level scheme, increasing capacity, whereas a remote far from the base station could use a low-level scheme, ensuring good BER performance, albeit at the expense of capacity. Once installed, however, FDMA systems are relatively inflexible in the allocation of capacity between the various remotes.

For systems employing FDMA in upstream direction, downstream transmission is usually via a continuous FDM modulated signal.

9.2.2 Time Division Multiple Access (TDMA)

In time division multiple access (TDMA), the total bandwidth available is used by the remote stations on a sequential basis. Each remote communicates by sending a burst of information in an assigned time slot. Each time a remote transmits a burst, the base station receiver must resynchronize to the new incoming signal. To accommodate this without loss of information data, the front end of every burst consists of a block of synchronization data,

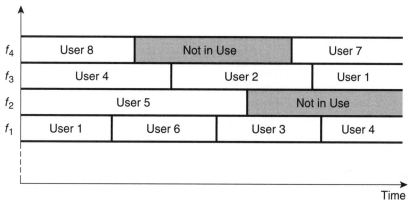

(a) Example of DAMA/FDMA Sub-Channel Allocation

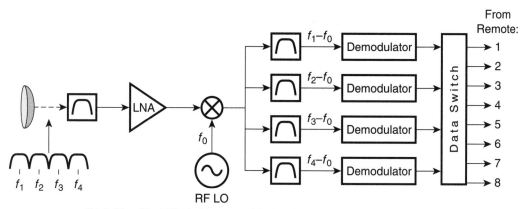

(b) A Simplified Block Diagram of DAMA/FDMA Base Station Receiver

FIGURE 9.1   DAMA/FDMA principles.

often referred to as the preamble. Like FDMA, TDMA systems can be made to function in either a PAMA mode or a DAMA mode.

In the PAMA mode, each remote is assigned the same time slot within a frame of constant repetition rate. Remotes are therefore assigned time slots on a synchronous basis. Note, however, that allocation of time slots between remotes need not be on an equal basis but can be chosen to best match the relative capacity needs of the remotes. Figure 9.2 shows an allocation of time slots between users at three remotes. Note the synchronous assignment, where user 1 is always assigned time slots 1 and 2, user 2 always assigned time slot 3, and user 3 is always assigned time slot 4.

In the DAMA mode, the time slots are assigned to users on a demand basis. Fig. 9.3(a) shows an example of a DAMA/TDMA system allocating four time slots in a synchronous fashion between users at eight remotes. In the first two frames, users 1, 5, 4, and 8 are assigned time slots 1, 2, 3, and 4, respectively. At the start of frame 3, however, user 1 no longer demands service but user 6 does, so time slot 1 is reassigned to user 6. At the start of frame 4 user 8 no longer demands service, and as no new remote demands service, time slot 4 becomes idle and no burst transmission takes place. Note the similarity with the DAMA/FDMA example given previously, the only difference now being that time slots replace subchannels.

In DAMA/TDMA systems, it is also possible, though more complex, to assign the time slots on an asynchronous basis. With such an assignment there is no repetitive allocation of a given time slot to a given user. Rather, time slots are allocated in what appears to be a random fashion but is in fact a sophisticated allocation based on individual throughput (often called in this case "bandwidth") and quality of service required. Figure 9.3(b) shows an example of asynchronous allocation of time slots. Such allocation allows a more versatile use of the total transmission capacity as compared to synchronous allocation. Channel assignment and throughput can be made highly dynamic, a feature well suited to most data communication needs.

Note that with TDMA only a single base station modem is required. This modem is quite complex, however, in that it has to process independent signal bursts. Note also that in TDMA systems, downstream transmission can either be continuous or burst.

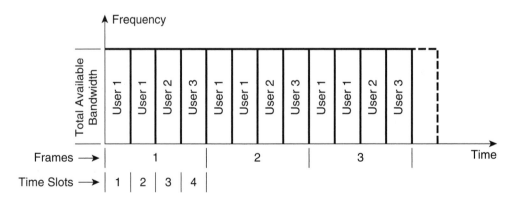

FIGURE 9.2   Example of PAMA/TDMA time slot allocation.

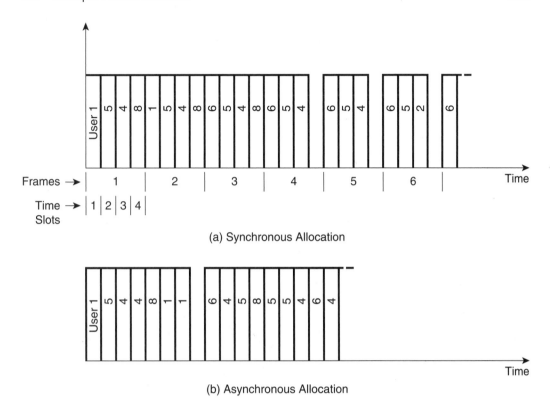

FIGURE 9.3   Examples of DAMA/TDMA time slot allocation.

9.2.3   Direct Sequence-Code Division Multiple Access (DS-CDMA)

In *direct sequence-code division multiple access* (*DS-CDMA*) systems all remotes can communicate simultaneously with the base station, each occupying the same bandwidth, without catastrophic interference effects. This is possible because users communicate with the base station via DSSS as described in Section 8.2.1. Each remote is assigned an individual and distinctive PN code, and these codes are almost uncorrelated with one another. If there are k active remotes, then the base station must despread and demodulate k independent DSSS messages. A simplified block diagram of a DS-CDMA base station receiver where the despreading occurs at IF is shown in Fig. 9.4. Note that like FDMA, multiple demodulators are required. For a DS-CDMA system with k active remotes, where

(a) the fundamental modulation is BPSK
(b) all remotes have the same bit rate f_b and the same chip rate f_C
(c) each signal from the k remotes presents the same power to the base station receiver, and

(d) thermal noise power is negligible compared to the total interfering noise power, it can be shown[1] that the probability of bit error P_{be} is given by

$$P_{be} = \frac{1}{2} erfc\left[2\left(\frac{1}{k-1}\right)\left(\frac{f_C}{f_b}\right)\right]^{\frac{1}{2}} \tag{9.1a}$$

$$= Q\left[4\left(\frac{1}{k-1}\right)\left(\frac{f_C}{f_b}\right)\right]^{\frac{1}{2}} \tag{9.1b}$$

When the received power from an unwanted user is much larger than that of the wanted user, a significant number of errors can occur. For remotes with the same output power this situation occurs when the wanted user is far from the base station and an unwanted user is near. This problem is therefore referred to as the *near-far problem*. It is normally addressed by using adaptive power control on all the remotes to ensure that the signal levels that they present at the base station receiver are essentially equal, regardless of their location and the fading environment.

In DS-CDMA systems, DS-CDMA can be used in the downstream as well as the upstream direction. When used in the downstream direction, signals from the base station with unique PN codes are transmitted to each active remote, and at each remote the receiver despreads and demodulates only that signal addressed to it. It should be noted that often DS-CDMA systems are simply referred to as CDMA systems.

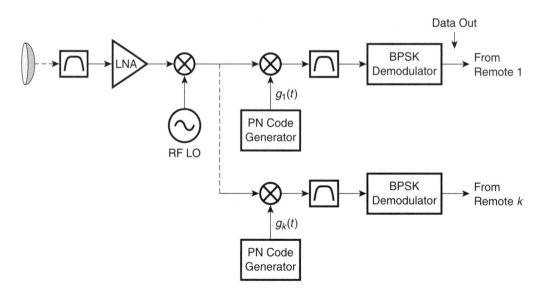

FIGURE 9.4   A simplified block diagram of a DS-CDMA base station receiver.

9.2.4 Frequency Hopping-Code Division Multiple Access (FH-CDMA)

In *frequency hopping-code division multiple access* (*FH-CDMA*) systems, remotes communicate with a base station via FHSS, as described in Section 8.2.3, each occupying the same bandwidth. Here, as in DS-CDMA systems, each remote is assigned an individual and distinctive PN code. If there are k active remotes, then the base station must independently dehop and demodulate the k received signals. Also as in DS-CDMA systems, in FH-CDMA systems, FH-CDMA can be used in the downstream as well as the upstream direction.

9.3 TRANSMISSION SIGNAL DUPLEXING

A transmission signal duplexing scheme is a method of accommodating the transmission and reception of signals on a two-way link. There are three schemes that are utilized in broadband wireless access networks—namely, *frequency division duplexing* (*FDD*), *frequency switched division duplexing* (*FSDD*), and *time division duplexing* (*TDD*).

9.3.1 Frequency Division Duplexing (FDD)

Frequency division duplexing (FDD) is the traditional form of duplexing and is discussed in detail in Section 7.5. In PMP systems utilizing FDD the downstream and upstream frequency channels are separate and thus all remote stations can transmit and receive simultaneously. The channels are usually, but not necessarily, of equal size. In systems employing FDD, the downstream transmission can be either continuous or bursty.

9.3.2 Frequency Switched Division Duplexing (FSDD)

Frequency switched division duplexing (FSDD), also known as *half duplex-frequency division duplexing* (*H-FDD*), is a duplexing scheme in which, like FDD, the downstream and upstream frequency channels are separate, but where some or all of the remote units cannot transmit and receive simultaneously but must do so sequentially. As a result, both downstream and upstream transmission is bursty. Those remotes that can operate simultaneously are said to operate in a *full-duplex* mode, whereas those that must operate sequentially are said to operate in a *half-duplex* mode. The design of remote terminals that operate in a half-duplex mode is simplified as the antenna coupling device now consists essentially of only a switch that switches between the transmitter output and receiver input instead of a frequency separating device as described in Section 7.5. An example of time allocation on a FSDD system is shown in Fig. 9.5.

9.3.3 Time Division Duplexing (TDD)

Time division duplexing (TDD) is a duplexing scheme where the downstream and upstream transmissions share the same frequency channel but do not transmit simultaneously. Thus, bursty transmission is required in both directions. However, because of the rapid speed of switching between these transmissions, "simultaneous" two-way communication

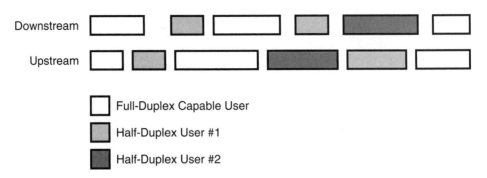

FIGURE 9.5   Example of FSDD time allocation.

is preserved. The allocation of time between downstream and upstream traffic is normally adaptive, making this technique highly attractive for situations where the ratio of downstream to upstream traffic is likely to be asymmetric and highly variable. One impediment to the deployment of TDD systems is that many global spectrum allocations dictate FDD and prohibit TDD because of the difficulty in coordinating it with FDD systems from an interference point of view. This difficulty arises because the transmission of the same frequency in both downstream and upstream directions can make the discrimination of a nearby antenna to these signals limited to nonexistent.

9.4   MEDIUM ACCESS CONTROL (MAC)

In the dowstream direction of a PMP system, all remotes receive the same transmission from the base station. Since the base station is the only transmitter operating in the downstream direction, it can transmit without having to coordinate with other transmitters in the system, except in the case of TDD, when it must allow periods for upstream transmission. In the upstream direction of a DAMA system, however, users share the transmission medium (i.e., the RF channel) be it via frequency division or via time division, on a demand basis. When such is the case, there must exist a mechanism to oversee this sharing in an efficient manner. This mechanism is referred to as the *medium access control (MAC)* of the system. The MAC coordinates and schedules transmission among competing remotes with the goal of low latency and good overall channel utilization. Further, it can be designed to allow significant flexibility in upstream access. For example, remotes can be granted varying classes of service. Class of service may be designed to vary from remote transmission only being allowed when the remote requests it and bandwidth is available to a continuing right over exclusively assigned bandwidth. There are many and varied ways to implement medium access control on broadband PMP DAMA/TDMA systems. Before reviewing techniques commonly in use in such systems, a very brief overview of a foundation laying access control system developed for satellite communications is in order.

Perhaps the first widely used "wireless" DAMA/TDMA MAC scheme was one implemented by the University of Hawaii to provide access via satellite to several university computers. The scheme, dubbed *ALOHA*, is extremely simple in its concept. Basically, with this scheme, all messages are packets of constant length, and any user can attempt to communicate at any time. After sending a packet the use listens for a positive acknowledgment (*ACK*) from the receiver. With receipt of an ACK the user knows successful transmission has been accomplished. However, because more than one user may transmit a packet at the same time, collisions can occur. When a collision occurs, the receiver is unable to decode the transmissions and broadcasts a negative acknowledgment (*NAK*). When a user who has just sent out a packet receives a NAK, it simply retransmits the packet. Now, however, it commences transmission after a random delay to minimize the probability of a collision with the other user(s) that it just collided with. This type of access scheme is referred to as a *contention* scheme, as the users contend for access. With lots of users the basic ALOHA quickly bogs down, so designers implemented improvements to it. One such improvement is called *slotted ALOHA (S-ALOHA)*. Here messages have to be sent in time slots defined by the base station, including retransmitted messages, which are transmitted after a random delay of an integer number of time slots. This change results in the rate of collisions being halved, since now only packets transmitted in the same slot can collide with each other. Another improvement, building on slotted ALOHA, is called the *reservation-ALOHA (R-ALOHA)*. Here, importantly, a time frame is established. This frame is divided into time slots, which, under certain circumstances, can be further divided into reservation subslots. If no users are requesting service, all time slots are divided into reservation subslots and these subslots are available to all users on a contention basis for the purpose of making a reservation. When a user wants to send a packet or packets, it sends a reservation request to the MAC, which grants it a specified time slot or slots. Once a time slot is granted, all time slots except the last in the frame are no longer divided into reservation subslots, but instead are made available for packet transmission. Though ALOHA and its evolution took place in a satellite communication environment, it is important to broadband TDMA communication because it put in place the building blocks for modern PMP DAMA/TDMA communication protocols.

Most state-of-the-art PMP DAMA/TDMA systems employ as their access scheme a derivative of R-ALOHA, employing a combination of randomly accessed contention slots for making reservations and reserved slots that are assigned by the base station based on many considerations, including the quality of service assigned to the requesting remote. Further, a polling mechanism is often overlaid in which the base station assigns a slot to each polled remote. This procedure is useful for ongoing interrogation of the health and status of remote units and for situations where most or all remotes have information to transmit, as can be the case if there is a momentary failure at the base station resulting in the need to reset parameters. For a PMP system employing a TDM downstream and a DAMA/TDMA upstream, Fig. 9.6(a) shows an example of a downstream frame structure, of period t_d, and Fig. 9.6(b) shows an example of an associated upstream frame structure, of period t_u. The downstream signal is continuous and at the beginning of each frame has control data for capacity assignment in the upstream channel, upstream frame sync purposes, polling, and so on. Following the control slot are information slots, which carry data packets that each consist of a header and a data field, the header containing a destination address and other control data. Each remote terminal receives full frames but outputs only the specific time slots addressed to it. The upstream frame is divided into two sections. The first contains random data slots, and these are used by the remotes on a contention basis to make a reservation. The second section contains reserved slots that can only be used by the remote to which it has been assigned. Each data slot, random or reserved, includes data that consists of a preamble for synchronization purposes, a header

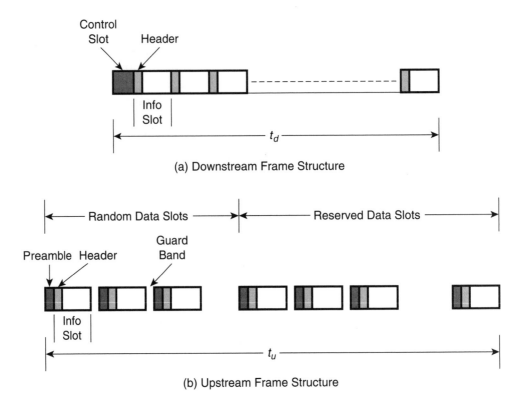

FIGURE 9.6   TDM/TDMA PMP system downstream and upstream frame structure examples.

for address and other purposes, and a data field. In order to eliminate collisions, adjacent data slots are separated from each other by a guard band.

9.5   DYNAMIC BANDWIDTH ALLOCATION (DBA)

Most sources of data, including voice, are bursty. Voice data, on average, requires a capacity-to-usage ratio of about 2:1 since in a given direction the speaker is only speaking about half the time. Computer and Internet derived data typically have capacity-to-usage ratios of 10:1 or more. As a result of such ratios, assigning a fixed bandwidth for data transmission from each source in a multisource system is inherently wasteful. As indicated previously, one of the goals of the MAC layer is to ensure the efficient use of the available bandwidth. A packet based DAMA/TDMA system is an excellent one for facilitating this goal because, in principle, the bandwidth can be allocated between users on a dynamic basis, thus minimizing any unused time. When so allocated, the system is said to employ *dynamic bandwidth allocation (DBA)*. With DBA, in a given downstream or upstream frame, packets, being individually addressed, can be assigned in any amount to any user. Knowing the total bandwidth available and the utilization at any instant in time, a DBA system can allocate any unused bandwidth to subscribers based on the quality of service assigned to them and their individual needs for more bandwidth. DBA allows the system to handle data traffic where the sum of the peak demand of each user significantly exceeds that of the channel. This is possible because of the very low probability that individual users require their peak capacity at the same time. Figure 9.7(a) shows a simple example of bandwidth allocation without DBA for a system that carries both voice and data traffic, and Fig. 9.7(b) shows bandwidth allocation for the same traffic but with DBA. Clearly, DBA has allowed the transmission of voice and data traffic that was restricted in the non-DBA case.

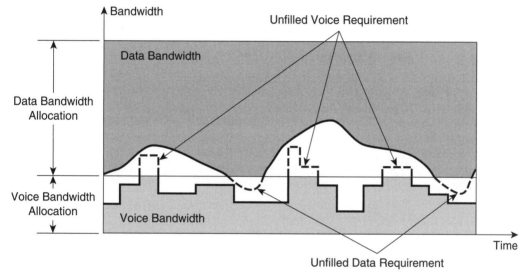

(a) Bandwidth Allocation without DBA

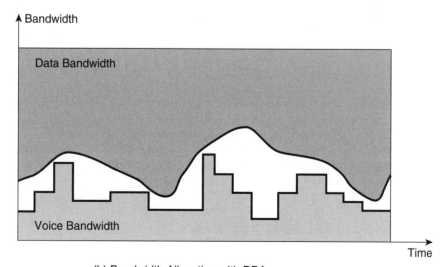

(b) Bandwidth Allocation with DBA

FIGURE 9.7   Example of DBA versus non-DBA.

9.6 ADAPTIVE MODULATION

In a PMP system layout, the further a remote station is from the base station the lower the average signal level it receives from the base station and the lower the average signal level the base station receives from it. In addition, the further the remote is from the base station, the larger the fading that is likely to occur on the remote/base station path, as fading is a function of path length. As we know, in TDMA systems, remotes communicate in bursts that occupy the same bandwidth, this being the entire assigned bandwidth. Thus, if a high-level modulation is used on one remote and a low level on another, the high-level modulation remote will be able to transmit more data in its burst than the low-level one. However, at the base station, the received level from the high-level modulation remote must be higher than that from the low-level modulation remote for similar BER performance. All of this would suggest that, for similar reliability on all individual remote/base station links, modulation type should be a function of path length, assuming equal transmitter power on all remotes. High-level modulation could be employed on short paths resulting in greater throughput per burst, and modulation of decreasing complexity employed as path length increases, with a corresponding decrease in throughput per burst. For this throughput maximization scheme to be truly effective, however, it must be implemented in the downstream direction as well. In TDMA type systems, having a different modulation on each remote transmitter is quite straightforward. However, having modulation at the base station that varies as a function of the particular remote being addressed at a particular time is more complex as it necessitates burst, not continuous transmission. Taking to the limit this concept of adjusting modulation per path as a function of received signal level, many manufacturers offer *adaptive modulation*. With such a scheme, modulation is adjusted automatically per base station/remote link, independently on the downstream and upstream directions, and in such a way as to optimize the trade-off between capacity and reach. Modulation can be adjusted on a burst-by-burst basis. In addition to maximizing throughput and coverage area, adaptive modulation can also help in combating co-channel interference. It does this by decreasing modulation complexity and hence increasing resistance to interference whenever performance starts to degrade as a result of such interference. Instantaneous measurements are made of signal-to-noise ratio (S/N) and carrier-to-interference ratio (C/I), and the modulation complexity varied dynamically. Thus, if, for example, a remote terminal is experiencing good S/N because its path is not experiencing a fade, a more complex downstream modulation can be used resulting in increased throughput per burst. If, on the other hand, the base station is simultaneously experiencing poor C/I because the upstream path, on a different frequency to the downstream path, is experiencing a fade, a less complex modulation can be used to maintain reliability. Typically, in systems employing adaptive modulation, modulation varies between 4-QAM and 16-QAM, and in some cases up to 64-QAM.

9.7 NON-LINE-OF-SIGHT (NLOS) TECHNIQUES

Broadband service to SOHO and residential customers is highly price sensitive. Thus, if PMP broadband wireless access is to be successful in this market arena, low hardware and installation costs as well as a high probability of establishing effective communication links is imperative. One way to make progress in this regard is to apply technology that permits *non-line-of-sight* (*NLOS*) communication between the base station and the remote terminals.[2] This is because, in the typical deployment scenario in an urban or suburban environment, as few as 30% of potential subscriber remote stations will actually have a LOS with the base station. Further, most of those that are fortunate enough to have LOS with adequate received signal level achieve this via the accurate alignment of a rooftop-mounted, highly directive antenna. Such mounting and alignment normally requires a skilled installer, is time consuming, and, as a result, costly. With NLOS transmission, on the other hand, less directive remote station antennas can be used, allowing installation by a less skilled worker or by the subscriber. Additionally, in some instances, it may be possible to install the antenna under the eaves of rooftops or even indoors. Frequencies above about 10 GHz essentially require line-of-sight for acceptable communication as even the slightest obstruction by buildings or trees imparts significant loss. As we move down in frequency from 10 GHz, however, the signal tends to diffract around or penetrate through obstacles and, as a result, obstruction loss becomes less and less. Further, a number of reflected signals tend to end up at the receiving antenna (i.e., multipath becomes more and more predominant). Taking advantage of these lower frequency phenomena, technologies have been and are being developed to permit effective NLOS communication, predominantly at frequencies between about 2 and 5 GHz. In a broad sense, these techniques can be grouped into two categories: specialized transmission techniques that perform well in a multipath environment, and spatial processing techniques, which seek to maximize signal strength relative to interference and noise. The application of a technique from one of these categories on its own, however, is hard pressed to provide true NLOS communication. Rather, it is the synergistic combination of techniques, one from each category, which can lead to effective NLOS links. Following is a high-level overview of two transmission techniques—namely *orthogonal frequency division multiplexing* (*OFDM*) and CDMA with *RAKE* receiver, as well as the more popular spatial processing techniques—that are applied in NLOS broadband fixed PMP systems.

9.7.1   Orthogonal Frequency Division Multiplexing (OFDM)

Orthogonal frequency division multiplexing (OFDM)[3,4] is not a modulation technique though it is often loosely referred to as such. Rather, it is a multicarrier transmission technique, which allows the transmission of data on multiple adjacent subcarriers, each subcarrier being modulated in a traditional manner with a linear modulation scheme such as QAM. In OFDM, the data for transmission is, via a serial to parallel converter, converted into several parallel streams and each stream used to modulate a separate subcarrier. Thus, only a small amount of the total data is transmitted via each subcarrier, in a subchannel a fraction of the width of the total channel. As a result, in a multipath fading environment, as a fade notch moves across the channel, the fading appears to each subchannel almost as a flat fade. It thus induces a significantly reduced amount of intersymbol interference compared to that which would be experienced by a single carrier modulated system with a spectrum extending across the entire band. Further, while those subchannels at or close to the notch may experience a deep fade and hence thermal noise and ISI induced burst errors, those removed from the notch won't. In the reconstructed original data stream, these burst errors are randomized, due to the interleaving that results from the parallel to serial process, and therefore more easily corrected with FEC. The robustness of OFDM to multipath interference is one of its most important properties. Figure 9.8 shows the spectrum of a standard four-channel FDM system. In such a scheme, modulated signals are stacked adjacent to each other with a guard band between each adjacent spectrum to ensure that there is no overlap between frequencies and to facilitate recovery via filtering of each signal at the receive end. While this approach works well, it suffers from the major drawback that, because of the spacing required between subcarriers, it wastes spectrum. As a result, it requires more bandwidth than would be required by a single carrier modulated by the original data stream, assuming that the same modulation is applied to the single carrier as to the subcarriers. With OFDM, however, the subcarriers are cleverly stacked close to each other. This results in overlapping spectra, which (1) eliminates the spectral utilization drawback without incurring an adjacent channel interference penalty, and (2) retains advantages in the multipath arena that accrue to parallel transmission of lower data rate streams.

OFDM achieves its close stacking property, without adjacent channel interference, by making the individual subcarrier frequencies *orthogonal* to each other (more on orthogonality later). This is accomplished by having each subcarrier frequency be an integer multiple of the symbol rate of the modulating symbols and each subcarrier separated from its nearest neighbor(s) by the symbol rate. Thus the multiple to generate each subcarrier is one integer different from those to generate its adjacent neighbors. Figure 9.9(a) shows an example of four subcarriers in the time domain over one symbol period, τ. Figure 9.9(b) shows these subcarriers in the frequency domain when each is modulated with symbols of period τ in the form of rectangular pulses. With such modulation, each subcarrier amplitude spectrum has the familiar sin x / x format, but note that, by the choice of subcarrier frequencies, the spectra overlap and each spectrum has a null at the center frequency of each of the other spectra in the system. Why this structure allows the individual modulated subcarriers to be demodulated with no interference from its neighbors is not intuitively obvious, at least not to the author, by simply looking at Fig. 9.9(b). Remember, we are looking at a frequency domain representation, not a time domain representation of overlapping pulses. If anything, this figure seems to represent the ultimate in adjacent channel interference. The explanation here, like the devil, is in the detail, the detail being orthogonality.

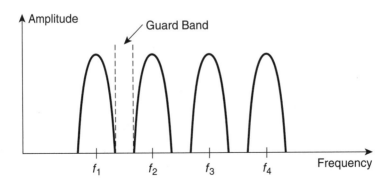

FIGURE 9.8   Standard FDM frequency spectrum.

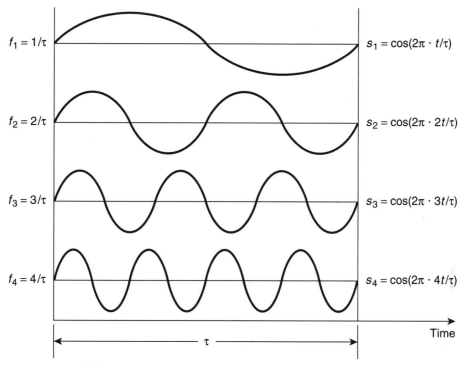

(a) Time Domain Representation of 4 OFDM Sub-Carriers

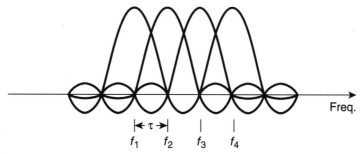

(b) Frequency Domain Representation of 4 OFDM Modulated Sub-Carriers

FIGURE 9.9   Time and frequency representations of OFDM subcarriers.

The term *orthogonal* refers to the total uncorrelation between variables. In the case of OFDM the term *orthogonal* is used in reference to the mathematical relationship be-tween subcarriers. For an OFDM system where the individual subcarrier modulating sym-bol period is τ, the symbol rate equals $1/\tau$. A cosine function derived subcarrier, S_n, of frequency n times the symbol rate is thus given by

$$s_n = \cos(2\pi n t / \tau) \tag{9.2}$$

and, when multiplied with itself and integrated over the period τ, we get

$$\int_0^\tau s_n \cdot s_n dt = \int_0^\tau \cos(2\pi nt/\tau)\cdot\cos(2\pi nt/\tau)\,dt$$

$$= \int_0^\tau \frac{1}{2}(1+\cos(4\pi nt/\tau))dt \qquad (9.3)$$

$$= \frac{\tau}{2}$$

(the latter term within the integral sign integrates to zero as the integration is over a whole number of cycles). When, however, two different subcarriers, of frequencies n and m times the symbol rate, are cross multiplied and integrated over the period τ, we get

$$\int_0^\tau s_n \cdot s_m dt = \int_0^\tau \cos(2\pi nt/\tau)\cdot\cos(2\pi mt/\tau)\,dt$$

$$= \int_0^\tau \frac{1}{2}\left(\cos\big(2\pi(n+m)t/\tau\big)+\cos\big(2\pi(n-m)t/\tau\big)\right)dt \qquad (9.4)$$

$$= 0$$

This latter result is because we are integrating both sinusoidal functions within the integral sign over a whole number of cycles. This is a direct consequence of our choice of subcarrier frequencies relative to the symbol rate $1/\tau$, and it is this relationship between cosine function subcarriers that make them orthogonal.

It can similarly be shown that with sin function derived subcarriers

$$\int_0^\tau \sin(2\pi nt/\tau)\cdot\sin(2\pi mt/\tau)dt = \frac{\tau}{2} \qquad (m=n) \qquad (9.5a)$$

$$= 0 \qquad (m \neq n) \qquad (9.5b)$$

and, further, that

$$\int_0^\tau \cos(2\pi nt/\tau)\cdot\sin(2\pi mt/\tau)dt = 0 \qquad \text{(for all } n \text{ and } m\text{)} \qquad (9.6)$$

To see how OFDM takes advantage of these orthogonal properties in the demodulation process, consider an OFDM system with N subcarriers, with subcarrier frequencies varying from 0 times the symbol rate to $N-1$ times the symbol rate. Further, assume QAM modulated subcarriers, with the modulated subcarrier of frequency n times the symbol rate being given by

$$s_{QAM,n} = a_n \cos(2\pi nt/\tau) + b_n \sin(2\pi nt/\tau) \qquad (9.7)$$

Then the total OFDM signal, S_{OFDM}, is the sum of all such subcarriers and given by

$$S_{OFDM} = \sum_{n=0}^{n=N-1} S_{QAM,n} \tag{9.8a}$$

$$= \sum_{n=0}^{n=N-1} \left\{ a_n \cos(2\pi nt / \tau) + b_n \sin(2\pi nt / \tau) \right\} \tag{9.8b}$$

At the receiver, to decipher the symbol information a_k, S_{OFDM} is simply multiplied by $\cos(2\pi kt / \tau)$ and the product integrated over the period τ. This follows since

$$\int_0^\tau S_{OFDM} \cdot \cos(2\pi kt / \tau)dt$$

$$= \sum_{n=0}^{n=N-1} \left\{ a_n \int_0^\tau \cos(2\pi nt / \tau) \cdot \cos(2\pi kt / \tau)dt + b_n \int_0^\tau \sin(2\pi nt / \tau) \cdot \cos(2\pi kt / \tau)dt \right\} \tag{9.9a}$$

and by applying Eq. (9.4) and (9.6) to Eq. (9.9a), we get

$$\int_0^\tau S_{OFDM} \cdot \cos(2\pi kt / \tau)dt = \frac{\tau}{2} a_k \tag{9.9b}$$

Similarly, to decipher the symbol information b_k, S_{OFDM} is multiplied by $\sin(2\pi kt / \tau)$ and the product integrated over the period τ, since

$$\int_0^\tau S_{OFDM} \cdot \sin(2\pi kt / \tau)dt$$

$$= \sum_{n=0}^{n=N-1} \left\{ a_n \int_0^\tau \cos(2\pi nt / \tau) \cdot \sin(2\pi kt / \tau)dt + b_n \int_0^\tau \sin(2\pi nt / \tau) \cdot \sin(2\pi kt / \tau)dt \right\} \tag{9.10a}$$

and by applying Eq. (9.5) and (9.6) to Eq. (9.10 a), we get

$$\int_0^\tau S_{OFDM} \cdot \sin(2\pi kt / \tau)dt = \frac{\tau}{2} b_k \tag{9.10b}$$

The OFDM signal S_{OFDM} is referred to as the baseband OFDM signal and in real systems is upconverted in the transmitter to the desired transmission frequency band. If the subcarriers are upconverted by f_c Hz, then the RF OFDM signal, $S_{OFDM,RF}$ is given by

$$S_{OFDM,RF} = \sum_{n=0}^{n=N-1} \left\{ a_n \cos\left(2\pi(f_c + n / \tau)t\right) + b_n \sin\left(2\pi(f_c + n / \tau)t\right) \right\} \tag{9.11}$$

In the receiver, the received signal is downconverted back to baseband prior to demodulation.

In order to avoid the construction of a large number of subchannel modulators in the transmitter and an equal number of filters and demodulators in the receiver, modern OFDM systems utilize *digital signal processing* (*DSP*) devices. In fact, it is the availability of such devices that has made the commercialization of OFDM possible. Directly as a consequence of the orthogonality of the OFDM signal structure, modulation is able to be performed, in part, by using DSP to carry out an *inverse discrete Fourier transform* (*IDFT*). Similarly, demodulation is able to be performed, in part, by using DSP to carry out a *discrete Fourier transform* (*DFT*). The Fourier transform allows events in the time domain to be related to events in the frequency domain, and vice versa for the inverse Fourier transform. The conventional transform, as discussed in Section 3.2.1, relates to continuous signals. However, digital signal processing is based on signal samples and so uses DFT and IDFT, which is a variant of the conventional transform. In fact, it is typically the *fast Fourier transform* (*FFT*) and the *inverse fast Fourier transform* (*IFFT*) that are normally applied, these being a rapid mathematical method for computer applications of DFT and IDFT, respectively.

Figure 9.10 shows the basic processes in an IFFT/FFT based OFDM system. The incoming serial data is first converted from serial to parallel in the S/P converter. If there are N subcarriers, N sets of parallel data streams are created. Each set contains a subset of parallel data streams, depending on the type of modulation. For example, if the modulation is 16-QAM, then each set contains four parallel data streams, the 4 bits in each symbol period of these streams being used to define a specific point in the 16-QAM constellation. The parallel data streams feed the mapper. For each subcarrier, the input data per symbol period is mapped into the complex number representing the amplitude and phase value of the subcarrier. For example, if the modulation is 16-QAM and the constellation diagram is as shown in Fig. 4.23(b), then 1110 is mapped to the complex number $1 + 3i$. The outputs of the mapper feed the IFFT processor. The IFFT knows the unmodulated subcarrier frequency associated with each input and uses this, along with the input, to define the modulated signal in the frequency domain. An IFFT is performed on this frequency representation, and the output is a set of time domain samples. The next process, the addition of a cyclic prefix, is optional and will be discussed later. The outputs of the cyclic prefix adder are multiplexed via the P/S converter to create a burst of serial samples per symbol. These samples are then transformed from discrete to analog (continuous) format via the D/A converter and low-pass filter. The output of the low-pass filter is the baseband OFDM signal and is upconverted to create the RF OFDM signal. At the receiver, the process is reversed. The received waveform is downconverted, digitized, and demultiplexed, and then converted back to the complex representation of the symbols by the FFT. These complex representations are then demapped to recreate the original parallel data streams, which are then transformed to the original serial stream by the P/S converter.

As indicated previously, OFDM is very robust in the face of multipath fading. Nonetheless, in the presence of such fading, a certain amount of ISI is unavoidable unless techniques are implemented to avoid it. Figure 9.11(a) is an illustration of how ISI can be incurred as a result of a delay signal. One technique for eliminating, if not significantly reducing, ISI is the adding of a *guard interval*, or *cyclic prefix*, of length τ_g to the beginning of each subcarrier transmitted useful symbol, of length τ_u, as shown in Fig. 9.11(b). This action allows time for the multipath signals from the previous symbol to die away before the information from the current symbol is processed over the unextended symbol period.

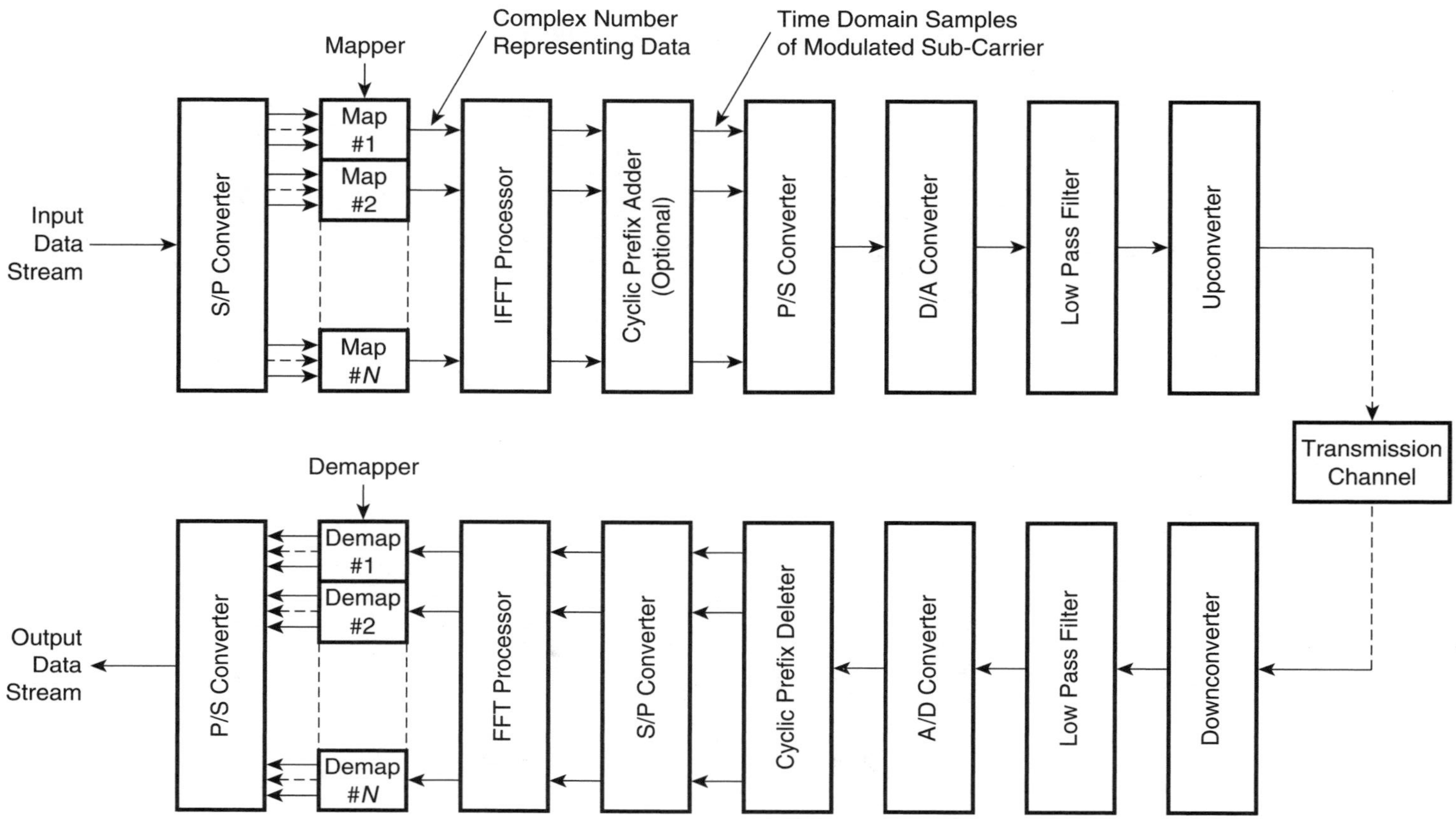

FIGURE 9.10   Basic processes in IFFT/FFT based OFDM system.

Figure 9.11(c) is an illustration of how the ISI, shown in Fig. 9.11(a), is avoided by the use of a cyclic prefix. Cyclic prefix addition is carried out while the symbol is still in the form of IFFT samples and is achieved by copying the last section of the symbol, typically 1/32 to 1/4 of it, and adding it to the front. In Fig. 9.10 it is shown taking place before the P/S converter, but it can also be done on the multiplexed signal after the P/S converter. With this addition, the symbol total duration is now $\tau_t = \tau_g + \tau_u$. Due to the periodic nature of the modulated subcarrier, the junction between the prefix and the start of the original burst is continuous. By adding the guard interval in this manner, the length of the symbol is extended while maintaining orthogonality between subcarriers. As long as the delayed signals from the previous symbol stay within the guard interval, then in the time τ_u there will be no ISI, only "interference" of the symbol by a delayed version of itself. This interference results in no distortion of the symbol, only in a change to its amplitude and a shift of its phase. In the guard interval there will be ISI, but the guard period is eliminated in the receive process and the received symbol is processed only over the period τ_u. Should the delayed signals from the previous symbol extend beyond the guard interval, however, ISI occurs, but is likely to be limited, as the strength of the delayed signals beyond the guard interval is likely to be small relative to the desired symbol. Nonetheless, should it be desired to minimize this residual ISI even further, this can be achieved by the addition of one-tap baseband transversal equalizers in the demappers. While adding a prefix eliminates or minimizes ISI, it is not without penalty. It reduces data throughput, since N symbols are now transmitted over the period τ_t instead of over the shorter period τ_u. For this reason, τ_g is usually limited to no more than about 1/4 of τ_u.

As indicated previously, the input to the transmitter's IFFT processor is a set of complex number representations of modulated subcarriers. In the processor, each of these frequency domain inputs is sampled once and processed, based on a symbol duration of τ_u, to produce a time domain sample. However, the IFFT process is usually carried out on a total number of samples of size 2^x, where x is a positive integer. In real realizations, therefore, for a system with N real subcarriers, the processor block size is chosen so that $2^x \geq N$. For example, if the number of real subcarriers is 200, then the smallest processor block size would be 256. In such a situation, there would in effect be 56 *null subcarriers*, i.e., "subcarriers" of zero value. It is the convention to designate an OFDM structure as the IFFT, and hence FFT, size or number of "points" which is the smallest power of two above the number of real subcarriers. Thus, the OFDM structure in the example above is designated as a "256 point FFT" one, even though there are only 200 real subcarriers.

In addition to being used to carry data, a few real subcarriers are sometimes used as pilot subcarriers. Such subcarriers can be employed for various purposes such as aiding in receiver and transmitter synchronization and in minimizing the effects of received signal distortion due to multipath fading. Null subcarriers, when present, are usually placed symmetrically above and below the real subcarriers. This way, if some or all of them fall within the channel bandwidth, they provide guard bands that enable the signal to decay naturally within this bandwidth.

Since, with cyclic prefix addition, the symbol duration out of the cyclic prefix adder is τ_t, then the period between useful symbol inputs into the adder, and hence between useful symbols out of the IFFT, must also equal τ_t. This in turn dictates the period between symbols into the IFFT to be τ_t, and hence the sampling rate of the representation of each sub-carrier to be $1 / \tau_t$. It is important to recall, however, that the symbol length used for

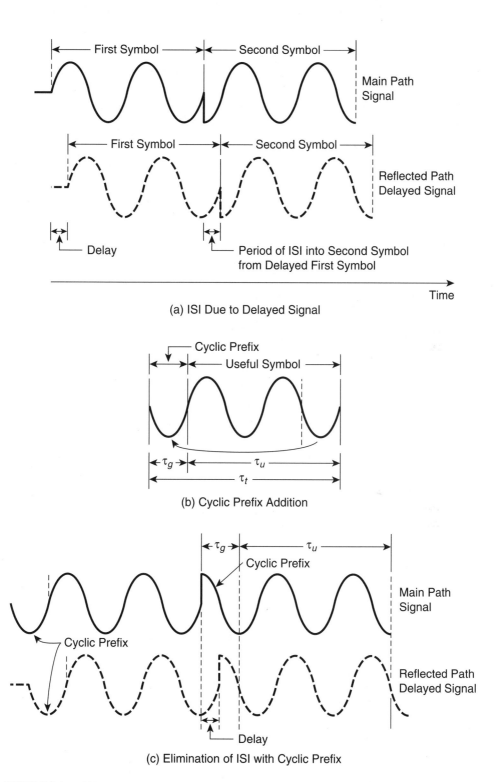

FIGURE 9.11   ISI generation and elimination.

IFFT processing is τ_u. In the absence of cyclic prefix addition, this sampling rate becomes $1 / \tau_u$. Thus, if the number of FFT points (FFT size) is N_{FFT}, then the total sampling rate, F_s, the rate representing the sampling of all possible subcarriers, is given by

$$F_s = \frac{N_{FFT}}{\tau_u} \qquad (9.12)$$

This sampling rate is often specified in OFDM systems, but it must be emphasized that it is the rate in the absence of cyclic prefix addition.

A significant problem with the practical implementation of OFDM is that it exhibits a high *peak-to-average ratio* (*PAR*) of signal power. As indicated previously, the OFDM signal is the sum of N separate sinusoidal signals. The amplitudes and phases of these sinusoids are uncorrelated, but in the normal course of operation certain input data sequences occur that cause all the sinusoids to add in phase leading to a signal with a very high peak relative to the average. For OFDM systems with a small number of subcarriers the PAR is normally about $10 \log_{10} N$ dB, and for a large number of subcarriers it is never lower than 13 dB.[5] As a result of this PAR characteristic, OFDM systems require very linear power amplifiers.

A number of techniques have been developed to minimize the PAR of OFDM systems, including coding that tries to avoid the data sequences that result in the constructive addition of the subcarriers. A discussion of such techniques, however, is outside the scope of this text.

OFDM is a multicarrier transmission technique that enables transmission between two points. However, it lends itself to form the basis of a multiaccess technique for upstream transmission, which is referred to as *orthogonal frequency division multiple access* (*OFDMA*).[6] In OFDMA, which is really a form of FDMA, the subcarriers are divided into sets called subchannels. The subcarriers in each subchannel are spread over the full spectrum to minimize the effects of multipath fading. The subchannels are assigned to users via MAC messages sent downstream. In the downstream direction, it is also possible to take advantage of OFDM by using subchannels. Here the subchannels can be addressed to different users, with the modulation, coding, and so on, on each subchannel tailored to conditions of the link to the user to which it is addressed.

9.7.2   DS-CDMA with RAKE Receiver

DS-CDMA as a multiple access scheme was discussed in Section 9.2.3. It was noted that, by using a unique PN code per subscriber, the DS-CDMA receiver is able to pick out intended information from all other received signals. In a NLOS environment where the receiver input consists of multipath signals, the DS-CDMA receiver can be modified to provide a measure of protection against multipath effects. This modified receiver is called a RAKE receiver, a simplified diagram of which is shown in Fig. 9.12. (Note that the word RAKE is not an acronym, but is derived from the garden rake appearance on the amplitude versus time plane of the multiple delayed signals processed by the receiver.) The modification consists of equipping n despreaders in the receiver instead of only one for each subscriber's data. At any given time, each despreader processes one of the n strongest multipath signals being received, by being fed with a recovered PN code that is synchronized to the particular

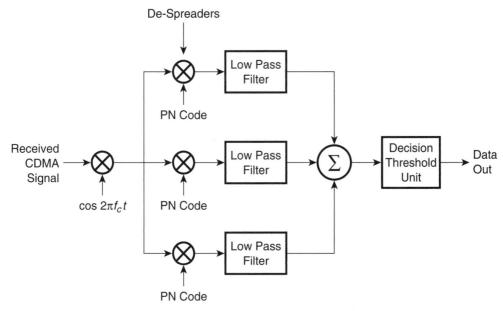

Note: CDMA signal is BPSK derived.

FIGURE 9.12   DS-CDMA RAKE receiver with three despreaders: Each of the three PN codes is synchronized to a different multipath signal.

multipath signal being processed. Since it's the same original transmitted signal being detected, the code fed to all descramblers is the same; only the timing differs. As indicated above, these time differences resemble the spacing of the prongs of a rake. The outputs of all the despreaders are added and fed to a decision threshold unit where a binary decision is made. In a typical realization, n is usually limited to about 4.

9.7.3   Spatial Processing Techniques

As important a role as transmission techniques such as OFDM can play in minimizing multipath induced problems, it cannot increase the strength of the processed received signal. The only way to accomplish this is to apply spatial processing techniques. There are many such techniques. Common to all is the use of multiple antennas at the receiver, at the transmitter, or both, together with intelligent signal processing and coding. Following we will review those techniques most commonly used with fixed broadband PMP systems, namely *spatial diversity (SD)*, *multiple-input, multiple-output (MIMO)*, and *adaptive beamforming*.

9.7.3.1   Spatial Diversity (SD)

Spatial diversity (SD) is enacted by combining the signals from multiple antennas; these antennas are either at the receive end or transmit end. When one transmitter antenna feeds multiple receiver antennas, the antenna system is referred to as a *single-input, multiple-output (SIMO)* one. When multiple transmitter antennas feed a single receiver antenna, the system is referred to as a *multiple-input, single-output (MISO)* one. The basic principle behind spatial diversity (also referred to as space diversity) is that antennas separated physically receive signals or transmit signals that travel over different paths and are thus uncorrelated or almost so regarding fading. As a result, these signals are unlikely to fade simultaneously. Therefore, by carefully combining them, the average signal to noise ratio or the average *signal-to-interference and noise ratio* (*SINR*), and hence BER performance, is improved relative to a *single-input, single-output* (*SISO*) system.

Spatial diversity at the receive end (SIMO) is the classical approach to spatial diversity and has already been discussed in Section 5.9.1. Measurements suggest that a separation between antennas of as little as 0.5 to 1 wavelength is enough to ensure a high SD improvement factor at the user end of a PMP link, with somewhat more spacing required at the base station side.[2] Cisco Systems, Inc. has developed a system that uses receiver space diversity in conjunction with OFDM and has labeled it *vector orthogonal frequency division multiplexing* (*VOFDM*).[7] In VOFDM, the receiver combines the signals from each of the two antennas on a subcarrier-by-subcarrier basis to create, per subcarrier, a single signal that maximizes the SINR and hence minimizes BER.

While spatial diversity at the transmit end (MISO) is a more recent approach, it is particularly attractive in PMP systems. This is because, with it, diversity has to only be installed at the base station to improve the quality of reception at all the remote units served by that base station. With SIMO, the receiver has "knowledge" of the individual paths and can use this knowledge to carry out optimum combining. With MISO, however, the transmitter has no knowledge of the paths if the system is an FDD one. On the other hand, if the system is a TDD one, then the transmitter can have knowledge of the paths if the antennas are also being used as receive antennas. Many transmit diversity schemes have been proposed. One of the simpler to implement that does not require channel knowledge at the transmitter is *delay diversity*.[8,9] With this approach, the signal transmitted from a second antenna is a delayed version of that sent from the first antenna. This delay between signals results in the signal at the receiver appearing as one that has undergone multipath distortion. With clever processing the receiver is able to resolve the multipath distortion and achieve diversity gain. Other more complex schemes that also do not require knowledge at the transmitter of the channel fall under the heading of *space-time coding* (*STC*). With STC the same data is transmitted via multiple antennas but coded differently. The received signal is the sum of these transmitted signals corrupted by noise, interference, and multipath effects. In the receiver, space-time decoding algorithms and channel estimation techniques are used to achieve both diversity and coding gain. Well-known STC codes are space-time trellis codes (STTCs)[10] and space-time block codes (STBCs).[11,12] Both these codes achieve the same diversity advantage as maximal ratio receive combining (Section 5.9.1). In addition to diversity gain, STTCs also provide a certain amount of coding gain. STBCs provide negligible coding gain but are much less complex in structure than STTCs.

9.7.3.2 Multiple Input, Multiple Output (MIMO) Antenna Technique

When multiple antennas are employed on both ends of the link the technique is referred to as a multiple input, multiple output (MIMO) one. MIMO based systems are very interesting because of their dual capability. They can be used to provide very robust spatial diversity by combining the features of MISO and SIMO systems. In addition to this, however, they can be used to increase throughput capacity with no additional transmit power or channel bandwidth compared to a SISO system.[13,14] Figure 9.13 shows a basic capacity-enhancing MIMO based one-way system, with two transmit antennas and three receive antennas. In such a scheme, if there are n transmit antennas, then the input data stream is divided via a serial to parallel converter into n substreams of bit rate $1/n$ that of the input stream. These substreams are then encoded and used to modulate n carriers within the same channel that feed the n transmit antennas. At the receive end, an m antenna array is equipped, where $m \geq n$. Thus, a "matrix" channel consisting of $n \times m$ spatial dimensions exists within the same assigned bandwidth. The MIMO-based system, therefore, increases throughput n times compared to a SISO, SIMO, or MISO one occupying the same channel bandwidth. Successful MIMO transmission requires a propagation environment that is rich in multipath resulting from signals bouncing and scattering of nearby objects such as buildings, trees, and cars. At the receiver, each of the m antennas picks up all the transmitted substreams and their multiple images, all superimposed on each other. However, because each substream is launched from a slightly different point in space, each one is scattered somewhat differently than the rest. It is these differences in scattering that are the key to successful multiple spatial transmission, since they allow substream recovery via sophisticated signal processing. Recovered substreams are decoded and recombined to re-create the original signal. Since successful transmission is so dependent on multipath scattering, not all remote stations will necessarily be able to benefit from this capacity-enhancing technique. Assuming strong scattering, remote stations close to the base station are likely to be able to use it. However, those far from the base station or those with near line-of-sight with the base station where scattering is weak are poor candidates for this technique.

For MIMO systems that offer capacity enhancement, one way of providing backup in the event of unsuitable channel conditions is to switch to spatial diversity whenever such conditions arise, the trade-off being a decrease in throughput in exchange for reliability.

9.7.3.3 Adaptive Beamforming

Adaptive beamforming[15] is a technique whereby an array of antennas is used adaptively for reception, transmission, or both in a way that seeks to optimize the transmission over the channel. In a PMP environment, the antenna array is usually located at the base station, but, unlike omnidirectional or sectorized antennas that cover an entire cell or

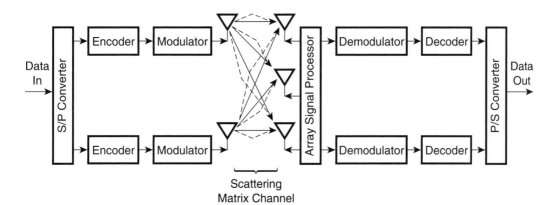

FIGURE 9.13   Capacity-enhancing MIMO based one-way system.

sector with almost equal energy, regardless of the location of the remotes, adaptive beam-forming targets the individual remote. It is able to do this by creating multiple beams si-multaneously, each beam directed to an individual remote. The shape of each beam can be dynamically controlled so that signal strength to and from a remote is maximized, by di-recting the main lobe in the direction of the strongest signal component, and sidelobes in the direction of multipath components. Further, it can simultaneously be made to mini-mize interference by signals that arrive at a different direction from the desired by locating nulls in the direction of the interference. Thus, this technique can be made to maximize the SINR. Figure 9.14 shows the beams of an adaptive beamforming antenna array com-municating with two remote stations in the presence of multipath and interference.

An *antenna array* consists of two or more individual antenna elements that are arranged in space and interconnected electrically via a feed network in such a fashion as to produce a directional radiation pattern. In an adaptive beamforming, the phases and amplitudes of the signals in each branch of the feed network are adaptively combined to optimize the SINR. To demonstrate the basic principles of an antenna array, consider the simple M-element *linear equally spaced* (*LES*) array shown on Fig. 9.15(a). This array is oriented on the x-axis with a spacing of Δx between adjacent elements and is shown receiving a plane wave from the direction (α, β). Fig. 9.15(b) shows the elements and its feed network and, to keep the analysis simple, a plane wave arriving in the x-y plane ($\beta = \pi / 2$). This results in the difference in length, Δl, between rays received by any two adjacent elements being

$$\Delta l = \Delta x \cos \alpha \qquad (9.13)$$

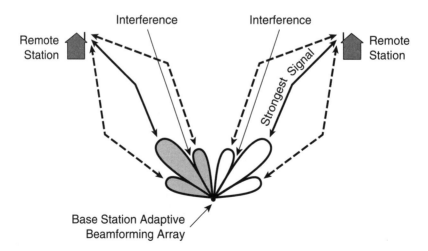

FIGURE 9.14   Beams from adaptive beamforming antenna array communicating with two remote stations.

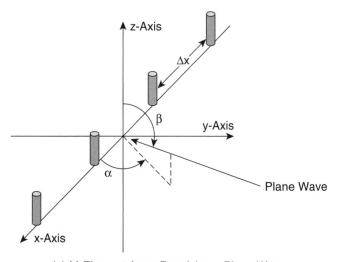

(a) M-Element Array Receiving a Plane Wave

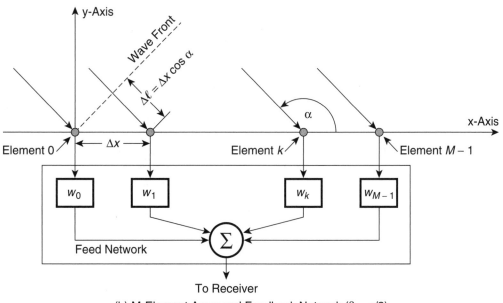

(b) M-Element Array and Feedback Network ($\beta = \pi/2$)

FIGURE 9.15   M-element linear equally spaced (LES) array.

and thus the difference in phase (i.e., phase shift), $\Delta\phi_k$, between the ray received by the kth element and the first element (where $k = 0$), being

$$\Delta\phi_k = 2\pi k \frac{\Delta l}{\lambda} \tag{9.14a}$$

$$= 2\pi k \frac{\Delta x}{\lambda} \cos\alpha \tag{9.14b}$$

where λ is the wavelength of the received wave.

To maximize reception of this wave, the despreading weighting elements w_k in the feed network are adjusted such that the signals from all elements sum coherently in phase. In effect, all signals other than that received on element $M-1$ are delayed so as to have the same phase as that on element $M-1$. Thus the signal on element 0, for example, is delayed by $\Delta\phi_M$. Note that since a wave arriving perpendicular to the x-axis would result in no phase shift between the received rays at the elements, then to maximize reception of such a wave requires no delay by any element. This example only shows phase being weighted, but in adaptive beamforming, as indicated previously, both phase and amplitude are normally adjusted. In a PMP system, all antenna elements receive signals from all active remotes simultaneously. These signals are then processed by sophisticated DSP circuitry to create the element weights necessary to simultaneously generate a separate antenna pattern for each remote.

The preceding example gave some insight into how adaptive beamforming works as a receiving antenna system. With the antenna at the base station this is good for upstream transmission, even if the signals arrive via NLOS routes. However, this is of little use if the downstream transmission is not similarly processed. To do this, information about the downstream paths is needed. With FDD, this can only be made available via feedback paths from the remotes to the base station, which adds complexity and consumes some of the upstream bandwidth. For this reason, most systems that employ adaptive beamforming use TDD as well. This way, since the downstream and upstream paths are now identical, information necessary for effective downstream transmission is available from upstream transmission. For transmission, the adaptive array operates in the reverse of its reception mode. The signal to be transmitted is fed through a feed network that splits it into M components and applies a spreading weight to each, prior to feeding an antenna element, with the spreading weight being the same as the despreading weight applied to that element. Thus, when these signals arrive at the intended remote, their weights are such that, even if arriving over a NLOS route, they combine with each other coherently.

Compared to standard omnidirectional or sector antennas, adaptive beamforming antennas, by their ability to focus beams on individual remotes, result in significantly increased coverage and capacity in both LOS and NLOS environments. As these improvement are achieved with all the "smarts" at the base station only, remote station cost and complexity is low compared to SIMO and MIMO systems, where multiple antennas and associated intelligence is required at each remote.

9.8   MESH NETWORKS

In mesh networks, also referred to as *multipoint-to-multipoint* (*MTM*) networks, all remote stations are interconnected with each other via a "mesh" arrangement. Interconnection to the public wired network is via an access point, which communicates with one or more remotes.

Each remote is able to communicate with one or more of the other remotes via PTP links, not only receiving and transmitting data for its own use, but also acting as a repeater for data destined elsewhere in the network. Data flows via one or more paths back and forth across the network in a manner analogous to Internet traffic flow. Figure 9.16(a) shows remote stations in a standard PMP arrangement with three potential remotes unable to join the network, two because of blockage and one because it is out of the range of the base station. Figure 9.16(b) shows the same remotes now all connected via a mesh network, including the three originally unable to access the network. These latter remotes are now able to join the network because they can communicate with other remotes. In this way, mesh networks solve to LOS problem associated with PMP systems. Note, however, that to start a mesh network, "seed nodes" have to be deployed so that initial subscribers can join the system. Also, to extend coverage into difficult areas additional seed nodes may have to be installed.

Mesh networks can be constructed via two basic antenna architectures. In the simpler approach, omnidirectional antennas are used at the remote stations. Networks operating in the 2- to 5-GHz range tend to use this approach. The other approach is to use directional antennas, this being the approach typically employed in networks operating at or above about 10 GHz.

Some of the advantages of a mesh network over a LOS PMP one are as follows:

- It solves the LOS problem.
- It extends the coverage area.
- It provides redundant paths that can route around remote station failures.
- It provides a possible improvement in overall BER quality as each link is likely to be quite short.
- It eliminates the high initial cost of a base station.
- It offers more scalability, with possibly lower initial deployment cost.
- When directional antennas are used, energy is only transmitted in the direction needed thus minimizing interference and facilitating frequency reuse.
- When omnidirectional antennas are used, the cost of the remote station antenna system is minimized.

However, against these advantages, the following disadvantages must be weighed:

- It requires highly sophisticated software control to manage the traffic flow through the network.
- Though it requires no base station, seed nodes are required to start the network.
- Its remote stations, by having to serve also as repeaters, are more complex and costly.
- Due to the possibility of a large number of links in series, it is susceptible to latency issues with delay sensitive service such as voice.

9.9   AIR INTERFACE STANDARDS

Successful deployment of metropolitan area fixed broadband wireless access systems depends on competitive offerings versus DSL and cable modem alternatives. One obvious route to lowered cost is the adoption of standards so that advantage can be taken of economy of scale. To this end, standards have been or are in the process of being developed. For worldwide application, the IEEE has developed an air interface standard. The first version, *IEEE 802.16*, covers systems operating between 10 and 66 GHz. An enhancement,

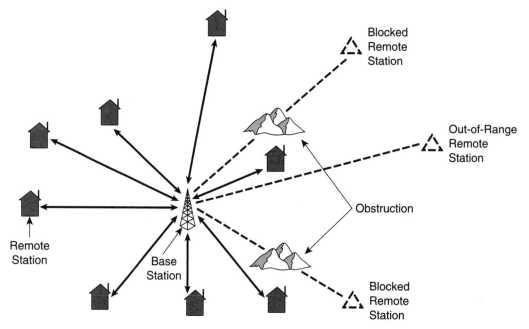

(a) PMP Network with 2 Blocked and 1 Out-of-Range Remote Station

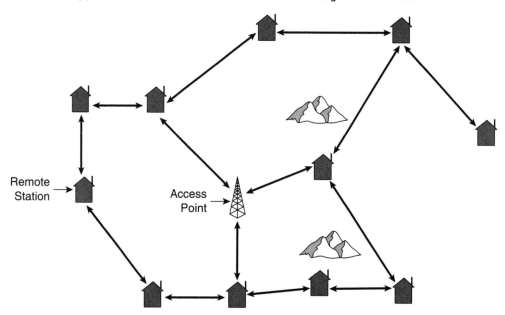

(b) Mesh Network with No Blocked Remote Stations

FIGURE 9.16   PMP and mesh networks.

published as *IEEE 802.16a*, expands the baseline standard to cover systems operating between 2 and 11 GHz. Further, currently under development is an amendment to 802.16a to permit combined fixed and mobile operations in licensed bands. This amendment will be released under the nomenclature *IEEE 802.16e*. In Europe, two standards are being developed by ETSI under a project titled *Broadband Radio Access Networks* (*BRAN*). The first, *HIPERACCESS*, applies to systems operating above 11 GHz and is very similar to IEEE 802.16. The second, *HIPERMAN*, applies to systems operating below 11 GHz and is very similar to IEEE 802.16a. Finally, we note that two IEEE standards, though developed for wireless LANs with maximum data rates of 54 Mb/s, can be utilized in external broadband wireless access systems by employing higher gain antennas than used in LAN applications. The first is *802.11a*, developed for systems operating in the U-NII 5-GHz bands. A metropolitan area system employing this technology is described later in this chapter. The second is *802.11g*, developed for systems operating in the ISM 2.4-GHz band.

A non-profit forum, *Worldwide Interoperability for Microwave Access*, normally referred to as *WiMAX*, was formed in 2001 to help promote deployment of broadband wireless access networks via global standardization and certification of compatibility and interoperability. It supports IEEE's 802.16/16a and ETSI's HIPERMAN standards. As a result, "WiMAX" is used loosely to imply 802.16/16a and HIPERMAN technologies, much the same way that Wi-Fi is used to imply 802.11a, b and g technologies, these latter technologies being supported and promoted by another nonprofit organization, the Wi-Fi Alliance. A key milestone in the introduction of WiMAX certified 802.16a equipment took place in July 2003, when Intel, the world's largest chip maker, announced a) its intention to develop chips based on the 802.16a standard and WiMAX certification and b) that it was working with Alvarion, a leading supplier of broadband wireless equipment, to have these chips incorporated into Alvarion's next generation of equipment.

Following is a high level summary of some of the salient features of the standards mentioned above.

9.9.1   IEEE 802.16 Standard, 10–66-GHz Systems (WiMAX Supported)

The IEEE 802.16 air interface standard[16,17] was approved in December 2001 and published in April 2002. Additional details were added in amendment 802.16c,[18] published in January 2003. It specifies the *physical layer* (*PHY*) and the MAC layer for fixed broadband wireless access systems. A simplified diagram of the 802.16 protocol stack is shown in Fig. 9.17. Some key physical layer features of this standard are as follows:

- Operating frequency: 10–66 GHz.
- PMP LOS operation.
- Transmission signal duplexing: Supports FDD, H-FDD, and TDD.
- Downlink (downstream) access: With TDD, subscriber access is via TDM. With FDD, subscriber access is via TDM for FDD subscribers, via TDMA for H-FDD subscribers.
- Uplink (upstream) access: Via TDMA.
- Frame duration: 1.0 ms.

- Each frame divided into physical slots of four QAM symbols each for the purpose of bandwidth allocation and for the demarcation of physical transitions; for example, transition from one modulation to another. Each frame consists of a downlink subframe and an uplink subframe. Figure 9.18 shows the TDD frame structure.

- The TDD downlink subframe structure is as shown in Fig. 9.19. The preamble is used by remotes for synchronization and equalization. The broadcast control section contains a downlink and an uplink map. The downlink map defines where in the downlink subframe bursts of user data are located and the profile of each of these bursts. A burst profile includes transmission parameters such as modulation and coding. Thus, a remote station, having read the downlink map, knows where in the received frame to look for a data burst addressed to it and the physical nature of that burst. The uplink map defines where in the next uplink subframe bursts of user data are to be located and the profile of each of these bursts. The final section of the downlink subframe, the TDM section, carries the user data bursts to remotes and is arranged in a contiguous fashion.

- The FDD downlink subframe structure is as shown in Fig. 9.20. It is essentially a TDD subframe structure with a TDMA portion added for transmission to H-FDD remotes that are scheduled to transmit earlier in the frame than they receive. In this added portion, which is also contiguous, each burst begins with a preamble for synchronization purposes.

- The uplink subframe, which is the same for TDD and FDD, accommodates three types of remote station bursts that may occur in any order within the subframe at the discretion of the MAC uplink scheduler. Figure 9.21 shows a structure for accommodating bursts in one particular order. In it, the first portion shown is for initial maintenance opportunities. These opportunities are time slots for bursts that are transmitted in contention to establish initial system access. The second portion is request contention opportunities, opportunities for bursts with bandwidth requests, also transmitted in contention, in response to polling to groups of remotes (multicast polling) or all remotes (broadcast polling). The third portion is for scheduled data. This consists of bursts of data from remote stations, each burst from a remote station being transmitted in a slot specifically granted to that remote by the MAC.

- Channel widths: 20, 25, and 28 MHz, with same width used for downlink and uplink.

- Modulation: Burst single carrier Gray coded QPSK, 16-QAM (optional for uplink), and 64-QAM (optional for both downlink and uplink).

- Pulse shaping: Root raised cosine with rolloff factor of 0.25.

- FEC: RS outer code, able to correct from 0 to 16 byte errors (with 0 correction, then no overhead). RS code used either alone, or (a) in the case of QPSK only, with rate 2/3 convolution inner code, or (b) with rate 8/9 parity check code. Optionally, stand alone turbo product code.

- Maximum throughput (64-QAM and no FEC overhead): 96 Mb/s in 20-MHz channel, 120 Mb/s in 25-MHz channel, and 134 Mb/s in 28-MHz channel.

- Converts multiple data units received from the MAC, referred to as MAC protocol data units (PDUs), into FEC blocks that form a single TDM or TDMA burst. The FEC blocks in a burst are of fixed length, except the final, which may be shortened to fit into the assigned burst length.

- Modulation and FEC adaptive on both downlink and uplink. Can be changed burst by burst, per remote station, as link conditions vary.

- Provides bandwidth on demand on a frame-by-frame basis.

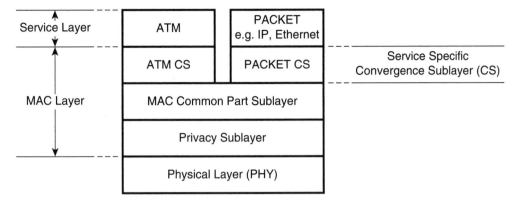

FIGURE 9.17   802.16 protocol stack.

Since the base station transmitter is the only one operating in the downlink direction, it can transmit without having to coordinate with the remote stations, except in the case of TDD operation, where it must time its transmission in intervals when remotes are not transmitting. In the uplink direction, however, where remotes share the same channel, their transmission must be controlled by a MAC protocol. The 802.16 MAC schedules remote station access via a combination of polling, contention procedures, unsolicited guaranteed bandwidth grants, and unsolicited bandwidth grants that are subject to the network load and thus not guaranteed. As can be seen in Fig. 9.17, the MAC protocol stack consists of three sublayers. At its top is the service-specific *convergence sublayer* (*CS*). Two versions of this sublayer are defined. One maps ATM data into the format required by the following layer, the other maps packet data, such as Ethernet and IP, into the format required by the following layer. The following layer is called the MAC *common part sublayer* (*CPS*). It performs the key MAC functions such as system access, bandwidth allocation, and QoS application. At the bottom is the *Privacy Sublayer*. Its

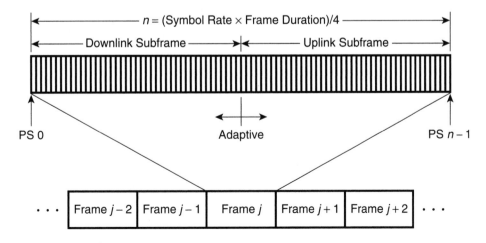

FIGURE 9.18   802.16 TDD frame structure. *(By permission from Ref. 16, ©2001 IEEE.)*

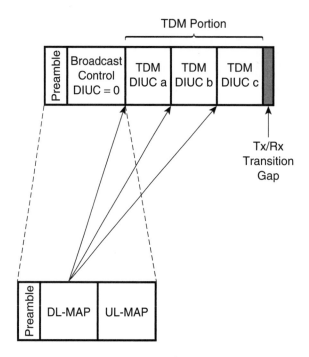

FIGURE 9.19   802.16 TDD downlink subframe structure. *(By permission from Ref. 16, ©2001 IEEE.)*

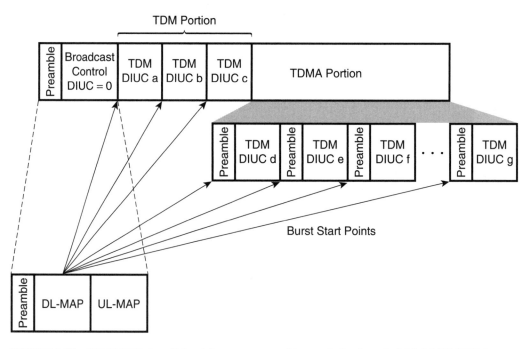

FIGURE 9.20   802.16 FDD downlink subframe structure. *(By permission from Ref. 16, ©2001 IEEE.).*

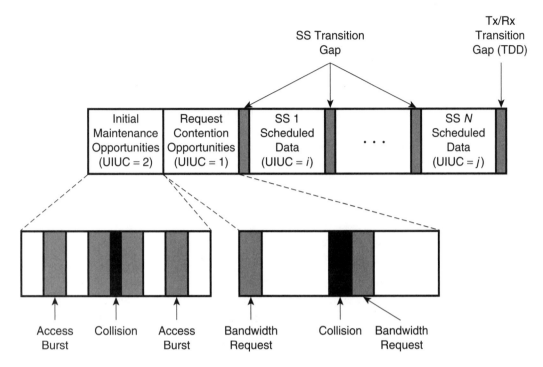

FIGURE 9.21 802.16 Uplink subframe structure. *(By permission from Ref. 16, ©2001 IEEE.).*

role is to increase the privacy of communication via encryption and authentication. Some key features of the MAC are as follows:

- Connection oriented: thus all services mapped to a connection. Note that a remote station may serve several connections.
- Supports continuous and burst traffic in the downlink direction and burst traffic in the uplink direction.
- Protocol independent, being able to support ATM, IP, Ethernet, and so on.
- Flexible QoS options.
- Provides authentication of each remote station and encryption of data to enhance security.
- Supports four classes of uplink service as follows:
  (1) *Unsolicited grant service* (*UGS*) for carrying real-time constant bit rate service such as T1/E1 on a periodic basis. With this class of service the connection does not have to request bandwidth on an ongoing basis as the base station schedules regularly a fixed amount of bandwidth negotiated at connection setup.
  (2) *Real-time polling service* (*rtPS*) for real time service of variable bit rate, such as voice over IP or MPEG video, on a periodic basis. Here the connection is polled individually (unicast polling) on a periodic basis, and offered opportunities in scheduled data slots to request bandwidth to meet its real-time needs.
  (3) *Non-real-time polling service* (*nrtPS*) for carrying, on a regular basis, non-real-time, variable bit rate service that can tolerate longer delay than rtPS traffic.

Here the connection is also polled individually, but less frequently than in the rtPS case. However, the connection may also use, for bandwidth requests, random access request contention opportunities as well as scheduled data slots offered via unsolicited bandwidth grants.

(4) *Best effort* (*BE*) service for providing service, as the name implies, on a best effort basis. It's the low man on the totem pole, providing no guarantee that a connection is able to transmit data. Bandwidth is requested using contention slots, polling, and scheduled data slots, offered via unsolicited bandwidth grants.

- Bandwidth requests to support QoS are always made by individual connections. However, bandwidth is granted to remote stations (not individual connections) which manage the bandwidth allocation among their user connections.
- The data units generated for the next lower level, the PDUs, are of variable length.

9.9.2   IEEE 802.16a Standard, 2–11 GHz Systems (WiMAX Supported)

The IEEE 802.16a air interface standard[19,6] was developed as an amendment to 802.16. It was approved in January 2003 and published in April 2003. It contains three air interface options, namely OFDM, OFDMA, and Single-Carrier. Some key physical layer features of this standard are as follows:

- Operating frequency: 2–11 GHz, in licensed and unlicensed bands, with unlicensed band operation designated as *WirelessHUMAN*.
- PMP LOS and NLOS operation.
- Mesh structure operation, operating in TDD mode only.
- OFDM air interface option:
  - 256-point FFT with 192 data subcarriers, 8 pilot subcarriers, and 56 null subcarriers, 55 of which are guard band subcarriers and one of which is located at the channel center frequency (see Fig. 9.22).
  - Cyclic prefix length/useful symbol length: adaptive; 1/4, 1/8, 1/16, 1/32.
  - Sampling rate/nominal channel bandwidth (F_s/BW): 8/7 for licensed channel bandwidths which are multiples of 1.75 MHz and licensed exempt channels. 7/6 for any other bandwidth.

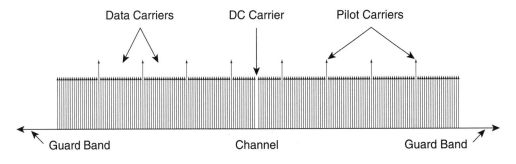

FIGURE 9.22   802.16a—OFDM frequency description. *(By permission from Ref. 19, ©2003 IEEE.)*

- QPSK, 16-QAM and 64-QAM adaptive OFDM sub-carrier modulation.
- FEC: RS outer code, convolution inner code, TPC and TCC optional.
- TDM downlink, TDMA uplink.
- OFDMA air interface option:
  - 2048-point FFT downlink carrier, with 1536 data subcarriers, 166 pilot subcarriers, and 346 null subcarriers.
  - 2048-point FFT uplink "carrier," with 1536 data subcarriers, 160 pilot subcarriers, and 352 null subcarriers.
  - Cyclic prefix length/useful symbol length: adaptive; 1/4, 1/8, 1/16, 1/32.
  - Sampling rate/nominal channel bandwidth (F_s/BW): 8/7.
  - QPSK, 16-QAM and 64-QAM adaptive OFDM sub-carrier modulation.
  - FEC: RS outer code, convolution inner code, TPC and TCC optional.
  - TDM downlink, OFDMA uplink.
- Single-Carrier air interface option:
  - BPSK, QPSK, 16-QAM, 64-QAM and 256-QAM adaptive modulation.
  - FEC: RS outer code, TCM inner code, TPC and TCC optional.
  - TDM downlink, TDMA uplink.
- Duplexing, licensed bands: FDD, H-FDD, TDD.
- Duplexing, unlicensed bands: TDD mandatory.
- Channel bandwidths, licensed bands: limited to regulatory bandwidth divided by any power of 2 but no less than 1.25 MHz.
- Channel bandwidths, 5 GHz unlicensed bands: 10 and 20 MHz.

Some key additions to the original 802.16 MAC in order to accommodate 802.16a are as follows:

- OFDM/OFDMA support.
- ARQ support. ARQ permits a more efficient transport of data in a high-BER environment.
- Support of dynamic frequency selection (DFS) in unlicensed bands. Allows the avoidance of channels already occupied by signals from other systems.
- Mesh topology support.
- Advanced antenna system support. Allows use of spatial diversity antenna systems.

The maximum data rate available with 802.16a technology is a function of the air interface option and the channel bandwidth. As the 5-GHz unlicensed system using OFDM is likely to be a popular configuration, let's investigate its maximum data rate. First, we note that this rate is achieved when the following parameters are operative:

Maximum channel bandwidth, BW = 20 MHz
Sub-carrier modulation: 64-QAM
Cyclic prefix length/useful symbol length, τ_g / τ_u, given by

$$\tau_g / \tau_u = 1/32 \tag{9.15}$$

Sampling rate/nominal channel bandwidth, F_s / BW, given by

$$F_s / BW = 8 / 7 \tag{9.16}$$

From Eq. (9.12) we have that the useful symbol duration, τ_u, given by

$$\tau_u = \frac{N_{FFT}}{F_s} \tag{9.17}$$

Substituting Eq. (9.16) into Eq. (9.17), we get

$$\tau_u = \frac{N_{FFT}}{BW} \times \frac{8}{7} = \frac{256}{20} \times \frac{8}{7} = 11.2 \ \mu \sec. \tag{9.18}$$

The symbol time after cyclic prefix addition, τ_t, is given by

$$\tau_t = \tau_u + \tau_g \tag{9.19}$$

Substituting Eq. (9.15) into Eq. (9.19), we get

$$\tau_t = \tau_u + \frac{\tau_u}{32} = \frac{33}{32} \tau_u = 11.55 \ \mu \sec. \tag{9.20}$$

Thus the symbol rate after cyclic prefix addition, F_t, is given by

$$F_t = \frac{1}{\tau_t} = 0.08658 \text{ Mega symbols / sec.} \tag{9.21}$$

Since this is the rate at which symbols are exiting the cyclic prefix adder, it must also be the rate at which they are entering it and hence the rate that sub-carrier symbols are being sampled by the IFFT processor. However, since we are using 64-QAM, every sub-carrier symbol represents 6 bits of data. Further, only 192 subcarriers carry data. Thus the date rate into the modulator, F_d, is given by

$$F_d = F_t \times 6 \times 192 = 99.74 \text{ Mb / s} \tag{9.22}$$

This is a very impressive rate for a 20 MHz channel, implying a spectral efficiency of 5 bits/s/Hz. Bear in mind however, that the data into the modulator is after FEC coding. A truer picture of the maximum throughput is the maximum rate into the FEC encoder. Using Reed-Solomon and convolution coding, the highest overall 802.11a coding rate for 64-QAM is 3/4, this resulting from an RS (120,108) code and a 5/6 convolution code. Thus the rate into this encoding structure, F_e, is given by

$$F_e = \frac{3}{4} F_d = 75 \text{ Mb / s} \tag{9.23}$$

Even this number, however, which implies a spectral efficiency of 3.75 bits/s/Hz, is some-what overstated. Recall that in the 5 GHz unlicensed bands, duplexing is constrained to TDD. As a result, this throughput is for downlink and uplink combined. It is further constrained by the need to provide guard intervals between transmission bursts.

9.9.3 IEEE 802.16e Planned Standard for Combined Fixed and Mobile Operation

Though this text is focused on fixed broadband wireless communications, much of the technology covered can be applied to portable and mobile broadband wireless communications. This point is best made by considering that the IEEE approved for development, on December 11, 2002, an amendment to its 802.16/802.16a standard. This amendment, which will be referred to as 802.16e, is for physical and medium access control layers for combined fixed and mobile operation in licensed bands. Some of its planned key features are as follows:

- Support of subscriber stations moving at vehicular speeds as well as fixed stations
- Operating frequency: 2–6 GHz in licensed bands
- Backward compatibility with fixed stations
- Packet-oriented architecture

9.9.4 ETSI/BRAN HIPERACCESS, Systems Above 11 GHz

The ETSI HIPERACCESS standard[20] is, architecturally, very similar to the IEEE 802.16 one, but not interoperable with it. A simplified diagram of its protocol stack is shown in Fig. 9.23. It consists of specifications for the physical (PHY) layer, the *data link control* (*DLC*) layer, and a convergence layer for mapping cell-based (ATM) and packet-based (IP, Ethernet, etc.) services into the format required by the DLC. Thus, the DLC layer is the equivalent of the common part sublayer and the privacy sublayer of the 802.16 MAC layer. The associated specifications are as follows:

- ETSI TS 101 999, the PHY protocol specification, published as version V1.1.1 in April 2002
- ETSI TS 102 000, the DLC protocol specification, published as version V1.3.1 in December 2002
- ETSI TS 102 115-1, part 1 of the Cell Based Convergence Layer specification, published as version V1.1.1 in September 2002
- ETSI TS 102 115-2, part 2 of the Cell Based Convergence Layer specification, published as version V1.1.1 in October 2002

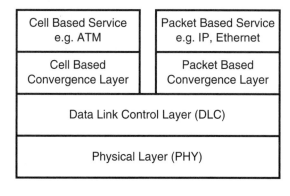

FIGURE 9.23   ETSI HIPERACCESS protocol stack.

- ETSI TS 102 117-1, part 1 of the Packet Based Convergence Layer specification, published as version V.1.1 in September 2002
- ETSI TS 102 117-2, part 2 of the Packet Based Convergence Layer specification, published as version V.1.1 in October 2002

Given that the HIPERACCESS standard is very similar to the 802.16 one, it is easier to list key features of HIPERACCESS that differ from 802.16. These features are as follows:

- Only one channel width supported: 28 MHz, for both downlink and uplink.
- Modulation: QPSK, 16-QAM and, optionally, 64-QAM, for downlink. QPSK and, optionally, 16-QAM, for uplink.
- Maximum throughput (64-QAM): 125 Mb/s.
- FEC: RS outer code, but in addition to a rate 2/3, a rate 5/6 and rate 7/8 block con-volution inner code, for critical communications. Turbo product code optional.
- Fixed MAC PDU length.
- ARQ support for uplink.

9.9.5   ETSI/BRAN HIPERMAN, Systems below 11 GHz (WiMAX Supported)

The ETSI/BRAN HIPERMAN standard,[21] when finalized, will have the following key features:
- Operating frequency: below 11 GHz.
- PMP LOS and NLOS operation.
- Mesh structure operation.
- A 256-point FFT OFDM mandatory mode identical to, and hence completely inter-operable with, IEEE's 802.16a mode (MAC and PHY).
- A 2048-point FFT OFDMA optional mode.
- No single-carrier modulation mode.

9.9.6   IEEE 802.11a Standard, 5 GHz Systems (Wi-Fi Supported)

The *IEEE 802.11a standard*[22] was developed for WLANs operating in the three 5-GHz unlicensed U-NII bands. It was released in 1999 and supports data rates of up to 54 Mb/s. Though developed for WLANs, its key physical layer features are reviewed here. This is because products employing this standard, and operating in the 5.8-GHz band, can be uti-lized for external PMP communications over several miles by the use of directive, rela-tively high-gain antennas. These key physical layer features are as follows:

- PMP operation.
- Operating frequency bands: 5.15–5.25, 5.25–5.35, and 5.725–5.825 GHz U-NII bands
- TDD operation
- TDM downlink, TDMA uplink
- OFDM air interface, with:
  - 52 OFDM subcarriers: 48 used for data transmission, 4 used as pilot subcarriers
  - OFDM symbol duration, $\tau_t = 4.0$ μs

- Guard interval, $\tau_g = 0.8\ \mu\mathrm{s}$
- BPSK, QPSK, 16-QAM and 64-QAM subcarrier modulation
- FEC convolution coding with a coding rate of 1/2 or 3/4 for BPSK, QPSK, and 16-QAM, and 2/3 or 3/4 for 64-QAM
- Data rates, depending on the coding rate, of 6 and 9 Mb/s with BPSK, 12 and 18 Mb/s with QPSK, 24 and 36 Mb/s with 16-QAM, and 48 and 54 Mb/s with 64-QAM

9.9.7 IEEE 802.11g Standard, 2.4 GHz Systems (Wi-Fi Supported)

The *IEEE 802.11g standard*, like 802.11b, is an amendment to the basic 802.11 standard for WLAN systems operating in the 2.4-GHz ISM band. It received final approval of the IEEE Standards Board in June 2003. Whereas 802.11b raises the original 802.11 maximum data rate from 2 Mb/s to 11 Mb/s by employing CCK, 802.11g raises it to 54 Mb/s by employing OFDM. Like 802.11a, its key physical layer features are reviewed here because products employing this standard can be utilized for external PMP communications over several miles by the use of directive, relatively high-gain antennas.

Like 802.11a, IEEE 802.11g uses TDD and, for downlink and uplink operation, TDM and TDMA, respectively. Also, it achieves its 54 Mb/s maximum data rate by using the same OFDM structure as used with 802.11a. However, in addition to achieving a high data rate, a mandatory requirement in the development of this standard was that it be backward compatible with 802.11b. This was specified so that 802.11g and 802.11b devices could coexist in the same network. To allow the backward compatibility with 802.11b, 802.11g uses CCK formatted request-to-send/clear-to-send (RTS/CTS) preamble communication prior to OFDM packet data transmission.[23] As 802.11b terminals support this RTS/CTS feature, both 802.11g and 802.11b users can monitor preamble communications and hence effectively share the network.

9.10 POPULAR FREQUENCY BANDS

Fixed broadband wireless access systems, like broadband point-to-point systems, operate in both licensed and unlicensed bands.

The unlicensed bands available in the United States are the same as those covered in Section 8.3, namely the 2.4- and 5.8-GHz ISM bands and the 5.3- and 5.8-GHz U-NII bands. The key specifications of PMP equipment operating in these bands are as those shown in Tables 8.4, 8.5 and 8.6 for PTP systems with the exception of "Max. Peak Output Power into Ant.," $P_{\max}$ say. For PMP systems, the regulations covering $P_{\max}$ that differ from those shown in Tables 8.4, 8.5 and 8.6 are as follows:

- For "digitally modulated" systems operating in both the 2.4- and 5.8-GHz ISM bands, if the radiating antenna's gain G is greater than 6 dBi, then $P_{\max}$ must be reduced below 30 dBm by the amount in dB that G exceeds 6 dBi.
- For FHSS systems operating in both the 2.4- and 5.8-GHz ISM bands, where the number of hopping channels ≥ 75, if the radiating antenna's gain G is greater than 6 dBi, then $P_{\max}$ must be reduced below 30 dBm by the amount in dB that G exceeds 6 dBi.

- For FHSS systems operating in the 2.4-GHz ISM band, where the number of hopping channels < 75, if the radiating antenna's gain G is greater than 6 dBi, then P_{max} must be reduced below 21 dBm by the amount in dB that G exceeds 6 dBi.
- For "radio frequency devices" operating in the 5.8-GHz U-NII band, if the radiating antenna's gain G is $\leq$ 6 dBi (not 23 dBi as is the case for PTP), then P_{max} must be the lesser of 30 or $17+10 \log B$ dBm.
- For "radio frequency devices" operating in the 5.8-GHz U-NII band, if the radiating antenna's gain G is > 6 dBi (not 23 dBi as is the case for PTP), then P_{max} must be reduced below the lesser of 30 or $17 + 10 \log B$ dBm by the amount in dB that G exceeds 6 dBi.

In summary, the P_{max} requirements for PMP systems are, depending on the transmitter antenna gain, the same or more stringent than for PTP systems.

Licensed U.S. frequency bands for FBWA and their corresponding channel/block bandwidths are shown in Table 9.1. In the provision of licensed spectrum for PTP systems, it

TABLE 9.1   Licensed U.S. Frequency Bands for FBWA Communications

Band Classification	FCC Part #	Freq. Range (MHz)	Ch./Block Bandwidth
MDS	21	2150–2162	6-MHz ch.
WCS	27	2305–2320	5, 10-MHz blocks
		2345–2360	5, 10-MHz blocks
ITFS	74	2500–2596	6-MHz ch.
		2644–2650	6-MHz ch.
		2656–2662	6-MHz ch.
		2668–2674	6-MHz ch.
		2680–2686	6-MHz ch.
MDS (MMDS)	21	2596–2644	6-MHz ch.
		2650–2656	6-MHz ch.
		2662–2668	6-MHz ch.
		2674–2680	6-MHz ch.
		2686–2690	4-MHz ch.
DEMS	101	24,250–24,450 / 25,050–25,250	40-MHz ch./40-MHz blocks
LMDS, A Block	101	27,500–28,350	850-MHz block
		29,100–29,250	150-MHz block
		31,075–31,225	150-MHz block
LMDS, B Block	101	31,000–31,075	75-MHz block
		31,225–31,300	75-MHz block
39-GHz Band	101	38,600–40,000	50-MHz ch./50-MHz blocks

Acronym	Full description
MDS	Multipoint distribution service
MMDS	Multichannel multipoint distribution service
LMDS	Local multipoint distribution service
WCS	Wireless communications service
ITFS	Instructional television fixed service
DEMS	Digital electronic message service

TABLE 9.2   CEPT-Recommended Licensed Frequency Bands for FBWA Communications

CEPT Recommendation	Freq. Range (GHz)	Ch./ Block Bandwidth	Comments
ERC Rec. 14-03 E	3.41–3.50	0.25–40 MHz ch.	FDD
	3.50–3.60	0.25–50.0 MHz ch.	FDD
	3.41–3.60	0.25–90 MHz ch.	FDD
ERC Rec. 12-08 E	3.6–3.7	0.25–50.0 MHz ch.	FDD
	3.7–3.8	0.25–50.0 MHz ch.	FDD
	3.6–3.8	0.25–100 MHz ch.	FDD
ERC Rec. 12-05 E	10.15–10.30 / 10.50–10.65	0.5–150 MHz ch.	FDD
ERC Rec. (00)05	24.5–26.5	28-MHz blocks	FDD/TDD
ERC Rec. (01)03	27.5–29.5	28-MHz blocks	FDD/TDD
ECC Rec. (02)02	31.0–31.3	3.5–28 MHz ch.	FDD/TDD
ERC Rec. (01)02	31.8–33.4	3.5–56 MHz ch.	FDD/TDD
ECC Rec. (01)04	40.5–43.5	Blocks up to 3 GHz	

is customary to assign channels of a specified bandwidth. However, in the case of FBWA, instead of assigning channels, authorities often assign blocks. A block is a bandwidth assignment that an operator may, within limits, partition as it sees fit. Thus, within an assigned block, an operator may deploy equipment with a variety of center frequencies and bandwidths. Note that included in the table are ITFS channels. Generally, such channels must be used for providing educational and instructional programming. However, they are included here because the regulations permit a licensee to lease excess capacity to third parties. Note also that the channels shown classified as MMDS are now generally lumped together with the original MDS channels (2150–2162 MHz) and all referred to as MDS channels.

CEPT-recommended licensed frequency bands for FBWA are shown in Table 9.2. However, CEPT has further recommended (ERC Rec. 13-04) that in the range between 3 and 29.5 GHz, the following bands be preferred:

- 3.4–3.6 GHz
- 10.15–10.3 / 10.5–10.65 GHz
- 24.5–26.5 GHz
- 27.5–29.5 GHz

Above 29.5 GHz, it is likely that the 40.5–43.5-GHz band will become a preferred one as it has been structured specifically for FBWA services.

9.11   TWO COMMERCIAL PMP SYSTEMS

In this the final section, two commercial PMP systems are briefly reviewed. Both are Alvarion Limited models, one being for operation in the United States' 5.8-GHz unlicensed band, the other for operation in the ERC recommended 26-GHz band. The former is a NLOS, OFDM TDM/TDMA-based system, the latter a LOS, multicarrier TDM/TDMA-based one. Key parameters of both systems are summarized in Table 9.3, and the receiver threshold at 10^{-6} BER of the 5.8-GHz system, as a function of data rate, modulation, and coding rate, is shown in Table 9.4.

TABLE 9.3   Key Parameters of PMP Equipment Reviewed

Parameter	Specification	
	BreezeACCESS VL	*WALKair 3000*
Operating frequency	5.275–5.850 GHz	24.5 – 26.5 GHz
Carrier channel bandwidth	20 MHz	14 MHz
Duplexing method	TDD	FDD
Downlink access	TDM	TDM
Uplink access	TDMA	TDMA
Transmission technique	OFDM	Multicarrier
Number of OFDM subcarriers	52	N.A.
Adaptive subcarrier modulation	BPSK, QPSK, 16-QAM, 64-QAM	N.A.
Adaptive carrier modulation	N.A.	QPSK, 16-QAM
FEC	Convolution	Reed-Solomon
Maximum number of carriers/ sector/polarization	1	6
Gross data rate per "carrier"	6–54 Mb/s, see Table 9.4	16 Mb/s, QPSK; 36 Mb/s, 16-QAM
Max. gross data rate/sector/ polarization	54 Mb/s	216 Mb/s
Number of remote stations per carrier	512	64
Sector angular dimension in degrees	60, 90, 120 (6, 4, 3 sectors/cell)	45, 90 (8, 4 sectors/cell)
Base station (BS) data interface	10/100BaseT	10/100BaseT, ATM over STM-1 or E3
Remote station (RS) data interface	10/100BaseT	4×100BaseT and $2 \times$ E1
Dynamic bandwidth allocation	Yes	Yes
ATPC, RS xmtr.	Yes	Yes
BS xmtr. output power	21 dBm	10 dBm per carrier
RS xmtr. output power	21 dBm	10 dBm
BS antenna gain, sector angle	16 dBi, 60; 16 dBi, 90; 15 dBi, 120	18 dBi, 45; 15 dBi, 90
BS antenna type	Flat panel	20-cm horn
RS antenna gain	21 dBi	33 dBi
RS antenna type	Flat panel	33-cm dish
BS/RS receiver threshold at 10^{-6} BER, typical (dBm)	See Table 9.4	Proprietary
Typical cell radius	8 km	3–4 km, 16-QAM; 4–5 km, QPSK

(Courtesy of Alvarion, Ltd.)

TABLE 9.4   BreezeACCESS VL Receiver Threshold as a Function of Data Rate, Modulation, and Coding Rate

Data rate (Mb/s)	Modulation	Coding Rate	Rcvr. Threshold at 10^{-6} BER (dBm)
6	BPSK	1/2	−88
9	BPSK	3/4	−87
12	QPSK	1/2	−86
18	QPSK	3/4	−84
24	16-QAM	1/2	−81
36	16-QAM	3/4	−77
48	64-QAM	2/3	−72
54	64-QAM	3/4	−70

(Courtesy of Alvarion, Ltd.)

9.11.1 The Alvarion BreezeACCESS VL, 54 Mb/s, 5.8 GHz, NLOS, OFDM, TDM/TDMA, PMP System

The Alvarion BreezeACCESS VL PMP system was designed by its manufacturer to address primarily last mile access to large businesses, campuses, and multitenant units. The VL model is one a number of BreezeACCESS models that operate in several frequency bands with varying modulation methods and capacity. It communicates in the downstream and upstream directions at *burst* rates of up to 54 Mb/s. However, as it uses TDD, 54 Mb/s is the total of the downstream and upstream rates averaged over time. Thus, the effective ongoing communication rate in a given direction is normally less than 54 Mb/s, with the actual number being dependent on the asymmetry of communication at the time. Its OFDM transmission technique is in fact based on the IEEE 802.11a standard (Sec. 9.9.6), developed for WLANs but adapted here for fixed point-to-multipoint outdoor communications via the use of relatively high gain, high directivity antennas, as permitted under FCC Part 15 regulations.

The system's base station consists of an access unit (AU) or access units and an indoor shelf. Each AU is comprised of an indoor unit (IDU) connected via CAT-5 cable to an outdoor unit (ODU), which in turn is connected to an antenna. The CAT-5 cable can be as long as 100 meters (328 feet). When only one AU is equipped, its IDU can be a standalone type. When more than one AU is equipped, then their IDUs are housed in the indoor shelf. This shelf can also support AUs of other BreezeACCESS models. The IDU acts as an Ethernet repeater. In the transmit direction it regenerates the incoming signal from the external network and sends it up the CAT-5 cable to the ODU. In the receive direction, it regenerates the incoming signal from the ODU and sends it on to the external network via a standard 10/100BaseT interface. In addition to its repeater functions, it sends/receives management and control signals to/from the ODU via the CAT-5 cable. Also, it generates 56 V DC, which it sends up the CAT-5 cable to power the ODU. Housed in the ODU is an OFDM transmitter/receiver and management and control circuitry. The right-hand side of Fig. 9.24 shows a base station IDU, ODU, and antenna.

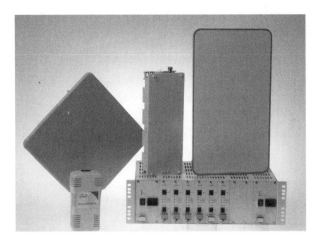

FIGURE 9.24   The Alvarion BreezeACCESS VL subscriber unit (on left) and base station (on right). *(Courtesy of Alvarion, Ltd.)*

The remote unit, referred to by the manufacturer as the subscriber unit (SU), consists of two units, a stand-alone IDU and an ODU with integrated antenna. The units are connected via a CAT-5 cable and their functions are similar to their counterparts at the base station. Two options of the SU are available: a 24-Mb/s model, upgradeable to 54 Mb/s, and a 54-Mb/s model. The SU connects to the subscriber's data equipment via a standard 10/100BaseT interface. The left-hand side of Fig. 9.24 shows a subscriber unit IDU and antenna (ODU unit behind antenna).

9.11.2   The Alvarion WALKair 3000, 26 GHz, LOS, Multicarrier TDM/TDMA PMP System

The Alvarion WALKair 3000 PMP system has been designed by its manufacturer for application by new and established carriers in the provision of last mile access to small and medium sized businesses and multidwelling units/multitenant units (MDUs/MTUs). It employs FDD, but, unlike the traditional single carrier per direction arrangement, it provides for multiple carriers in each direction. The maximum net data rate per carrier is 36 Mb/s. Up to six downstream carriers can be equipped per sector via a single ODU; thus the maximum downstream net data rate per sector per ODU is 216 Mb/s. The system employs 16-QAM and QPSK, switching between the two adaptively. The maximum net carrier data rate of 36 Mb/s is achieved with 16-QAM. When QPSK is employed, this rate drops to 16 Mb/s. Each carrier pair, employing TDM downstream, TDMA upstream, can support communications with up to 64 remote stations, allowing these remote stations to communicate at a maximum collective downstream rate of 36 Mb/s and similarly upstream.

A photograph of the base station (BS) is shown in Fig. 9.25(a). It consists of an indoor unit (IDU) and outdoor unit (ODU), connected together by a single coaxial cable that supports both upstream and downstream IF signals as well as providing DC power to the ODU.

In the BS IDU, modules are housed in a shelf, and the main building blocks of this shelf are modem basic unit (BS-BU) cards with Ethernet ports for IP backbone connectivity,

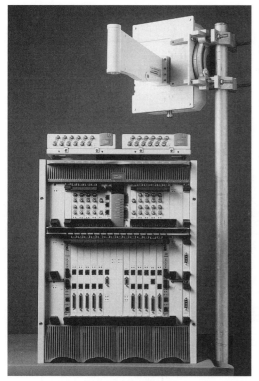

(a) Base station

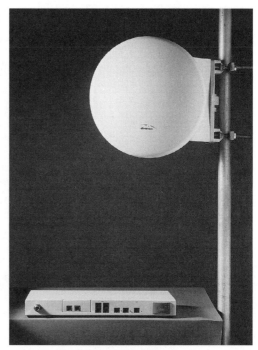

(b) Terminal station

FIGURE 9.25   The Alvarion WALKair 3000 base station and terminal station. *(Courtesy of Alvarion, Ltd.)*

network interface unit (NIU) cards for ATM backbone connectivity, and the IF mux module. An NIU card performs ATM to IP and IP to ATM transformations. If a particular external network interface is an IP one, connection between the BS and the external network is via the Ethernet port of a BU. However, if the external network interface is an ATM one, then the connection is via a NIU card that in turn interfaces with a BU. Each BU card converts incoming IP data to an IF signal and its incoming IF signal to IP data. In the transmit section of the IF mux, up to six IF signals from BUs are multiplexed and combined with 48 V DC for coaxial cable transmission to the ODU. In the receive section, IF signals from the ODU are de-multiplexed and fed to individual BUs.

The BS ODU consists of an RF unit and an antenna. In the transmit section of the RF unit, the multiplexed IF signal from the BUs is upconverted to RF, amplified, and fed to the antenna via a duplexer. In the receive section, the incoming signals from the antenna duplexer are downconverted to IF for cable transmission to the IF mux.

Like the base station, the remote terminal, referred to by the manufacturer as a terminal station (TS), consists of an IDU and ODU connected via a single coaxial cable that transmits IF signals in both directions and 48 V DC to the ODU. The IDU is equipped with four 10/100BaseT and two E1 interfaces and, optionally, an additional interface card comprising E1 and Ethernet ports. A photograph of the terminal station is shown in Fig. 9.25(b).

REFERENCES

1. Taub, H., and Schilling, D. L., *Principles of Communication Systems, 2nd ed.,* Mc-Graw Hill, 1986.
2. Gesbert, D, et al, "Technologies and Performance for Non-Line-of-Sight Broadband Wireless Access Networks," *IEEE Communications Magazine*, Vol. 40, No. 4, April 2002, pp. 86–95.
3. Burr, A., *Modulation and Coding for Wireless Communications*, Pearson Education/ Prentice Hall, 2001.
4. Bingham, J. A. C., "Multicarrier Modulation for Data Transmission: An Idea Whose Time Has Come," *IEEE Communications Magazine*, Vol. 28, No. 5, May 1990, pp. 5–14.
5. Merchan, S., Armada, A. G., and Garcia, J. L., "OFDM Performance in Amplifier Nonlinearity," *IEEE Trans. on Broadcasting*, Vol. 44, No. 1, March 1998, pp. 104–114.
6. Koffman, I., and Roman, V., "Broadband and Wireless Access Solutions Based on OFDM Access in IEEE 802.16," *IEEE Communications Magazine*, Vol. 40, No. 4, April 2002, pp. 96–103.
7. Ayanoglu, E, et al, "VOFDM Broadband Wireless Transmission and Its Advantages over Single Carrier Modulation," *Proc. IEEE 2001 ICC*, Helsinki, June 2001.
8. Seshadri, N., and Winters, J. H., "Two Signaling Schemes for Improving the Error Performance of FDD Transmission Systems Using Transmitter Antenna Diversity," *Proc. 1993 IEEE Vehicular Technology Conf.*, May 1993, pp. 508–511.
9. Winters, J. H., "The Diversity Gain of Transmit Diversity in Wireless Systems with Rayleigh Fading," *Proc. 1994 ICC/SUPERCOM*, New Orleans, May 1994, Vol. 2, pp. 1121–1125.
10. Tarohk, V., Seshadri, N., and Calderbank, A. R., "Space-Time Codes for High Data Rates Wireless Communications: Performance Criterion and Code Construction," *IEEE Trans. Inform. Theory*, Vol. 44, March 1998, pp. 744–765.
11. Alamouti, S. M., "A Simple Transmit Diversity Technique for Wireless Communications," *IEEE J. on Select Areas in Commun.*, Vol. 16, No. 8, Oct. 1998, pp. 1451–1458.
12. Tarokh, V., Jafarkhani, H., and Calderbank, A. R., "Space-Time Block Codes from Orthogonal Designs," *IEEE Trans. Inform. Theory*, Vol. 45, July 1999, pp. 1456–1467.
13. Foschini, G. J., and Gans, M. J., "On Limits of Wireless Communications in a Fading Environment when Using Multiple Antennas," *Wireless Personal Commun.*, Vol. 6, March 1998, pp. 311–335.
14. Foschini, G. J., Golden, G. D., Valenzuela, R. A., and Wolniansky, P. W., "Simplified Processing of Wireless Communication at High Spectral Efficiency," *IEEE J. Select Areas Commun.* (Wireless Communications Series), Vol. 17, Nov. 1999, pp. 1841–1852.
15. Monzingo, R. A., and Miller, T. W., *Introduction to Adaptive Arrays*, Wiley, 1980.
16. *IEEE Std 802.16-2001,™ IEEE Standard for Local and Metropolitan Area Networks—Part 16: Air Interface for Fixed Wireless Access Systems*, The Institute of Electrical and Electronic Engineers, Inc., April 8, 2002.

17. Eklund, C., Marks, R., and Stanwood, K., "IEEE Standard 802.16: A Technical Overview of the WirelessMAN™ Air Interface for Broadband Wireless Access," *IEEE Communications Magazine*, Vol. 40, No. 6, June 2002, pp. 98–107.

18. *IEEE Std 802.16c, IEEE Standard for Local and Metropolitan Area Networks—Part 16: Air Interface for Fixed Wireless Access Systems–Amendment 1: Detailed System Profiles for 10–66 GHz,* The Institute of Electrical and Electronic Engineers, Inc., January 15, 2003.

19. *IEEE Std 802.16a™-2003, IEEE Standard for Local and Metropolitan Area Networks—Part 16: Air Interface for Fixed Wireless Access Systems–Amendment 2: Medium Access Control Modifications and Additional Physical Layer Specifications for 2–11 GHZ,* The Institute of Electrical and Electronic Engineers, Inc., April 1, 2003.

20. *ETSI Technical Report ETSI TR 102 003 V.1.1.1 (2002-03), Broadband Radio Access Networks (BRAN);HIPERACCESS; System Overview*, ETSI, France, March 2002.

21. *ETSI Technical Report ETSI TR 101 856 V.1.1.1 (2001-03), Broadband Radio Access Networks (BRAN); Functional Requirement for Fixed Wireless Access Systems below 11 GHz: HIPERMAN*, ETSI, France, March 2002.

22. *IEEE Std 802.11a-1999, Part 11: Wireless LAN Medium Access Control (MAC) and Physical Layer (PHY) Specifications, High-Speed Physical Layer in the 5 GHz Band,* Institute of Electrical and Electronic Engineers, 1999.

23. Zyren, J., Enders, E., and Edmondson, T*., IEEE 802.11g Offers Higher Data Rates and Longer Range*, Intersil Corporation, December 2002.

APPENDIX

HELPFUL MATHEMATICAL IDENTITIES

TRIGONOMETRIC IDENTITIES

$$\sin(x \pm y) = \sin x \sin y \pm \cos x \sin y$$

$$\cos(x \pm y) = \cos x \cos y \mp \sin x \sin y$$

$$\sin x \sin y = \frac{1}{2}\cos(x - y) - \frac{1}{2}\cos(x + y)$$

$$\cos x \cos y = \frac{1}{2}\cos(x + y) + \frac{1}{2}\cos(x - y)$$

$$\sin x \cos y = \frac{1}{2}\sin(x + y) + \frac{1}{2}\sin(x - y)$$

$$\cos x \sin y = \frac{1}{2}\sin(x + y) - \frac{1}{2}\sin(x - y)$$

$$\sin^2 x = \frac{1}{2}(1 - \cos 2x)$$

$$\cos^2 x = \frac{1}{2}(1 + \cos 2x)$$

$$\sin x = \frac{e^{jx} - e^{-jx}}{2j}$$

$$\cos x = \frac{e^{jx} + e^{-jx}}{2}$$

$$e^{jx} = \cos x + j\sin x$$

STANDARD INTEGRALS

Where a, b, and c are constants,

$$\int \sin(ax + b)\, dx = -\frac{1}{a}\cos(ax + b) + c$$

$$\int \cos(ax + b)\, dx = \frac{1}{a}\sin(ax + b) + c$$

$$\int a\, dx = ax + b$$

$$\int (ax + b)^n\, dx = \frac{1}{a(n+1)}(ax + b)^{n+1} \qquad n \neq -1$$

$$\int \frac{1}{ax + b}\, dx = \frac{1}{a}\ln|ax + b| + c$$

$$\int e^{ax+b}\, dx = \frac{1}{a}e^{ax+b} + c$$

ABBREVIATIONS
AND ACRONYMS

A/D	Analog to Digital Converter
AAL	ATM Adaptation Layer
ABPSK	Associated Binary Phase Shift Keying
AC	Alternating Current
ADSB	Associated Double-Sideband Supressed Carrier
ADSL	Assymetrical Digital Subscriber Line
AM	Amplitude Modulation
AMI	Alternative Mark Inversion
ANSI	American National Standards Institute
ARQ	Automatic Request for Repeat
ATM	Asynchronous Transfer Mode
ATPC	Automatic Transmitter Power Control
b/s	Bits per Second
BCH	Bose-Chaudhuri-Hocquenghem (block codes)
BE	Best Effort (service)
BER	Bit Error Rate
bits/s/Hz	Bits per Second per Hertz
BPF	Bandpass Filter
BPSK	Binary Phase Shift Keying
BRAN	Broadband Radio Access Networks (ETSI)
C/I	Carrier-to-Interference Ratio
C/N	Carrier-to-Noise Ratio
CB	Center Bit
CCK	Complementary Code Keying
CDF	Cumulative Distribution Function
CDMA	Code Division Multiple Access
CEPT	European Conference of Postal and Telecommunications Administrations
CFM	Composite Fade Margin
CMA	Constant-Modulus Algorithm
CMI	Coded Mark Inversion
CO	Central Office
CPFSK	Continuous Phase Frequency Shift Keying
CPS	Commom Part sublayer
CS	Convergence Sublayer
CSMA/CD	Carrier-Sense Multiple Access/ Collision Detection
D/A	Digital to Analog converter
DACS	Digital Access and Cross-Connect System
DAMA	Demand Assigned Multiple Access

DBA	Dynamic Bandwidth Allocation
DC	Direct Current
DEMS	Digital Electronic Message Service
DFE	Decision Feedback Equalizer
DFM	Dispersive Fade Margin
DLC	Data Link Control (layer)
DPSK	Differential Phase Shift Keying
DQPSK	Differential Quadrature Phase Shift Keying
DSB	Double Sideband
DSBSC	Double-Sideband Supressed Carrier
DS-CDMA	Direct Sequence-Code Division Multiple Access
DSL	Digital Subscriber Line
DSSS	Direct Sequence Spread Spectrum
ECC	Electronic Communications Committee (CEPT)
EGC	Equal Gain Combiner
EIA	Electronics Industry Association (U.S.)
EIRP	Equivalent Isotropically Radiated Power
ERC	European Radiocommunications Committee (CEPT)
ETSI	European Telecommunications Standards Institute
FBWA	Fixed Broadband Wireless Access
FCC	Federal Communications Commission (U.S.)
FDD	Frequency Division Duplexing
FDE	Frequency Domain Equalizer
FDM	Frequency Division Multiplexing/Multiplexed
FDMA	Frequency Division Multiple Access
FEC	Forward Error Correction
FFM	Flat Fade Margin
FH-CDMA	Frequency Hopping-Code Division Multiple Access
FHSS	Frequency Hopping Spread Spectrum
FM	Frequency Modulation/Modulated
FSDD	Frequency Switched Division Duplexing
FSK	Frequency Shift Keying
Gb/s	Gigabits per Second
GFSK	Gaussian Frequency Shift Keying
GHz	Gigahertz
HDSL	High-Speed Digital Subscriber Line
HDTV	High-Definition Television
H-FDD	Half-Duplex-Frequency Division Duplexing
Hz	Hertz
IDFT	Inverse Discrete Fourier Transform
IDU	Indoor Unit
IEE	Institution of Electrical Engineers (U.K.)
IEEE	Institute of Electrical and Electronic Engineers (U.S.)
IF	Intermediate Frequency
IFFT	Inverse Fast Fourier Transform
IFM	Interference Fade Margin

IP	Internet Protocol
ISI	Intersymbol Interference
ISM	Industrial, Scientific, and Medical (frequency band, U.S.)
ITFS	Instructional Television Fixed Service (U.S.)
ITU	International Telecommunications Union
kb/s	Kilobits per Second
KHz	Kilohertz
ks/s	Kilosamples per Second
LAN	Local Area Network
LES	Linear Equally Spaced
LMDS	Local Multipoint Distribution Service (U.S.)
LMS	Least Mean Square (equalizer)
LO	Local Oscillator
LOS	Line-of-Sight
LPF	Low-Pass Filter
LSB	Least Significant Bit
LTE	Line Terminating Equipment
MAC	Medium Access Control
Mb/s	Megabits per Second
MDS	Multipoint Distribution Service (U.S.)
M-FSK	Multilevel (>2) Frequency Shift Keying
MHz	Megahertz
MIMO	Multiple-Input, Multiple-Output
MISO	Multiple-Input, Single-Output
ML	Maximal Length (sequence)
MLSD	Maximum Likelihood Sequence Detection
MMA	Multimodulus Algorithm
MMDS	Multichannel Multipoint Distribution Service (U.S.)
MMSE	Minimum Mean Square Error (equalizer)
MPC	Maxium Power Combiner
MPEG	Moving Picture Expert Group
MRC	Maximal Ratio Combiner
MSB	Most Significant Bit
MTM	Multipoint-to-Multipoint
MTSO	Mobile Telephone Switching Office
MTU	Maximum Transfer Unit
NDFM	Nondispersive Fade Margin
NLOS	Non-Line-of-Sight
nrtPS	Non-Real-Time Polling Service
NRZ	Nonreturn to Zero
NTIA	National Telecommunications and Information Administration (U.S.)
OC	Optical Carrier Level
ODU	Outdoor Unit
OFDM	Orthogonal Frequency Division Multiplexing
OFDMA	Orthogonal Frequency Division Multiple Access
OQPSK	Offset Quadrature Phase Shift Keying

P/S	Parallel to Serial (converter)
PAM	Pulse Amplitude Modulation/Modulated
PAMA	Pre-Assigned Multiple Access
PAR	Peak-to-Average Ratio
PBX	Private Branch Exchange
PCM	Pulse Code Modulation/Modulated
PDF	Probability Distribution Function
PDH	Plesiochronous Digital Hierarchy
PHY	Physical (layer)
PLL	Phase Lock Loop
PMD	Physical Media Dependent
PMP	Point-to-Multipoint
PN	Pseudonoise (sequence)
PSK	Phase Shift Keying
PSTN	Public Switched Telephone Network
PTP	Point-to-Point
PVC	Permanent Virtual Connection
QAM	Quadrature Amplitude Modulation
QoS	Quality of Service
QPSK	Quaternary/Quadrature Phase Shift Keying
RBER	Residual Bit Error Rate
RCA	Reduced Constellation Algorithm
RCVR	Receiver
RF	Radio Frequency
RS	Reed-Solomon (block codes)
rtPS	Real-Time Polling Service
RTS/CTS	Request-to-Send/Clear-to-Send
S/P	Serial to Parallel (converter)
SD	Spatial Diversity
SDH	Synchronous Digital Hierarchy
SDM	Synchronous Digital Multiplex
SEAL	Simple and Efficient Adaptation Layer
SES	Severely Errored Second
SIMO	Single-Input, Multiple-Output
SINR	Signal-to-Interference and Noise Ratio
SISO	Soft-Input, Soft-Output (decoder)
SISO	Single-Input, Single-Output (antenna system)
SNR	Signal-to-Noise Ratio
SOHO	Small Office/Home Office
SONET	Synchronous Optical Network
SPE	Synchronous Payload Envelope
SRSP	Standard Radio System Plan (Canada)
SSB	Single Side Band
STBC	Space-Time Block Codes
STC	Space-Time Coding/Code
STE	Section Terminating Equipment

STM	Synchronous Transport Module
STS	Synchronous Transport Signal
STTC	Space-Time Trellis Code
SVC	Switched Virtual Connection
T/I	Threshold to Interference Ratio
TC	Transmission Convergence
TCC	Turbo Convolution Code
TCM	Trellis Coded Modulation
TCP	Transmission Control Protocol
TDD	Time Division Duplexing
TDE	Time Domain Equalizer
TDM	Time Division Multiplexing/Multiplexed
TDMA	Time Division Multiple Access
TFM	Thermal Fade Margin
TIA	Telecommunications Industry Association (U.S.)
TPC	Turbo Product Code
T-TCM	Turbo Trellis Coded Modulation
UDP	User Datagram Protocol
UGS	Unsolicited Grant Services
UHF	Ultra High Frequency
U-NII	Unlicensed National Information Infrastructure (frequency band, U.S.)
VC	Virtual Connection/Channel/Circuit (ATM)
VC	Voice Channel
VCO	Voltage Controlled Oscillator
VOFDM	Vector Orthogonal Frequency Division Multiplexing
VP	Virtual Path
WAN	Wide Area Network
WCS	Wireless Communications Service
Wi-Fi	Wireless Fidelity
WiMAX	Worldwide Interoperability for Microwave Access
WLAN	Wireless Local Area Network
XMTR	Transmitter
XPD	Cross-Polarization Discrimination
XPIC	Cross-Polarization Interference Cancellation/Canceller
ZF	Zero Forcing (equalizer)

Index

A

Adaptive equalizer/equalization, 225–36
Adaptive beamforming, 313, 315–18
Adaptive modulation, 302
Air interface standards, 319–31
ALOHA, 298–99
  reservation (R-ALOHA), 299
  slotted (S-ALOHA), 299
Alternative mark inversion (AMI), 17, 252, 289
Altium MX 311, 252, 261–64
Alvarion, 321, 333–37
Amplitude spectral density, 36–38,51,67
Analog channel plan, 243, 244
Angle diversity, 178, 182–83
Antenna array, 316–18
Antenna combiner, 248
Antenna coupler, 248
Antenna duplexer, 248–49
Antennas, 128–34
  beamwidth, 129
  cross-polarization discrimination, 130
  dual reflector/Cassegrain, 131
  flat plane/planar array, 128, 133
  front-to-back ratio, 129–30
  gain, 128–29
  half-wave dipole, 133
  horn reflector, 128, 132
  lens-corrected horn, 128, 132
  parabolic, 128, 130–32
  polarization, 130
  radome, 130
    shield, 130
    stacked dipole/collinear array, 128, 132–33
Application layer, 25
Array codes, 222–23
Asymmetrical DSL (ADSL), 5
Asymptotic coding gain, 190, 191, 216
Asynchronous digital multiplexing, 19
Asynchronous transfer mode (ATM), 24, 29–33
ATM adaptation layer (AAL), 31
Atmospheric absorption, 142, 145–46
Atmospheric multipath fading, 155–57
Atmospheric reflection, 142
Autocorrelation, 275–76
Automatic request for repeat (ARQ), 190, 327, 330
Automatic transmitter power control (ATPC), 237–38, 252
Availability, 163

B

B3ZS, 18, 252
B6ZS, 18
B8ZS, 18, 289
Balanced discriminator, 114–15, 116
Baseband filtering, 51–58
Baseband signal, 35
Beam bending, 154
Best effort (BE) service, 326
Binary complementary codes, 278
Binary phase shift keying (BPSK), 72–75, 99, 100–101, 266

Bit-by-bit interleaving, 19
Blind time domain equalization, 235–36, 262
   constant-modulus algorithm (CMA), 235–36
   multi-modulus algorithm (MMA), 235
   reduced constellation algorithm (RCA), 235
Block codes/coding, 191–95
   Bose-Chaudhuri-Hocquenghem (BCH), 193
   cyclic, 192–93
   Reed-Solomon (RS), 193–95, 252, 289
Block interleaving, 218–19
Bose-Chaudhuri-Hocquenghem (BCH)
   codes, 193
BreezeACCESS VL, 334, 335–36
Broadband Radio Access Networks
   (BRAN), 321
Broadband signals, 1
Burst errored second, 163
Byte, 15

C

Cable modem technology, 6
Carrier recovery, 98–105
   Costas loop method, 100–103
   decision directed method, 104–5
   multiply-filter-divide method, 98–100
Carrier-sense multiple access, 27
Carson's rule, 118–19
Cassegrain antenna, 131
Central office, 5
Channel plans, 243
Checks on checks, 222, 223
Chip rate, 268
Circuit switching, 9
Climate factor, 173
Climate-terrain factor, 173
Co-channel plan, 243, 244
Code division multiple access (CDMA), 292,
   295–97
   direct sequence (DS-CDMA), 295–96
   frequency hopping (FH-CDMA), 297
   RAKE receiver, 303, 312–13
Code interleaving, 218–20, 222–24
   block interleaving, 218–19
   de-interleaver, 219, 220, 222–24
   interleaver, 219–20, 222–24
   interleaving depth, 218–219

Code of Federal Regulations, 242
Code tree, 198–99, 200
CODEC, 33
Coded mark inversion (CMI), 18
Coding gain, 190–91
   asymptotic, 190, 191, 216
Coherent detection, 68
Collinear array antenna, 128, 132–33
Collision detection, 27
Commom part sublayer (CPS), 323
Commutator, 13, 14
Companding, 12
Complementary code keying (CCK), 277–82,
   288, 289
   binary complementary codes, 278
   complementary codes, 277–79
   complex complementary codes, 278–79
   jamming margin, 281–82
   polyphase complementary codes, 278–79
Complementary codes, 277–79
Complementary error function (*erfc*), 45
Complex complementary codes, 278–79
Composite fade margin (CFM), 170–72
Compressor, 12, 13
Concatenated codes/coding, 191, 219–25
   iterative decoding, 220–21, 222, 223–24
   serial, 219–20
   turbo codes, 220–25
   turbo convolution codes (TCCs), 221–22
   turbo product codes (TPCs), 221, 222–25
Connectionless, 26
Connection-oriented, 25
Constant-modulus algorithm (CMA), 235–36
Constellation/signal space diagram, 72, 73,
   77, 82, 85, 91, 96, 260–61
Constellation™ 3xDS3, 252, 257–60
Constraint length, 195–96
Contention scheme, 298–99
Continuous-phase FSK (CPFSK), 113, 117,
   118, 252
Convergence sublayer (CS), 323
Convolution codes/coding, 191, 195–207
   code tree, 198–99, 200
   constraint length, 195–96
   Euclidean distance, 203–6
   hard decision decoding, 203

linear addition, 198
linear codes, 191
maximum likelihood sequence detection (MLSD), 199–201
puncturing/punctured, 206–207
sequential decoding, 199
soft decision decoding, 203–5
survivor path, 199–201
systematic codes, 191
truncation window, 202
Viterbi decoding, 199–202, 204–6
Coordinated transmit power, 237
Costas loop carrier recovery, 100–103
Crosscorrelation, 275–76
Cross-polarization discrimination, 130, 161–62
Cross-polarization interference canceller/ cancellation (XPIC), 162, 236–37
Cumulative distribution function (CDF), 41–43
Cyclic codes, 192–93
Cyclic prefix, 308–12, 326

D
Data link layer, 27
Decision directed carrier recovery, 104–5
Decision directed equalizer, 231
Decision feedback equalizer, 231–33
De-interleaver, 219, 220, 222–24
Delay and multiply timing recovery method, 105
Delay diversity, 314
Demand assigned multiple access (DAMA), 292, 294
Descrambler/descrambling, 97–98
Differential discrimination, 114, 115–17
Differential encoding/decoding, 100, 105–7
Differential phase shift keying (DPSK), 107–9
Differential quadrature phase shift keying (DQPSK), 109–12
Differential/product detection, 114, 115–17, 253
Diffraction, 149–50, 151
Digital access and cross-connect system (DACS), 22
Digital signal processing (DSP), 308
Digital subscriber line (DSL), 5
Digitized video, 33

Direct sequence code division multiple access (DS-CDMA), 295–96
Direct sequence spread spectrum (DSSS), 266–77
BPSK derived, 266–70
interference, 270–73
jamming margin, 272–73
processing gain, 271–72
QPSK derived, 273–75
Discrete Fourier transform (DFT), 308
Discrimination detection, 114
Dispersion signature, 160
Dispersion, 52
Dispersive fade margin (DFM), 171, 172, 252
Distributor, 13
Diversity baseband switch, 180
Diversity improvement factor, 178, 181, 184
Double-sideband suppressed carrier (DSBSC), 66–72
Double-sideband, 67
Downconverter, 122, 123
Downlink, 292
Downstream, 292
Dribbling errored second, 163
Dual reflector antenna, 131
Ducting, 140–41, 154
Dynamic bandwidth allocation (DBA), 300–301

E
E1 carrier, 4, 5,15–16, 19–20
Earth radius factor, 138–40, 150–53
Encapsulation, 25, 26
Energy density, 40
Equal gain combiner, 178–80
Equivalent isotropically radiated power (EIRP), 130
Error free state, 163
Error function (*erf*), 45
Ethernet, 24, 27–29
Euclidean distance, 203–6
European Conference of Postal and Telecommunications Administrations (CEPT), 242
European Telecommunications Standards Institute (ETSI), 242
Expander, 12, 13
Expectation, 43

Extrinsic information, 223, 224
Eye diagram/pattern, 56, 255–56

F
Fade margin, 137
    composite (CFM), 170–72
    dispersive (DFM), 171, 172, 252
    flat (FFM), 171, 172
    interference (IFM), 171, 172
    nondispersive (NDFM), 181, 182, 184, 185
    thermal (TFM), 137, 171
Fading, 127, 137–62
    atmospheric absorption, 142, 145–46
    atmospheric multipath/Rayleigh, 155–57
    atmospheric reflection, 142
    beam bending, 154
    diffraction, 149–50, 151
    ducting, 140–41, 154
    flat, 154
    frequency selective, 154–57
    minimum phase, 159, 160, 161, 172, 226, 227
    multipath channel model, 157–61
    multiple refractive paths, 141
    nonminimum phase, 159, 160, 161, 172, 226, 227
    rain attenuation, 142–44, 154
    scintillation, 141–42
    terrain/ground reflection, 146–47, 154–55
Fast Fourier transform (FFT), 308–12
Fast frequency hopping, 283
Federal Communications Commission (FCC), 242
Feedforward transversal equalizer, 228–31
Fiber optic technology/transmission, 2, 5
Fixed broadband wireless access (FBWA), 291–337
Fixed wireless systems, 1
Flat fade margin (FFM), 171, 172
Flat fading, 154
Flat plane antenna, 128, 133
Forward error correction (FEC), 190–225
Four-frequency plan, 246–47
Fourier transform, 36–38
Fragmentation, 26
Frames/framing, 13,15, 16, 323, 324, 325

Free Euclidean distance, 208
Free space loss, 134–36
Frequency bands, 243, 244, 245, 247–48, 284–87, 331–33
Frequency diversity, 178, 183–85
Frequency division duplexing (FDD), 297, 321, 322
Frequency division multiple access (FDMA), 292–93
Frequency division multiplexed/multiplexing (FDM), 1, 304
Frequency domain equalizer/equalization, 225–27, 252
Frequency hopping code division multiple access (FH-CDMA), 297
Frequency hopping spread spectrum (FHSS), 266, 282–84
    fast frequency hopping, 283
    interference, 284
    processing gain, 283
    slow frequency hopping, 283
Frequency modulated/modulation (FM), 1, 114
Frequency selective fading, 154–57
Frequency shift keying (FSK), 112–21
    balanced discriminator, 114–15, 116
    continuous-phase FSK (CPFSK), 113, 117, 118, 252
    differential discrimination, 114, 115–17
    differential/product detection, 114, 115–17, 253
    discrimination detection, 114
    Gaussian FSK (GFSK) , 117
    quadrature modulation, 113–14, 253
Frequency switched division duplexing (FSDD), 297
Fresnel zones, 147–49
Front-to-back ratio, 129–30

G
Gaussian CDF, 45
Gaussian FSK (GFSK) , 117
Gaussian PDF, 44
Gray coding, 86, 89, 94, 95, 96
Ground reflection, 146–47, 154–55
Guard interval, 308–12, 326

H

Half duplex-frequency division duplexing (H-FDD), 297, 321, 322
Half-wave dipole antenna, 133
Hamming distance, 192
Hard decision decoding, 203
Harris Corporation, 251, 252, 257–60
HDB3, 17–18, 252, 289
Header, 25, 26
High-definition television (HDTV), 33
High-speed DSL (HDSL), 5
HIPERACCESS, 321, 329–30
HIPERMAN, 321, 330
Horn reflector antenna, 128, 132
Huygens' principle, 149

I

IEEE 802.11a, 321, 330–31
IEEE 802.11b, 277–82, 321
IEEE 802.11g, 321, 331
IEEE 802.16, 321–26
IEEE 802.16a, 321, 326–28
IEEE 802.16e, 321, 329
Image reject filter, 122, 123
Industrial, Scientific and Medical (ISM) bands, 285–87, 331–32
Industry Canada, 242
Intel, 321
Interference:
    in DSSS, 270–73
    external, 162–63
    fade margin (IFM), 171, 172
    in FHSS, 284
    intersymbol (ISI), 56
    postcursor ISI, 227
    precursor ISI, 227
    threshold to interference ratio (T/I), 162, 252
Interleaved channel plan, 243, 245
Interleaver, 219–20, 222–24
Interleaving depth, 218–19, 220, 221
Intermediate frequency, 121, 123
International Telecommunications Union (ITU), 241
Internet layer, 25–27
Internet protocol (IP), 25–27
Intersymbol interference (ISI), 56

Intrinsic information, 223
Inverse discrete Fourier transform (IDFT), 308
Inverse fast Fourier transform (IFFT), 308–12
Iterative decoding, 220–21, 222, 223–24
    extrinsic information, 223, 224
    intrinsic information, 223
    log-likelihood ratio, 223
    soft-input soft-output (SISO), 223

J

Jamming margin, 272–73, 281–82
Justification, 19

K

Karnaugh map, 214, 215

L

Least mean square (LMS) equalizer, 231, 232, 262
Lens-corrected horn antenna, 128, 132
Licensed bands, 241, 247–48, 331, 332–33
Line codes, 16–18
Line layer, 22
Line terminating equipment (LTE), 22
Linear codes, 191
Linear system, 38–39
Linear transversal equalizer, 228–31
Local area network (LAN), 29
Local oscillator, 122, 123
Log-likelihood ratio, 223
Long-haul, 2

M

Manchester coding, 29
Maximal length (ML) PN sequences, 276
Maximal ratio combiner, 180
Maximum likelihood sequence detection (MLSD), 199–202, 204–206
Maximum power combiner, 178–80
Maximum transfer unit (MTU), 26
MDR 2400/5800 SR, 288–90
Medium access control (MAC), 298–300, 321, 322, 323–26, 327
Mesh networks, 7, 291, 318–19, 320, 327
Millimeter wave, 2
Minimum dispersion combiner, 180

Minimum distance, 192, 193–94
Minimum Euclidean distance, 208
Minimum mean square error (MMSE)
    equalizer, 231
Minimum phase fade, 159, 160, 161, 172,
    226, 227
Mobile telephone switching office (MTSO), 3
Modulo-2 addition, 191
Monitored hot standby protection, 249–50
Motion picture expert group (MPEG), 33
Multidimensional TCM, 217–18, 252, 257
Multi-modulus algorithm (MMA), 235
Multipath channel model, 157–61
Multipath fading, 155–57
Multiple refractive paths, 141
Multiple-input, multiple-output (MIMO), 313,
    314–15
Multiple-input, single-output (MISO), 313, 314
Multiply-filter-divide carrier recovery, 98–100
Multipoint-to-multipoint, 318

N
Narrowband noise representation, 48, 49, 70
National Telecommunications and Informa-
    tion Administration (NTIA), 242
Near-far problem, 296
Noise bandwidth, 48–50
Noise figure, 124–25, 252, 289
Nondispersive fade margin (NDFM), 181,
    182, 184, 185
Non-line-of-sight (NLOS) techniques, 302–18
Nonminimum phase fade, 159, 160, 161, 172,
    226, 227
Nonperiodic function, 35
Non-real-time polling service (nrtPS), 325–26
Non-return to zero (NRZ), 16
Normalized energy, 39
Normalized power, 39
Null sub-carriers, 310
Nyquist:
    bandwidth, 52
    criterion, 52

O
Offset quadrature phase shift keying
    (OQPSK), 81, 82
Optical carrier (OC), 22, 24, 260, 262

Orderwire, 34
Orthogonal frequency division multiple
    access (OFDMA), 312, 326
Orthogonal frequency division multiplexing
    (OFDM), 303–12
    cyclic prefix, 308–12, 326
    guard interval, 308–12, 326
    null sub-carriers, 310
    orthogonal sub-carriers, 305–7
    peak-to-average ratio (PAR), 312
    pilot sub-carriers, 310, 326, 327
Orthogonal sub-carriers, 305–7
Outage, 163–77
Overshoot radiation, 162, 246–47

P
Packet switching, 9, 24–33
Parabolic antenna, 128, 130–32
Parseval's theorem, 40
Path layer, 22
Path reliability, 164
Path terminating equipment (PTE), 22
Peak-to-average ratio (PAR), 124, 312
Permanent virtual connection (PVC), 31
Phase modulated/modulation (PM), 1
Phase shift keying (PSK), 2, 95–96, 210–16
Photonic layer, 22
Physical media dependent (PMD) sublayer,
    32–33
Pilot sub-carriers, 310, 326, 327
Planar array antenna, 128, 133
Plesiochronous digital hierarchy (PDH), 19–21
Plesiochronous, 19
Plessey Broadband Wireless, 288–90
Point-to-multipoint (PMP), 7, 8, 291–318,
    321, 326, 330
Point-to-point (PTP), 6–7, 241, 265, 291
Polyphase complementary codes, 278–79
Postcursor ISI, 227
Power spectral density, 40, 74, 78, 83, 118,
    119, 255–56, 262, 263
Pre-assigned multiple access (PAMA), 292, 294
Precursor ISI, 227
Predistorter, 122, 124
Privacy sublayer, 323–24
Private branch exchange (PBX), 5, 291
Probability density function (PDF), 41–43

Processing gain, 271–72, 283
Product codes, 222–23
Protocol data unit (PUD), 31, 32
Pseudonoise (PN) sequences, 266, 275–76
    maximal length, 276
Pseudorandom, 97
Public switched telephone network (PSTN), 1, 3
Pulse amplitude modulation (PAM), 11, 42–43,
    58–72
Pulse code modulation (PCM), 2, 9–13,
    15–16, 19
Pulse stuffing, 19
Punctured codes, 206–207

Q

Quadrature amplitude modulation (QAM), 2,
    51, 75–95, 124, 216–17, 233–35, 252
Quadrature crosstalk, 233
Quadrature FSK modulation, 113–14, 253
Quadrature phase shift keying (QPSK), 2, 75,
    76–80, 102–3, 215–16
Quality of service (QoS), 30, 326
Quantization, 11–13
    error, 11
Quaternary phase shift keying.
    See Quadrature phase shift keying

R

Radome, 130
Rain attenuation, 142–44, 154
Raised cosine filter/filtering, 52–54
RAKE receiver, 303, 312–13
Random variable, 41
Rayleigh CDF, 46, 47
Rayleigh fading, 155–57
Rayleigh PDF, 46, 47
Real-time polling service (rtPS), 325
Receive backside reception, 246
Received signal level, 136–37
Receiver front end, 124–25
Receiver low-noise amplifier, 124
Receiver noise figure, 124–25, 252, 289
Receiver sensitivity, 137
Receiver threshold, 137, 252, 289
Reduced constellation algorithm (RCA), 235
Reed-Solomon (RS) codes, 193–95, 252, 289
Reflection coefficient, 147

Refractive index, 138
Requantization, 13
Residual bit error rate (RBER), 163, 190, 252,
    289
Roll-off factor, 52–53
Root-raised cosine filter/filtering, 57
Roughness factor, 173

S

Sampling function, 36
Sampling theorem, 9
Satellite Internet transmission, 6
Scintillation, 141–42
Scrambler/scrambling, 97–98
Section layer, 22
Section terminating equipment (STE), 22
Sequential byte interleaving, 15
Sequential decoding, 199
Service channel, 34
Set partitioning (in TCM), 209–11
Severely errored second (SES), 163
Sheet, 142
Short-haul, 2
Signal space/constellation diagram, 72, 73,
    77, 82, 85, 91, 96, 260–61
Signal to interference and noise ratio
    (SINR), 314
Signaling bits, 13, 15, 16
Simple and efficient adaptation layer
    (SEAL), 32
Single sideband amplitude modulation
    (SSB -AM), 2
Single-input, multiple-output (SIMO), 313,
    314
Single-sided, 62
Slow frequency hopping, 283
Soft decision decoding, 203–5
Soft-input soft-output (SISO), 223
Space diversity, 178–82, 313–14
Space-time block codes (STBCs), 314
Space-time codes/coding (STC), 314
    space-time block codes (STBCs), 314
    space-time trellis codes (STTCs), 314
Spatial diversity, 178–82, 313–14
Spectral efficiency, 74, 245, 252
Spread spectrum techniques, 265–84
    chip rate, 268

Spread spectrum techniques, *(cont.)*
  complementary code keying (CCK),
    277–82, 288, 289
  direct sequence spread spectrum
    (DSSS), 266–77
  frequency hopping spread spectrum
    (FHSS), 266, 282–84
  processing gain, 271–72, 283
Square and filter timing recovery method, 105
Stacked dipole antenna, 128, 132–33
Staggered QPSK. *See* Offset quadrature phase
  shift keying
Standard deviation, 43
Standard Radio System Plans (SRSP), 242
Stochastic gradient algorithm (SGA), 230
Stratex Networks, 251, 252, 261–64, 290
Substandard/subrefractive atmosphere, 139
Superstandard/superrefractive atmosphere, 139
Suppressed carrier, 67
Survivor path, 199–201
Switched virtual connection (SVC), 31
Synchronous digital hierarchy (SDH), 19,
  21–24
Synchronous digital multiplexing (SDM), 19,
  21–24
Synchronous optical network (SONET), 19, 22
Synchronous payload envelope (SPE), 22, 23
Synchronous transport signal (STS), 22, 24
Synchronous transport module (STM), 22, 24
System gain, 137, 252, 289
Systematic codes, 191

T

T1 carrier, 4, 5, 15, 19–20
Tapped delay-line equalizer, 228–33
TD-2 system, 1
Terrain reflection, 146–47, 154–55
Terrain roughness, 173
Thermal fade margin (TFM), 137, 171
Thermal noise, 46–48, 125
Threshold to interference ratio (T/I), 162, 252
Time division duplexing (TDD), 297–98
Time division multiple access (TDMA), 292,
  293–95
Time division multiplexing (TDM), 2, 13–14
Time domain equalizer initialization methods,

235–36, 262
Time domain equalizer/equalization,
  227–236, 252
  least mean square (LMS) equalizer, 230,
    231, 262
  zero-forcing (ZF) equalizer, 230, 231
Timing jitter, 56
Timing recovery, 105
  delay and multiply method, 105
  square and filter method, 105
Transmission control protocol (TCP), 25
Transmission control protocol/Internet
  protocol (TCP/IP), 24, 25–27
Transmission convergence (TC) sublayer, 32,
  33
Transmit backside radiation, 246
Transmitted spectrum, 255–56, 262–63
Transmitter emission limits, 243–45, 255–56,
  262–63
Transmitter power amplifier, 123–24
Transport layer, 26
Transversal equalizer (TVE), 228–33
Trellis coded modulation (TCM), 207–18
  free/minimum Euclidean distance, 208
  Karnaugh map, 214, 215
  multidimensional, 217–18, 252, 257
  parallel branches, 210, 212
  set partitioning, 209–11
  turbo (T-TCM), 222
Trellis diagram, 199, 201
Truncation window, 202
Turbo (T-TCM), 222
Turbo codes, 191, 220–25
Turbo convolution codes (TCCs), 221–22
Turbo product codes, 221, 222–25
Two-frequency plan, 246–47
Two-sided, 36

U

Ultra high frequency (UHF), 2
Unacceptable performance, 163
Unavailability, 163
Unipolar, 16
Unlicensed bands, 265, 284–87, 331–33
  Industrial, Scientific and Medical (ISM),
    285–87, 331–32

Unlicensed National Information
  Infrastructure (U-NII), 285–87, 331–32
Unsolicited grant service (UGS), 325
Upconverter, 122, 123
Uplink, 292
Upstream, 292

V

Variance, 43
Vector orthogonal frequency division
  multiplexing (VOFDM), 314
Vigants multipath fade occurrence factor, 173
Virtual connection (VC), 30, 31
Virtual path (VP), 30, 31
Viterbi decoding, 199–202, 204–6

W

WALKair 3000, 334, 336–37
Wayside channel, 34
Wide area network (WAN), 29
Wi-Fi, 277, 321, 330–31
WiMAX, 321, 326, 330
Wireless local area network (WLAN), 330, 331
Wireless path, 127
WirelessHUMAN, 326
Witcom Limited, 251, 252–57
WitLink-2000, 252–57
Word-by-word interleaving, 15

Z

Zero-forcing (ZF) equalizer, 230, 231

ABOUT THE AUTHOR

DOUGLAS H. MORAIS

Doug Morais has over 30 years of experience in wireless design, engineering management, and executive management. Before starting Adroit Wireless Strategies (*www.adroitwireless.com*), a wireless consulting and training company, in 1999, he held executive management positions at Harris Corporation, Digital Microwave Corporation (now Stratex Networks), California Microwave, Inc., and Ortel Corporation. He spent 24 years at Harris, where, during his first 11 years, he held microwave radio design and engineering management positions, playing a key role in the introduction of digital microwave systems.

He holds a B.Sc. from the University of Edinburgh, Scotland, and a M.Sc. from the University of California, Berkeley, both in electrical engineering. He received a Ph.D., also in electrical engineering, from the University of Ottawa, Canada, his thesis being on various aspects of digital modulation for wireless communications. He has authored several papers on digital microwave communications and holds a U.S. patent that addresses point-to-multipoint microwave radio communications. He is a graduate of the AEA/Stanford Executive Institute, Stanford University, California, is a senior member of the IEEE, and a member of the IEEE Communications Society.

A short course based on the material in and of the same title as this text is given by Dr. Morais via Besser Associates (*www.besserassociates.com*), a leading provider of RF and wireless training.